Photovoltaic Modules

The book provides a comprehensive review of developments in all aspects of solar photovoltaic technology in a single volume. It discusses maximum power point tracking (MPPT) control for achieving maximum possible power, robust control to maintain stable operation under varying internal as well as the ambient environment, inverter control for constant frequency operation, and automating the maintenance of photovoltaic solar plants.

This book:

- Presents modeling methods based on mathematical and physical principles for solar photovoltaic cells, power quality analysis of rooftop grid-connected PV, and PV generation analyzed by bidirectional long short-term memory networks (BiLSTM) to evaluate the performance reliability of the bifacial module and the control system of the synchronous reference CCVSI for active power injection.
- Provides an overview of SPECS control, various control loops, control algorithms, controllers, and their impact on the prosumer and the smart grid and discusses instantaneous power theory (pq theory).
- Covers control techniques of power electronic converters, optimization techniques, and management of the grid-connected solar PV arrays, qualification testing of bifacial modules as per IEC- 61215: 2021 and IEC 61730, including analytical approach elaborated for the performance of a building-integrated solar PV/T system.
- Discusses and comprehensively reviews degradation mechanisms, characterization techniques, and occurrence frequencies based on field testing, long-term analyses of PV installations, harmonic compensation, and the enhancement of Power Quality for the entire system, a novel approach of developing an effective and systematic brownout procedure and a novel game theory auctioning framework for trading energy in smart grids and explains Gbest-guided Artificial Bee Colony (GABC) optimization.
- Includes real-life case studies.

It will serve as an ideal reference text for senior undergraduate, graduate students, and academic researchers in fields including electrical engineering, electronics and communications engineering, environmental engineering, and renewable energy.

Photovoltaic Modules

Fundamentals, Modeling, Performance Analysis and Control

Edited by
Dhiraj Magare, Prasiddh Trivedi
and Kedar V. Khandeparkar

CRC Press
Taylor & Francis Group
Boca Raton London New York

CRC Press is an imprint of the
Taylor & Francis Group, an **informa** business

Designed cover image: Shutterstock

First edition published 2026
by CRC Press
2385 NW Executive Center Drive, Suite 320, Boca Raton FL 33431

and by CRC Press
4 Park Square, Milton Park, Abingdon, Oxon, OX14 4RN

CRC Press is an imprint of Taylor & Francis Group, LLC

ISBN: 978-1-032-39304-9 (hbk)
ISBN: 978-1-041-03276-2 (pbk)
ISBN: 978-1-003-62316-8 (ebk)

DOI: 10.1201/9781003623168

Typeset in Sabon
by Deanta Global Publishing Services, Chennai, India

Contents

Preface

Photovoltaic power generation is rapidly growing in popularity as a sustainable and reliable source of clean energy. As the demand for photovoltaic systems increases, so does the need for a comprehensive understanding of the fundamentals, modeling, performance analysis, and control of these systems.

This book, *Photovoltaic Modules: Fundamentals, Modeling, Performance Analysis, and Control*, covers all aspects of photovoltaic systems in a structured and systematic manner. The primary goal of this book is to provide readers with a deep understanding of the theory and practical applications of photovoltaic systems.

The book begins by introducing the fundamentals of photovoltaic technology, including the physics of solar cells, different types of solar panels, and the design of photovoltaic systems. It then delves into the modeling of photovoltaic systems, covering topics such as system sizing, performance prediction, and energy yield estimation.

The book also explores the performance analysis of photovoltaic systems, including techniques for assessing the efficiency, reliability, and degradation of solar panels. In addition, it provides an overview of the various control strategies used to optimize the operation of photovoltaic systems, such as maximum power point tracking and grid integration.

Throughout the book, practical examples and case studies are used to illustrate key concepts and principles. The book is intended for students, researchers, and professionals in the fields of renewable energy, electrical engineering, and sustainability, who are seeking to enhance their knowledge and skills in photovoltaic systems.

We hope that this book will serve as a valuable resource for anyone interested in learning more about photovoltaic systems and their applications in the modern world. We believe that a solid understanding of photovoltaic technology is crucial for shaping a sustainable future, and we invite you to join us on this journey of exploration and discovery.

Dhiraj Magare,
Prasiddh Trivedi,
Kedar Khandeparkar

About the editors

Dhiraj Magare is an Associate Professor and Smart Grid Lab in-charge in the Department of Electronics and Communication Engineering, D.Y. Patil School of Engineering, Ramrao Adik Institute of Engineering, Nerul, Navi Mumbai. He obtained his B.E. in Instrumentation Engineering from the Government College of Engineering, Jalgaon, Maharashtra, India, and did his M.Tech. in Instrumentation Engineering from Shri Guru Gobind Singhji Institute of Engineering and Technology, Nanded, Maharashtra, India. He worked in embedded industrial automation and industrial control systems. He has around 15 years of industrial/teaching experience. He has contributed a Scilab-based global optimization toolbox for worldwide use; downloads reached up to 9000. He has published more than 30 international journal and conference papers in the area of estimation of environmental effects on different PV technology modules, thermal modeling, and simulation. In addition, it analyzes the PV module performance of different technologies. He actively contributed to the India-UK joint project "Stability and Performance of Photovoltaics," under which he visited Loughborough University, UK, for part of his research and also visited CREST, Loughborough University, UK, as part of his research activity of India–UK project. Recently, this project received the British Council-CII Award for Excellence in Research. Received the Best Innovator award for the mentoring project presented at a regional meeting by the Ministry of Education's Institution Innovation Council in 2024. Completed Foundation and Advance Level certification as Innovation Ambassador from Ministry of Education Innovation Cell & AICTE during 2022–2023.

Prasiddh Trivedi is Assistant Professor and Program Coordinator of Electrical Engineering at Gati Shakti Vishwavidyalaya, Vadodara. He obtained PhD from IIT Bombay and MTech from IIT-BHU. He has been a visiting research scholar at Leicester University as part of UKIERI initiative. Dr Trivedi has authored over 20 international journal and conference papers in the area of nonlinear, robust, and sliding mode control. His current research interests include robust and nonlinear control, e-mobility, and

Digital twins. He is also a co-founder of a tech start-up manufacturing heavy-duty industrial fans.

Kedar V. Khandeparkar is Assistant Professor in the Department of Computer Science and Engineering at the Indian Institute of Technology, Dharwad, India. His scientific research area interests include Cyber-Physical Systems, Data Science and Cyber Security in Smart Grids, and Applied Machine Learning.

List of contributors

Alexandre Hugo da Silveira
Programa de Pós-Graduação em Engenharia Elétrica – PPGEE
Universidade Federal do Rio Grande do Sul – UFRGS
Rio Grande do Sul, Brazil

Amandeep Singh Makhija
Research Scholar
Gujarat Technological University
Gujarat
India

Anshul Agarwal
Faculty of Department of Computer Science and Engineering
Visvesvaraya National Institute of Technology (VNIT) Nagpur
Nagpur
India

Anuradha
Department of Electrical Engineering
IIMT University
Meerut
India

Arturo Conde Enríquez
Faculty of Mechanical and Electrical Engineering
Autonomous University of Nuevo León
San Nicolás de los Garza
Mexico

Arup Dhar
National Institute of Solar Energy
Gurgaon
India

Birinchi Bora
National Institute of Solar Energy
Gurgaon
India

Deepak Yadav
National Institute of Solar Energy
Gurgaon
India

Dharmendra Kumar Dheer
Electrical Engineering Department
National Institute of Technology
Patna
India

Eric Lacerda Silva
Programa de Pós-Graduação em Engenharia Elétrica – PPGEE
Universidade Federal do Rio Grande do Sul – UFRGS
Rio Grande do Sul
Brazil

Fausto Bastos Líbano
Programa de Pós-Graduação em Engenharia Elétrica – PPGEE
Universidade Federal do Rio Grande do Sul – UFRGS
Rio Grande do Sul
Brazil

Reinier Herrera Casanova
Faculty of Mechanical and Electrical Engineering
Autonomous University of Nuevo León
San Nicolás de los Garza
Mexico

Roberto Chouhy Leborgne
Programa de Pós-Graduação em Engenharia Elétrica – PPGEE
Universidade Federal do Rio Grande do Sul – UFRGS
Rio Grande do Sul
Brazil

Maicon Jaderson Silveira Ramos
Programa de Pós-Graduação em Engenharia Elétrica – PPGEE
Universidade Federal do Rio Grande do Sul – UFRGS
Rio Grande do Sul
Brazil

Mugala Naveen Kumar
National Institute of Solar Energy
Gurgaon
India

Mukta Singh
Electrical Engineering Department
Madhav Institute of Technology and Science
Gwalior
India

Mukul Dixit
Department of Electrical & Electronics Engineering
COER University
Roorkee
India

Nitin Chand
Department of Electrical & Electronics Engineering
COER University
Roorkee
India

Saurabh Kumar Rajput
Centre for Internet of Things
Madhav Institute of Technology and Science
Gwalior
India

Shabbir S. Bohra
Electrical Engineering Department
Sarvajanik College of Engineering & Technology
Surat
India

Subhra Das
Solar Engineering Department
Amity School of Engineering and Technology
Amity University Haryana
Gurugram
India

Sudhir Kumar
Solar Energy Consultant and Trainer
Pune
India

T. Sathiyanarayanan
Researcher
Erode
India

Section I

Solar photovoltaic fundamentals

Chapter 1

Solar photovoltaic cell

History, working principle, and landscape

Sudhir Kumar

1.1 PHOTOVOLTAIC CONCEPT

Photovoltaic (PV) essentially means "voltage" created by light "photons". Photons are minute energy packets of electromagnetic radiation, also called "light quantum". The concept originated (1905) in Albert Einstein's explanation of the photoelectric effect, in which he proposed the existence of discrete energy packets during the transmission of light. A photon is assumed to be a subatomic particle, having energy and momentum but no mass or electric charge. The definition of photovoltaic is derived from the Greek words' "photo", meaning "light", and "voltaic", meaning "voltage".

Photovoltaic cells convert artificial light or sunlight into electricity. They are more commonly known as SPV (solar photovoltaic) cells due to their ability to utilize the free energy source of solar energy to generate marketable electricity. A photovoltaic (PV) cell, or solar cell, is a nonmechanical device that instantly converts sunlight directly into electricity, functioning as a silent electric generator. Sunlight is composed of photons, or energy packets, of solar energy. Corresponding to the different wavelengths of the solar spectrum, photons contain varying amounts of energy. A PV cell is made of semiconductor material. When photons strike a PV cell, the portion of absorbed photons (say 20%) provides energy to generate electricity, while the rest either are reflected or pass through.

Solar radiation has three ranges or bands in its spectrum: ultraviolet, visible, and infrared (Figure 1.1). Of the light that reaches Earth's surface, infrared radiation makes up 49.4%, while visible light contributes 42.3%. Ultraviolet radiation makes up just over 8% of the total solar radiation (Sze et. al. Chapter 1). Most infrared photons correspond to higher wavelengths with lower energy and may not excite electrons in solar cells to produce electricity. UV photons have lower wavelengths and high energy and will excite too many electrons that will get lost before generating electrical current. It is the visible or near-infrared photons with optimum energy that are mainly responsible for generating electricity through solar cells.

DOI: 10.1201/9781003623168-2

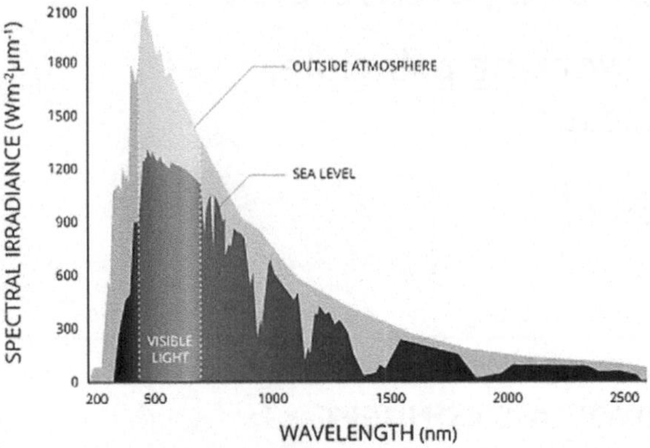

Figure 1.1 Solar spectrum in outer atmosphere and at sea level of the Earth. (Courtesy: Refer Annexure A)

1.2 HISTORICAL DEVELOPMENT

Literature beautifully documenting historical development in a concise manner is available (Elizabeth et. al.). Recalling history as an event on this day is presented in an interesting way [1]. However, a detailed chronological account has been presented elsewhere [3]. Fraas has divided the historical developments into six time periods, beginning with the discovery in 1839.

a. 1800–1904: Discovery Years
b. 1905–1950: Scientific Foundation
c. 1950–1959: First Practical Device Demonstration
d. 1960–1980: Enthusiastic Progress and New PV Devices
e. 1980–2000: Sluggish Development Phase
f. 2000–Present: International Support and New Opportunities

The story starts from the early observation of the photovoltaic effect in 1839. French physicist Alexandre Edmond Becquerel was working with metal electrodes in an electrolyte solution when he noticed that small electric currents were produced when the metals were exposed to light. He just reported the effect but could not give any explanation. As we know today, semiconductors, rather than metals, can show a photovoltaic effect. It appears that the metal electrodes were unknowingly converted to semiconductors due to inherent impurities. A few decades later, French mathematician Augustin Mouchot was inspired by the physicist's work. He began registering patents for solar-powered engines in the 1860s. In 1873, Willoughby Smith discovered that selenium had photoconductive potential, leading to William Grylls Adams' and Richard Evans Day's 1876 discovery that selenium creates electricity when exposed to sunlight. For the first time, the real solar cell was demonstrated in 1883 when New York inventor Charles Fritts created the first solar cell by coating selenium with a thin layer of gold. Fritts reported that the selenium module produced a current "that is continuous, constant, and of considerable force". This cell achieved an energy conversion rate of 1 to 2% (compared to 20% of today's solar cells). Following this, Wilhelm Hallwachs made a semiconductor-junction solar cell (copper and copper oxide) in 1904.

Now that the occurrence of photovoltaic effect was confirmed, proven, and established in laboratory experiments, it was time to uncover the science behind it. It took 45 years (from 1905 to 1950) to establish a solid scientific foundation for photovoltaic theory. It started, in 1905, with Albert Einstein's explanation of photoelectric effect based on quantum theory. Later, Jan Czochralski, a Polish scientist, invented a method in 1918 to grow single crystal of metal, which 50 years later served as the foundation for the commercial production of single crystal silicon. However, the production of single crystals created immense interest in the study of their ordered structure with its electrical and electronic properties based on the new quantum theory. This led F. Bloch in 1928 to develop band theory based on the single crystal periodic lattice. Following this, A. H. Wilson developed, in 1931, the theory of high-purity semiconductors. The next major advancement in solar cell technology came in 1940 from Russell Shoemaker Ohl, a semiconductor researcher at Bell Labs. He was investigating some silicon samples, one of which had a crack in the middle. He noticed that in this particular sample, current flowed through this crack when it was exposed to light. This crack, which had probably formed when the sample was made, actually marked the boundary between regions containing different levels of impurities, so one side was positively doped and the other side negatively doped. Ohl had inadvertently made a p-n junction, the basis of a solar cell. Excess positive charge builds up on one side of the p-n barrier, and excess negative charge builds up on the other side of the barrier, creating an electric field. When the cell is hooked up in a circuit, an incoming photon that hits the cell can then give an electron a kick, starting the flow of current. Ohl patented his solar cell, which had an efficiency of about one percent. In 1948, Gordon Teal and John Little adapted the Czochralski method of crystal growth to produce single-crystalline germanium and, later, silicon. The concept of deliberate doping of pure semiconductors with impurities (dopant atoms) to increase electrical conductivity was beginning to take shape.

The real foundation for modern silicon solar cells was laid down during 1950–1959 when the first practical device was demonstrated. The first practical silicon solar cell was created by a team of scientists working together at Bell Labs. In 1953, the Bell Labs engineer Daryl Chapin tried selenium solar cells, but found them very inefficient. Meanwhile, his other colleagues, Calvin Fuller, a chemist, and Gerald Pearson, a physicist, were working on controlling the properties of semiconductors by introducing impurities. They also created a silicon p-n junction that provided encouraging photoresponse. Soon they realized that silicon was a better material for solar cells. Chaplin also joined them. The three then worked for several months on improving the properties of their silicon solar cells. After making some other improvements to the design, they linked several solar cells together to create what they called a "solar battery". Bell Labs announced the invention on April 25, 1954, in Murray Hill, New Jersey. They demonstrated their solar panel by using it to power a small toy Ferris wheel and a solar-powered radio transmitter. These cells had about 6% efficiency. The *New York Times* forecasted that solar cells would eventually lead to a source of "limitless energy of the Sun". The confidence was so high that solar cells were used in outer space where solar energy was used to power satellites. In 1958, the Vanguard I satellite used a tiny 1-watt panel to power its radios. Later that year Vanguard II, Explorer III, and Sputnik-3 were all launched with PV technology on board. In 1958, T. Mandelkorn, US Signal Corps Laboratories, created n-on-p silicon solar cells, which were more resistant to radiation damage and were better suited for space. Hoffman Electronics created a 10% efficient commercial solar cell in 1959 that opened the vast door of market.

The fourth phase of 1960–1980 saw an encouraging response from all over the world to this sector, with huge amounts of funds pouring into research projects and field demonstrations. A strong feeling was prevailing all around that solar cells were the solution to the energy crisis. Pollution-free technology was envisaged as the savior of the Earth. In 1960, Hoffman Electronics created a 14% efficient solar cell. The Telstar communications satellite was powered by solar cells in 1962. In 1964, NASA launched the Nimbus spacecraft, the first satellite capable of running entirely on a 470-watt solar array. In 1966, NASA launched the world's first Orbiting Astronomical Observatory, powered by a 1-kilowatt array. Soyuz 1, in 1967, was the first human-crewed spacecraft to be powered by solar cells. Due to its suitability, using solar cells for space applications became routine. The first highly efficient GaAs heterostructure solar cells were created in 1970 by Zhores Alferov and his team in the USSR. In 1972, Hovel and Woodall at IBM demonstrated AlGaAs/GaAs solar cells with 18%–20% efficiency. For the first time a pathbreaking project to transition silicon PV from space to terrestrial applications was undertaken by Jet Propulsion Laboratory, California Institute of Technology, in 1975. The world production of photovoltaic cells exceeded 500 kW by 1977. L. Fraas and R. Knechtli presented the InGaP/GaInAs/Ge triple junction concentrator cell, predicting 40% efficiency at the concentration of 300 suns in 1978. In short, the platform was ready for solar cells' use in terrestrial applications soon.

The momentum gained in the fourth phase was dampened in the fifth phase of 1980–2000, and the progress was not as expected due to a lack of political will from world leaders. Investments and research project fundings were reduced in the first half of this phase. The second half witnessed some recovery. Still, the damage was done and the dream of making solar cells as a common commodity remained unfulfilled. The best part was that technology matured with high reliability. With a very low degradation rate, the life of solar cells was predicted to be a minimum of 25 years with high confidence. The only hitch was the high cost. Now the main focus was to reduce the production cost with innovative technological improvements. The world was also looking at some innovative financial and commercial instruments to drastically reduce production costs such as scaling up, using cheaper products or services, market-stimulating policies, and improvements to module efficiency. Meanwhile, there was slow but steady progress toward commercialization through incentivization policies. The first Concentrating PV system using Fresnel Lenses went into operation with 350 kW funded by the US and Saudi Arabia SOLERAS project in 1981. By 1983, worldwide photovoltaic production exceeded 21.3 megawatts, and sales exceeded $250 million. A 20% efficient silicon cell was created by the Centre for Photovoltaic Engineering at the University of New South Wales in 1985. L. Fraas et al. (1990) reported a 35% efficient two-chip stack GaAs/GaSb concentrator solar cell. A premier institute NREL, US was created in 1993. Kuryla, Fraas, and Bigger (1992) reported 25% efficient CPV module using GaAs/GaSb stacked cell circuit. The first thermophotovoltaic stove with heat and electricity co-generation, Midnight Sun, was demonstrated by JX Crystals Inc. in 1998. Despite all the hindrances, it appeared that the solar PV market was ready to expand with a big bang anytime.

As stated earlier, the fifth phase was struggling mainly with high costs. Soon the sixth phase, from 2000 to the present, broke all the barriers. Suntech Power was founded in China in 2001. SunPower Corp's first manufacturing facility, which makes 20% A-300 cells, came online in the Philippines, and the company's first utility-scale power plant came online in Bavaria in 2004.

In 2007, a 15 MW Nellis Solar Power Plant was installed using SunPower Corp modules. Polysilicon use in photovoltaics exceeded all other polysilicon uses for the first time. The most important development in the history of solar cells was the construction of fast-growing factories in China, pushing the manufacturing costs from $40 per watt in 1980 down to about $1.25 per watt for silicon photovoltaic modules. As a result, installations doubled worldwide, with cumulative worldwide solar PV installations crossing over 100 GW in 2013.

Finally, the journey of around two centuries (1889 – 2022) of solar cells has culminated into commoditization and democratization of environmentally benign technology for the benefit of all. Earlier a little less efficient polycrystalline silicon solar cells were dominating the market due to some cost advantages over more efficient mono-crystalline silicon. Innovative manufacturing processes have brought parity of cost between both types of silicon solar cells. Naturally, the recent market is dominated by monocrystalline silicon solar cells. Solar cells are now available to everybody, everywhere, and at affordable cost. What a story to narrate to our future generations! Today, the cost of solar modules is just \$0.26 per watt (October, 2022). The total installed capacity in the world (off grid + grid) is around 1000 GW, and the figure is still growing at a faster rate. This is a mind-boggling achievement, considering the hurdles that the solar sector has faced. Nobody can deny the great contribution of China, who single handedly has brought down the cost to an unbelievable scale due to cheaper labor, dedicated manufacturing hubs, and innovative inventory management. This is like the happy ending story of a Hollywood film with a win-win situation for all the players. Are we not moving toward a 100% renewable energy and cleaner world?

1.3 SOLAR CELL DEVELOPMENT

The solar cell is an electronic device which converts solar energy directly into electrical energy through the photovoltaic effect. It is a typical semiconductor p-n junction device. When the light falls on the device, the light photons of certain wavelengths are absorbed by the semiconducting material and electrical charge carriers, electrons, and holes are generated. These carriers diffuse to the junction where a strong electric field exists. The electrons and holes are separated by this field and produce an electric current in the external circuit (Figure 1.2). A typical silicon p-n junction solar cell has a p-type wafer with a very thin coating of n-type silicon over it, with a back electrical contact of metal paint and a front electrical contact of a very fine metal grid.

An ideal solar cell material must be a semiconductor in the solid state. It should be responsive to the visible range of the solar spectrum. Since it will be used under rough environmental conditions for a long time, the material should be stable under outdoor use with no or minimum annual degradation. Abundant availability of the material in nature is highly desirable. For wider and common use, it should have affordable cost compared to contemporary energy-producing materials. The most critical criterion of all is that the solid semiconductor material must have a suitable bandgap. This can be easily explained by the band theory of semiconductors, as explained below.

The band theory of semiconductors has been the most suitable and convenient explanation in the history of material physics in a simplistic manner (Sze et. al. Chapter 1). Let us start with the simplest atom of hydrogen, which has the electronic configuration $1s^1$ with just one electron in the *s* atomic orbital around the hydrogen nucleus. Once two hydrogen atoms (H) approach each other, they make a hydrogen molecule H_2 with each

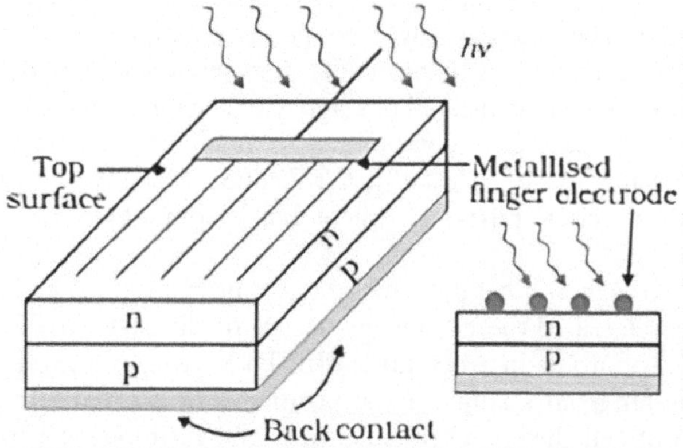

Figure 1.2 Typical p-n junction solar cell. (Courtesy: Refer Annexure A)

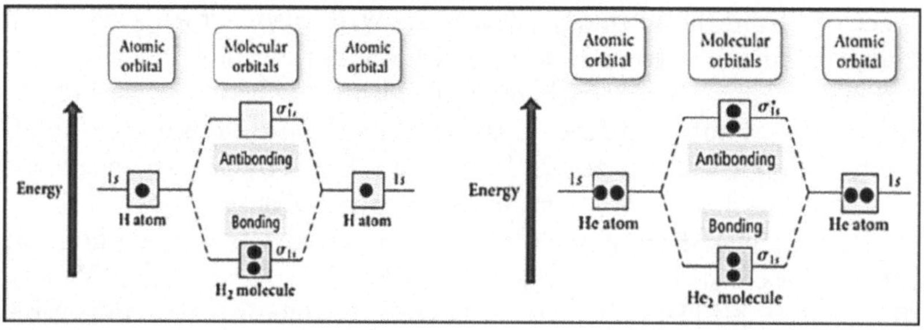

Figure 1.3 Molecular orbital diagram of hydrogen and helium. (Courtesy: Refer Annexure A)

electron, combining with each other to form a molecular bond (MO). Two atomic bonds combine to make two molecular bonds: the lower one, called bonding MO and the upper one antibonding MO. Both electrons settle in bonding MO, and the antibonding MO remains empty, similar to water filling the lower level first. Similarly, if we take the example of the helium molecule (He_2), the helium atom has molecular configuration $1s^2$. Thus, the He_2 molecule has a total of 4 electrons: 2 electrons filling bonding MO and 2 electrons filling antibonding MO (Figure 1.3).

As stated above, an ideal solar cell material must be a semiconductor in the solid state. Before looking at the band theory of solid semiconductors, let us first look at the available materials. Based on electrical conductivities, solid-state materials are categorized into three classes: insulators, semiconductors, and conductors. Insulators have the least conductivity (10^{-18}–10^{-8} S/cm). Semiconductors have mid-value conductivities (10^{-8}–10^{3} S/cm). Conductors have high conductivities (10^{3}–10^{8} S/cm). The conductivities are presented as an international system of units (S/cm = siemens/cm, S = A/V). Examples are insulators (glass, SiO_2, Diamond), semiconductors (Si, Ge, GaAs, CdS), and conductors (silver, copper, gold). Si, Ge, and Sn are called element semiconductors. A binary compound semiconductor is a combination of two elements such as GaAs and CdS. Semiconductors of ternary compounds and quaternary compounds are made of alloys of binary compounds such as $CuInSe_2$, Cu_2SnS_3, Cu_2ZnSnS_4, GaAsSb, and InGaAsSb. However, for practical purposes, silicon has become the most sought-after semiconductor.

Unlike hydrogen and helium, silicon is a solid material. The band theory of silicon is related to the crystalline nature of the material. The periodic arrangement of atoms in a crystal is called a lattice. In a crystal, an atom never strays far from a single, fixed position. For a given semiconductor, there is a unit cell that is representative of the entire lattice; by repeating the unit cell throughout the crystal, one can generate the entire lattice [4].

A silicon atom has electronic configuration $1s^2\ 2s^2\ 2p^6\ 3s^2\ 3p^2$. The 2 electrons from the first *s* orbital, 2 electrons from the second *s* orbital, and 6 electrons from the second *p* orbitals are well fixed in their orbitals and stabilized. These electrons (a total of 10) are not available for either conduction or forming bonds with any other element. The 4 electrons (2 in 3s and 2 in 3p orbitals) available in the outermost third orbital are available for interaction with other atoms. In the crystal lattice structure of silicon, a unit cell has a silicon atom surrounded by four silicon atoms, each sharing one electron. This sharing of electrons is known as covalent bonding; each electron pair constitutes a strong covalent bond. Each unit has a 3D tetrahedron structure. The same can be shown as a simplified two-dimensional bonding diagram for the tetrahedron (Figure 1.4).

Now we can conveniently develop the band theory of solid semiconductor crystals. Similar to the examples of hydrogen and helium, when two identical Si atoms are brought closer, the two atomic energy levels will split into two molecular energy levels by the interaction between the atoms. But in Si crystal, a large number of isolated atoms are brought together to form a solid and the orbits of the outer electrons of different atoms overlap and interact with each other. In such interaction, instead of two levels, a large number of separate but closely spaced levels are formed. This results in an essentially continuous band of energy.

The electrons can no longer be treated as belonging to their parent atoms. They belong to the crystal as a whole. The topmost band is called the Conduction Band and the lower one is called the Valence Band. A generalized concept is that the conduction band is devoid of electrons and the valence band is filled with bonded electrons. The energy gap between the lowest energy level of the conduction band and highest energy level of the valence band is called the Forbidden Band, or more commonly known as the Bandgap, as shown in Figure 1.5.

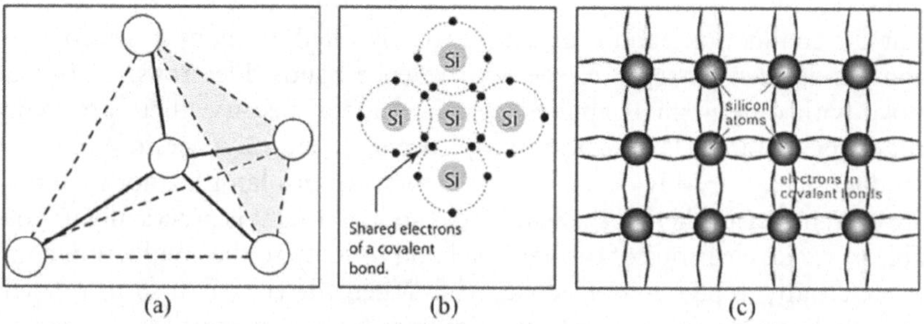

Figure 1.4 Bonding diagram for the (a) single unit of silicon 3D tetrahedron (b) two-dimensional representation of a tetrahedron bond, and (c) simplified sketch of the actual crystal structure. (Courtesy: Refer Annexure A)

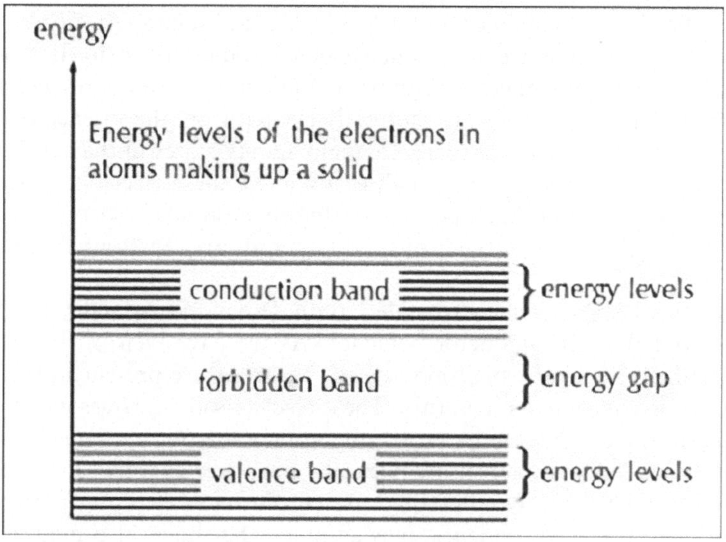

Figure 1.5 Formation of conduction band, valence band, and bandgap in a solid semiconductor. (Courtesy: Refer Annexure A)

This bandgap has very high significance while deciding the electrical nature of the solid material, such as whether it is a conductor or an insulator or a semiconductor. Conduction in solid crystal occurs when tightly held electrons in the valence band are pushed or excited to jump to the conduction band by externally applying an electric field with energy more than the bandgap energy. The empty energy levels in the conduction band allow the excited electrons to move freely throughout the crystal, enhancing conductivity. The lower the bandgap, the higher the possibility of valence band electrons moving toward the conduction band in larger numbers.

Interestingly, the bandgap in a conductor, also called metal, is so close that the conduction band is already partially filled. In many cases conduction bands even partially merge with valence bands. Electrons are free to move with only a small applied field in a metal because there are many unoccupied states close to the occupied energy states. Therefore, current conduction can readily occur in conductors. An insulator is another opposite extreme case wherein the bandgap is so high that it is practically impossible to excite electrons of the valence band to jump to the conduction band by externally applying an electric field. When the electric field is applied, so few electrons actually occupy conduction band states that the overall contribution to electrical conductivity is very small, resulting in a very high resistivity. Therefore, an insulator cannot conduct current. Insulators usually have bandgap energy of more than 3 eV.

Now is the interesting case of semiconductors. Semiconductors are reported to have bandgap energy 0.6–3.0 eV. At absolute zero-degree temperature, all electrons are in the valence band, and there are no electrons in the conduction band, making them poor conductors. At room temperature, good numbers of electrons are thermally excited from the valence band to the conduction band. Since there are many empty states in the conduction band, a small applied potential can easily move these electrons, resulting in a moderate current. Such behavior is highly desirable in electronics and solar applications. Three types of solid materials are represented in Figure 1.6.

Scientists have introduced the electron-hole concept to make it easy to understand the semiconductor behavior. As we have seen in the case of semiconductors, a very small number of electrons are present in the conduction band at room temperature. These electrons are the ones which have jumped to the conduction band, leaving vacant space in the valence band, which are called holes. Holes are naturally positively charged. The number of holes in the valence band is equal to that of excited electrons in the conduction band. For convenience, it is assumed that holes also migrate like electrons, but in the opposite direction. A semiconductor with an equal number of holes in the valence band and electrons in the conduction band is called an Intrinsic Semiconductor. It has such low conductivity that it is not suitable for practical purposes.

Intrinsic semiconductors need to be converted to extrinsic semiconductors wherein either electrons are in excess of holes or holes are in excess of electrons to attain an optimum conductivity for useful purposes. This is desirable in electronic and solar photovoltaic applications. Extrinsic properties are achieved simply by incorporating an additional atom in the intrinsic semiconductor crystal, replacing a Si atom. The additional atom may be having an extra electron compared to silicon that is donated to conduction band; hence it is called a donor element. On the other hand, if the additional atom is devoid of one electron compared to the silicon atom, it

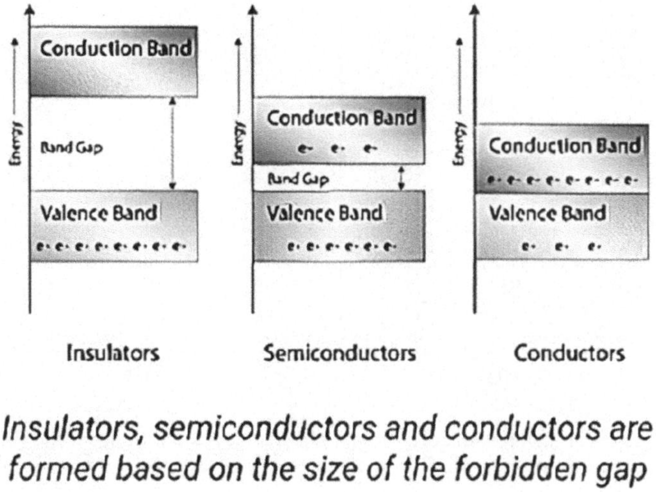

Figure 1.6 Energy band diagram of insulator, semiconductor, and conductor based on bandgap energy. (Courtesy: Refer Annexure A)

accepts an additional electron from the valence band and generates a hole there; hence, it is called an acceptor element.

Figure 1.7 shows the silicon lattice with phosphorus donor and boron acceptor. A silicon atom with atomic number 14 is replaced (or substituted) by a phosphorus atom. Phosphorus with atomic number 15 has the electronic configuration $1s^2$, $2s^2$, $2p^6$, $3s^2 3p^3$ with five valence electrons. The phosphorus atom forms covalent bonds with its four neighboring silicon atoms. The fifth electron has a relatively small binding energy to its host phosphorus atom and can be "ionized" to become a conduction electron at a moderate temperature. The silicon is called an n-type semiconductor with an additional negative charge carrier. Similarly, a silicon atom with atomic number 14 is replaced (or substituted) by a boron atom. Boron, with atomic number 5, has the electronic configuration $1s^2$, $2s^2 2p^1$ with three valence electrons. When a boron atom with three valence electrons substitutes for a silicon atom, an additional electron is "accepted" to form four covalent bonds around the boron, and a positively charged "hole" is created in the valence band. The silicon is now called a p-type semiconductor with an additional positive charge carrier.

The addition of impurities in the form of donor or acceptor atoms is in extremely small proportion e.g., 1 atom B or P in 10,00,000 Si. The n-type silicon is rich in electrons, while the p-type is rich in holes. The ability of the extrinsic Si crystal to conduct electricity is comparatively increased. The impurity atoms are imperfections and interrupt the perfect periodicity of the lattice. The impurity atoms will introduce the creation of energy

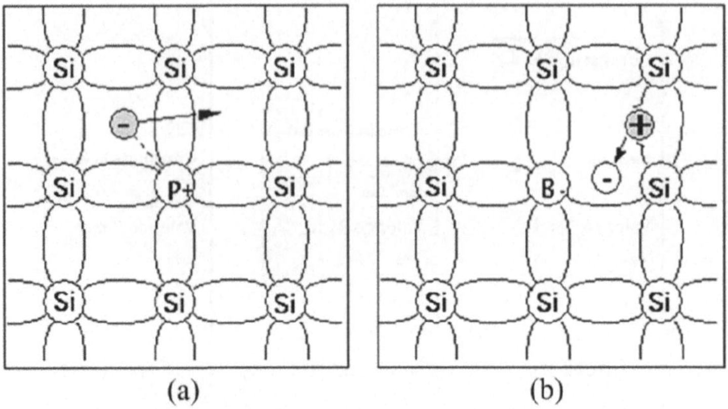

Figure 1.7 Extrinsic semiconductor with (a) phosphorus as donor atom and (b) boron as acceptor atom in the Si crystal lattice. (Courtesy: Refer Annexure A)

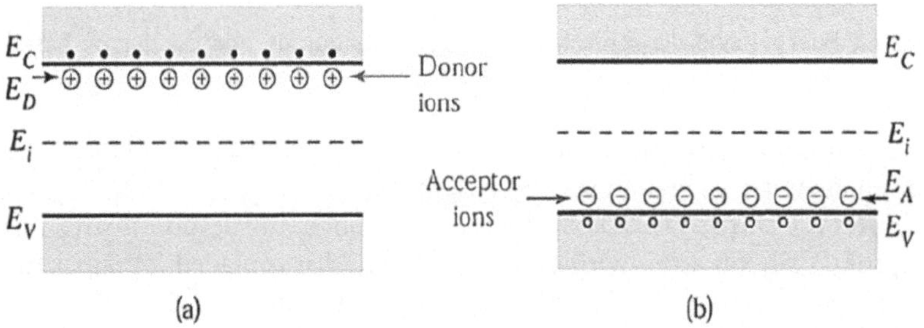

Figure 1.8 Creation of energy levels E_D (donor) and E_A (acceptor) in the bandgap due to impurity atoms. (Courtesy: Refer Annexure A)

levels E_D (donor) or E_A (acceptor) in the bandgap (Figure 1.8). For the sake of comparison, a reference level E_i termed "intrinsic level" is defined as the location of the Fermi level when the number of electrons is the same as that of holes. The Fermi level is defined as the highest energy level that an electron can occupy at the absolute zero temperature. The Fermi level lies between the valence band and the conduction band because at absolute zero temperature, the electrons are all in the lowest energy state. The Fermi level determines the probability of electron occupancy at different energy levels. The closer the Fermi level is to the conduction band energy, the easier it will be for electrons in the valence band to transition into the conduction band.

A typical Si solar cell works with a junction of p- and n-type semiconductors by depositing a very thin layer of 300 nm of n-type on 0.3 mm p-type wafer, thus creating a p-n junction. When the p- and n-type semiconductors combine, holes from the p-side diffuse into the n-side, and electrons from the n-side diffuse into the p-side. Consequently, a negative space charge forms near the p-side of the junction and a positive space charge forms near the n-side. This space charge region creates an electric field that is directed from the positive charge toward the negative charge (Figure 1.9). The width of the space charge region is 500–1000 nm. This is the p-n junction at thermal equilibrium without external bias. If we apply a positive voltage to the p-side with respect to the n-side, the p-n junction becomes forward-biased. The total electrostatic potential across the junction decreases and the forward bias reduces the depletion layer width. By contrast, if we apply positive voltage to the n-side with respect to the p-side, the p-n junction now becomes reverse-biased. The total electrostatic potential across the junction increases and reverse bias increases the depletion layer width. Needless to say, reverse biasing is favorable if we want quicker charge separation before the electron-holes recombine. That is what happens in a p-n junction solar cell.

A p-n junction solar cell schematic is represented in Figure 1.10. It consists of (a) a shallow p-n junction formed on the surface, (b) a front ohmic electrical contact strip and fingers, (c) a back ohmic contact that covers the entire back surface, and (d) an antireflection coating on the front surface. When the cell is exposed to the solar spectrum, a photon that has an energy less than the bandgap E_g makes no contribution to the cell output. A photon that has energy same as E_g contributes an energy E_g to the cell output. Energy greater than E_g is wasted as heat. When electron-hole pairs are created in the depletion layer, they are separated by the built-in electric field. If a wire is connected from the cathode (n-type silicon) to the anode (p-type

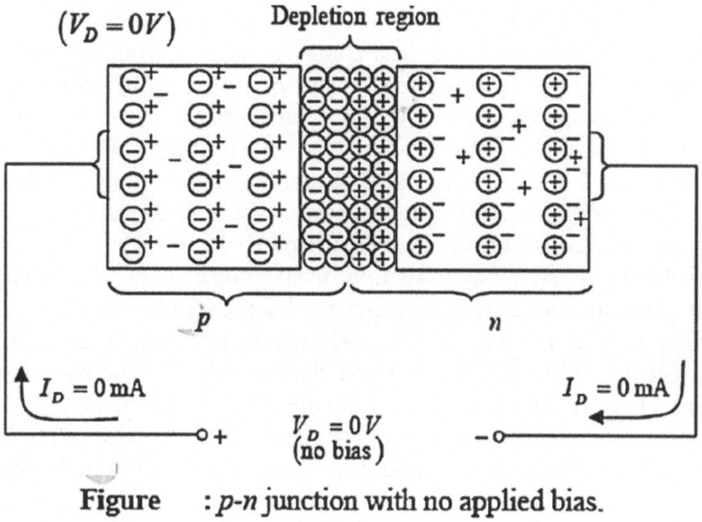

Figure 1.9 A p-n junction at thermal equilibrium with no applied bias. (Courtesy: Refer Annexure A)

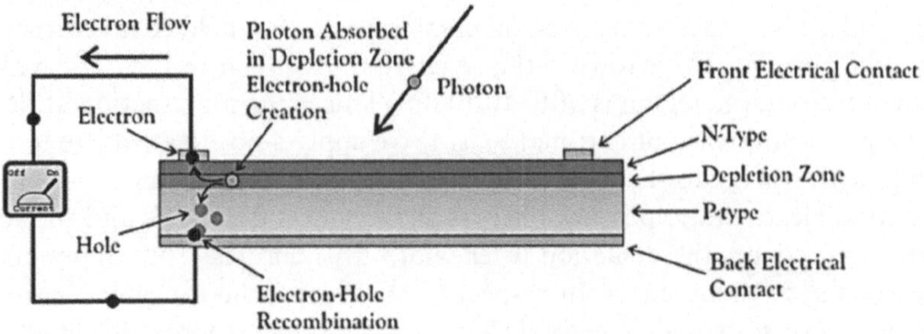

Figure 1.10 A p-n junction solar cell schematic. (Courtesy: Refer Annexure A)

silicon) electrons will flow through the wire. The electron is attracted to the positive charge of the p-type material and travels through the external load (meter), creating a flow of electric current. The hole created by the dislodged electron is attracted to the negative charge of n-type material and migrates to the back electrical contact. As the electron enters the p-type silicon from the back electrical contact, it combines with the hole, restoring the electrical neutrality.

A typical crystalline silicon solar cell is made from a p-type silicon wafer of thickness 0.3 mm, width 6" and length 6". A very thin film (300 nm) of n-type silicon is deposited on the p-type silicon wafer. Backside contact is a thin coating of silver that covers the entire back surface. The entire front surface cannot be coated since the n-type layer with a shallow p-n junction needs to be exposed to light for electron-hole pair generation. However, since we need to collect the photogenerated electrons for useful purposes, ohmic contact is necessary. The best compromise is to print very thin (0.05 mm) silver grid fingers on the front surface using the technology of photolithography. Vertical bus bars of 0.2 mm flat strips of copper or aluminum are then fixed on the grid fingers (Figure 1.11). Busbars with tapered ends have lower losses than a busbar of constant width. The fingers collect the generated electrons and deliver them to the busbars. A comparatively larger surface area of busbars allows heat to dissipate.

Under light exposure, electrons flow from p to n; hence, the n-type side is the negative terminal (cathode) and the back contact of the p-type side is the positive terminal (anode). The photocurrent is actually a reverse bias current because electrons flow toward the cathode and the holes flow to the anode. For creating a solar module, the negative terminals of bus bars on n-type Si of one cell are connected to the positive terminal of the p-type back surface paint of another cell and so on in series. Traditionally 2 busbar Si cells were manufactured in the 1980s and were replaced by 3 busbar cells in the 1990s. Thereafter, practically all solar cells have 5 busbars (Figure 1.12).

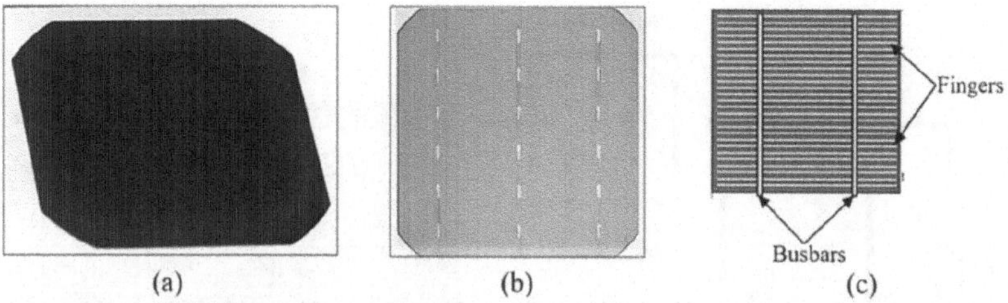

Figure 1.11 A single Si solar cell (a) 0.3 mm thick p-Si wafer, (b) entire back surface silver metal thin coating, and (c) front surface thin (0.05 mm) silver grid fingers and vertical bus bars of 0.2 mm flat strips of copper or aluminum. (Courtesy: Refer Annexure A)

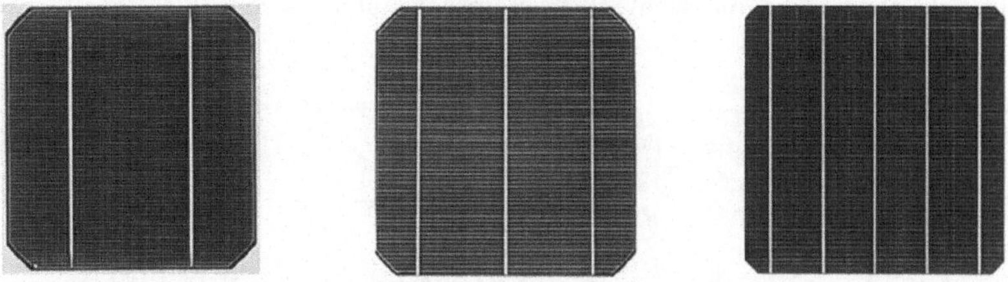

Figure 1.12 Silicon solar cells with 2, 3, and 5 busbars. (Courtesy: Refer Annexure A)

The increase in the number of busbars, although desirable for collecting maximum number of electrons, reduces the cell area, causing loss of productive energy. Therefore, having 5 bus bars is a good compromise.

1.4 CURRENT-VOLTAGE CHARACTERISTICS OF SOLAR CELL

Current-Voltage or I-V characteristics are essential for characterization of a solar cell to know its efficiency, V_{oc}, I_{sc}, and power rating. [5]. These solar cell characteristics are determined by connecting a variable load resistor to the cell and plotting the resulting currents and voltages at different loads, as indicated in Figure 1.13.

The characteristics of a solar cell without any irradiance (dark characteristics) correspond to the characteristics of a diode [6]. When the solar cell is illuminated, these characteristics shift by the amount of the photocurrent in the blocking direction (Figure 1.14).

For the sake of convenience, the I-V curve is more generally represented by Figure 1.14 (b), which is an inversion of Figure 1.14 (a) about the voltage axis. By choosing a proper load, close to 80% of the product I_{sc} and V_{oc} can be extracted, where I_{sc} is the short-circuit current and V_{oc} is the open-circuit

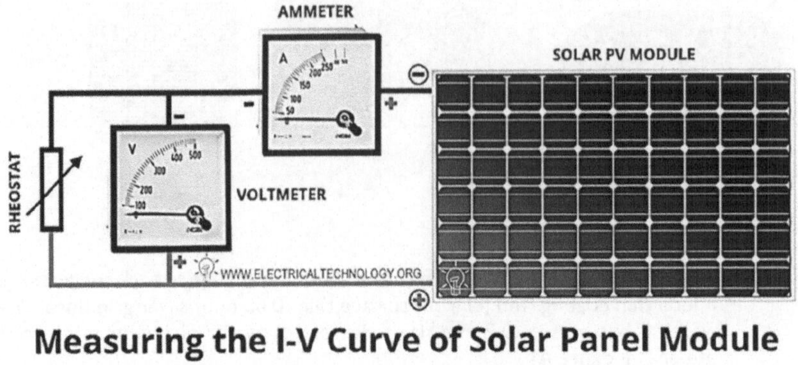

Figure 1.13 Devices to measure I-V characteristics and efficiency. (Courtesy: Refer Annexure A)

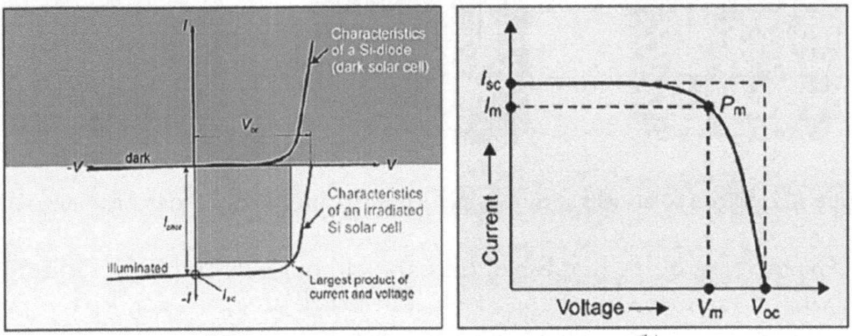

(a) (b)

Figure 1.14 I-V characteristics of solar cells in (a) dark and under illumination and (b) its inverted representation. (Courtesy: Refer Annexure A)

voltage of the cell. The theoretically attainable (optimal) power P_{op}, which could be taken from the terminal, is the product of short-circuit current I_{sc} and open-circuit voltage V_{oc}:

$$P_{op} = I_{sc} \times V_{oc}$$

The attainable power P_{mp} is defined by the greatest possible product of V and I at an operating point:

$$P_{mp} = I_{mp} \times V_{mp}$$

The ratio of P_{mp} to P_{op} is called the fill factor or FF. It describes the "rectangular-ness" of I-V characteristic. For a good photovoltaic cell, the FF of a minimum of 0.7 is desired.

The photovoltaic conversion efficiency is defined by the ratio of the PV electrical power output to the irradiated power on a solar cell. This depends on irradiance and spectrum. The conversion efficiency is determined under standard test conditions (STC): an irradiance of 1,000 W/m^2 perpendicular onto the front surface, a cell temperature of 25 °C and a spectral distribution according to solar irradiance passing at an elevation angle of 41.8° through the atmosphere (air mass 1.5). For example, let us assume a solar panel of 600 W with size 2.4 m × 1.3 m = 3.12 m^2. The solar radiation incident on the solar panel at STC will be 3.12 m^2 × 1,000 W/m^2 = 3120 W. The module's efficiency will be power output divided by solar power received in the solar cell area i.e. (600 W/3120 W) × 100 = 19.23%.

1.5 EFFICIENCY LIMITATION

High values of I_{mp} and V_{mp} are desirable to get maximum P_{mp}. However, solar cells suffer from many loss mechanisms. Around 31% of incident light is reflected and is not available for conversion to electrical energy in a silicon solar cell. Photons with less energy than the bandgap E_g make no contribution to the cell output. Photons with energy equivalent to the bandgap E_g contribute to the cell output. Photons with energy higher than the bandgap E_g transfer the surplus energy as lattice vibrations, i.e., as heat. Silicon is a so-called indirect bandgap semiconductor that makes the absorption of a photon dependent on the occurrence of a phonon (lattice vibration); hence its absorption coefficient is comparatively low. This is not the case with direct bandgap solar cells such as GaAs. Details of direct and indirect bandgaps are beyond the scope of this topic. In addition, shadowing losses caused by the front contacts, losses by non-absorbed (transmitted) irradiance, ohmic losses from series resistors, recombination losses, and additional diode losses contribute to lowering the efficiency. The practical efficiency of Si crystalline solar cells has been obtained at a maximum of 21%.

Efforts are on to improve efficiency. However, the theoretical limit of efficiency of semiconductors based on their bandgaps has been calculated and is known as the Shockley–Queisser limit [7], which is the plot of bandgap vs. maximum efficiency (Figure 1.15). Of all the power contained in sunlight (about 1000 W/m^2) falling on an ideal single-junction solar cell, only 33.7% of that could ever be turned into electricity. This corresponds to the bandgap of 1.34 eV. The most popular solar cell material, silicon, has a less favorable bandgap of 1.1 eV, resulting in a maximum efficiency of about 32%. Modern commercial monocrystalline solar cells produce a maximum

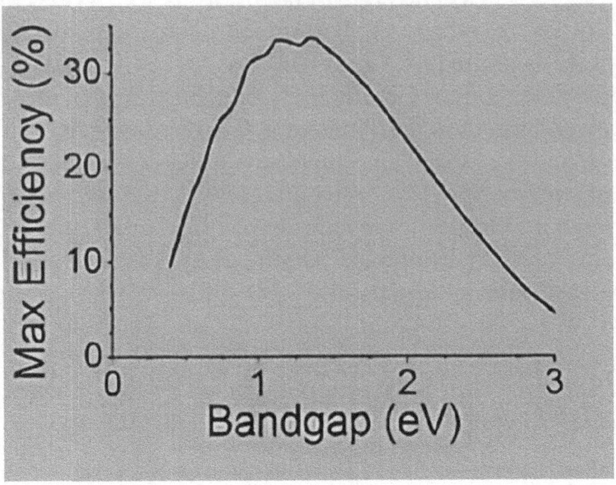

Figure 1.15 Shockley–Queisser theoretical efficiency limit single p-n junction solar cells. (Courtesy: Refer Annexure A)

of about 24% conversion efficiency. This theoretical limit is for single p-n junction solar cells. Multi-junction solar cells may have a theoretical limit of up to 80%.

1.6 SILICON MATERIAL PREDOMINANCE

As stated earlier, an ideal solar cell material must be a semiconductor in the solid state, responsive to the visible range of the solar spectrum, be stable under outdoor use with no or minimum annual degradation, be abundant in nature, and have an affordable cost and a suitable bandgap. We can see some semiconductor materials in Table 1.1.

As per table, GaAs, CdTe, and InP are the ideal materials since their band gaps are very close to the ideal bandgap of 1.34 eV. Si hardly comes closer to the ideal bandgap value. Still Si is the most used material in the world today for solar cells. GaAs could have been the best material if it could be affordable. It is stable in outdoor conditions, has an ideal bandgap, has a direct bandgap, has the least annual degradation, and has high efficiency. That is why it is being used mostly in space applications where cost is a secondary consideration. CdTe is again a costly rare earth material. All other qualities match GaAs. It must be used in thin-film form to save material and to reduce cost. But thin films have their own limitations such as lower efficiency and higher degradation rates [8]. The most critical fact is that Cd is highly poisonous and needs to be avoided, considering the potential health hazards during manufacturing and after-life disposal. InP is not a stable

Table 1.1 Bandgaps of some solar cell materials

S.N.	*Semiconductor Material*	*Chemical Formula*	*Bandgap (eV)*
1	Silicon	Si	1.11
2	Aluminum Antimonide	AlSb	1.6
3	Gallium Phosphide	GaP	2.26
4	Gallium Arsenide	GaAs	1.43
5	Gallium Sulfide	GaS	2.5
6	Indium Phosphide	InP	1.35
7	Cadmium Sulfide	CdS	2.42
8	Cadmium Selenide	CdSe	1.73
9	Cadmium Telluride	CdTe	1.49
10	Copper Oxide	Cu_2O	2.17

material in outdoor conditions and is yet to be produced commercially. Silicon has become a practical semiconductor material for electronics and solar cells. The main reasons for the use of silicon in vogue are that it exhibits better properties at room temperature, and high-quality or high-purity silicon can be grown thermally with ease. A device-grade silicon costs much less than any other semiconductor material. Silicon in the form of silica and silicates comprises 25% of the Earth's crust, and silicon is second only to oxygen in abundance. Even with its non-ideal bandgap and low absorption coefficient, crystalline silicon is the most sought-after material, mainly due to its stability under outdoor rugged conditions. Its sustained performance till the useful life of 25–30 years has been proven due to a comparatively low degradation rate (0.3%–0.4% per annum).

1.7 EFFECT OF TEMPERATURE AND RADIATION

Open-circuit voltage practically remains constant (very little increase) with an increase in radiation level. The short-circuit current is directly proportional to the irradiance because the current is equivalent to the number of electron-hole pairs generated by the absorbed photons. Consequently, the output power of a PV module increases with an increase in irradiance (Figure 1.16).

An increase in cell temperature at constant radiation causes a reduction of the open-circuit voltage while short-circuit current remains practically constant (Figure 1.17). A solar cell with a V_{oc} of 0.6 V at 25°C reaches a value of 0.45 V at 75°C. This is a considerable reduction, which can be about 25% in practice. The short-circuit current increases with increasing temperatures at a rate of about 0.1% for each degree centigrade rise in temperature. A solar cell with a short-circuit current of 2.0 A at 25 ^{0}C

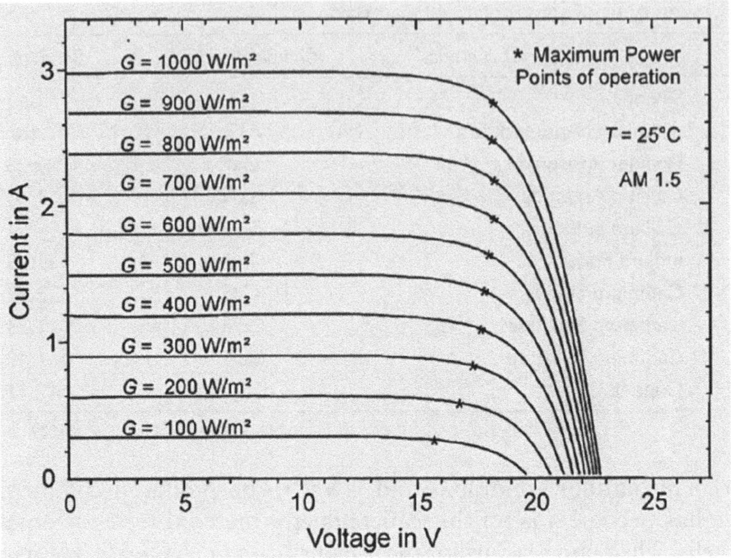

Figure 1.16 Current-voltage characteristics of a multi-crystalline silicon PV module at different radiation levels. (Courtesy: Refer Annexure A)

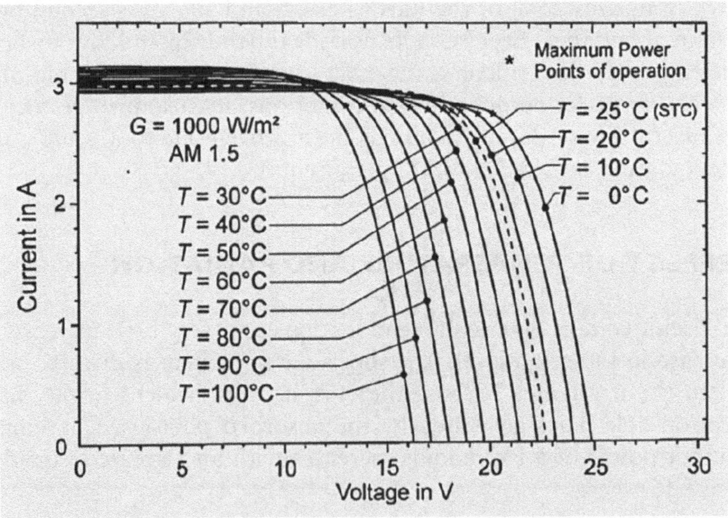

Figure 1.17 Current-voltage characteristics of a multi-crystalline silicon PV module at different temperatures. (Courtesy: Refer Annexure A)

reaches a value of 2.1 A at 75°C. This means an increase of 5%. Due to the combined effect of current and voltage, the output power decreases by -0.4%/°C to -0.5%/°C for crystalline silicon solar cells, while the values for thin-film solar cells are –0.25%/°C to –0.35%/°C. That is why output of solar power projects have been reported to suffer loss of 10%–20% in summer and enjoy 10%–20% gain in winter. This is significant for large solar power projects.

1.8 TYPES OF SOLAR CELLS

Silicon is the most abundant element available on the Earth, and the earliest solar cell was made of crystalline silicon. Today, silicon is used in monocrystalline, polycrystalline and amorphous forms for fabrication of solar cells (Figure 1.18). The majority of the commercial cells use monocrystalline silicon. The polycrystalline and amorphous types of silicon are a result of the engineering of silicon material where lattice perfection is sacrificed to achieve a significant reduction in the cost of silicon technology for solar cell applications. Many other semiconducting compounds have also been investigated [9]. Solar cells based on gallium arsenide, cadmium telluride, and copper indium di-selenide are now commercially available. Thus, based on the formation of material, solar cells are grouped into three major categories: a) crystalline, b) thin film, and c) emerging technologies. Each type of solar cell is briefly described below.

1.8.1 Crystalline solar cells

Under the crystalline category, there are two types of technologies, namely, monocrystalline and polycrystalline technology.

1.8.1.1 Monocrystalline silicon solar cells

Most of the solar cells manufactured all over the world today are fabricated using monocrystalline silicon chips or wafers as the base. The input

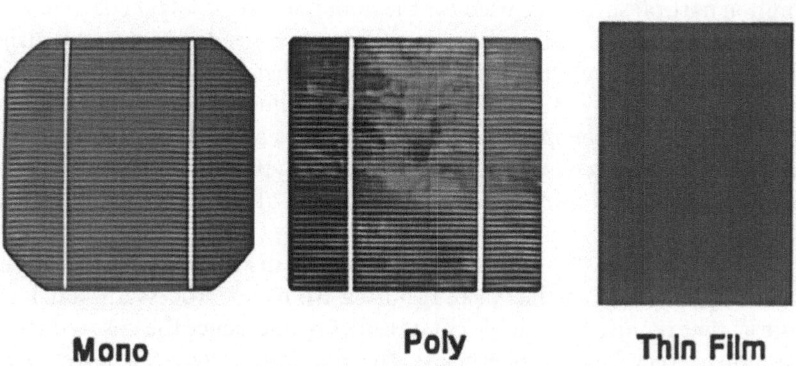

Figure 1.18 Mono-crystalline, polycrystalline, and thin-film solar cells. (Courtesy: Refer Annexure A)

material is SiO_2, available either in the form of quartz sand or as natural crystalline quartz converted into metallurgical grade silicon in a furnace through a reduction process using coal. Metallurgical silicon, after purification, is converted into rods of polycrystalline nature, i.e., many small crystals ordered close to each other. These polycrystalline rods are melted in a crucible to produce pure silicon ingots of monocrystalline nature, i.e., completely structured single crystal lattice, prepared by the crystal pulling method. The principle of the Czocharalski process is predominantly used for single crystal growth. In this process, a seed of single crystal silicon contacts with the melt of high-purity silicon, and as the seed is slowly raised, the atoms of the molten silicon adapt to the pattern of the single crystal seed as it cools and solidifies into a single crystal structure. This rod can be up to 2 m long. It is made of its own crystals and therefore the name "monocrystalline". From this single-crystalline ingot, wafers are sliced and textured to improve solar energy absorption. These wafers are then fabricated into p-n junctions by high temperature diffusion of dopants, mainly boron and phosphorous, that modify the surface layer composition.

Monocrystalline silicon solar cells have proven their credibility as a source of reliable electric power both on land and in space. Practical efficiencies in the range of 17–21% have been measured for the monocrystalline silicon cells. Recently on December 12, 2022, a Chinese solar cell company JinkoSolar has announced a new record, achieving a maximum solar conversion efficiency of 26.4% for its 182 mm and above large-size monocrystalline silicon TOPCon solar cell. This result has been independently confirmed by the National Institute of Metrology, China (https://www.jinkosolar.com/en/site/newsdetail/1827). The lab research product is soon expected to come into the market for commercial applications.

1.8.1.2 Polycrystalline silicon solar cells

After monocrystalline silicon, polycrystalline silicon is the second most common natural substance used for the manufacture of solar cells. For solar cells, one does not require silicon as pure as one needs for manufacturing semiconductors for electronics applications. Therefore, another approach to silicon technology is to prepare polycrystalline silicon blocks with no lattice perfection. The manufacturing process is simpler and cheaper. The most popular commercial process is the "casting process" wherein the molten silicon is poured into rectangular mold and allowed to solidify into an ingot.

The ingot is then sliced into wafers. This means that the process of cutting and polishing and the waste resulting from this process are much the same as that required for single crystal silicon, and hence the cost reduction may not be very large. Nonetheless, the poly technology has lowered the costs of PV technology because the "casting" process is relatively cheaper and less sophisticated compared to the Czochralski process. Due to the presence of structural imperfections, mostly grain boundaries, the efficiencies are slightly lower, around 15%–18%.

Both monocrystalline and polycrystalline silicon solar cell wafers are available in the world market with thicknesses of 200–300 μm and sizes 150 mm × 150 mm to 210 mm × 210 mm.

1.8.2 Thin-film solar cells (TFSC)

A thin film is a material created ab initio by the random nucleation and growth processes of individually condensing/reacting atomic/ionic/molecular species on a substrate. The structural, chemical, metallurgical, and physical properties of such a material are strongly dependent on a large number of deposition parameters and may also be thickness dependent [10].

Pure crystalline silicon has been conventionally an expensive material. So, continuous efforts have been made to produce cells with very little quantity of solar cell material. This type of cell technology is usually known as thin-film technology. Thin-film modules are made by depositing very thin layers of photosensitive materials onto a low-cost backing. Typical inexpensive substrates used for the purpose are made of glass, stainless steel, or even plastic. Three types of thin-film modules are commercially available at present.

1. Amorphous silicon
2. Cadmium telluride
3. Copper indium di-selenide

1.8.3 Amorphous silicon

In amorphous Si (a-Si), the atoms are arranged in a haphazard manner. Before 1975, it was the usual opinion that amorphous silicon could not be used to produce solar cells. Later, it became possible for the first time to dope a-Si, in which an alloy of silicon and hydrogen from the gaseous form of silane could be separated. This product was named a-Si:H (hydrogenated amorphous silicon). It could be doped during the process of separation. There are several methods today which can be used to deposit a-Si layers on a base substance. Already in the laboratory, efficiencies of about 12.5%, which are still very much lower than those of monocrystalline silicon PV, have been obtained.

Amorphous silicon is widely accepted as a thin-film solar cell material because: (a) it is abundant and non-toxic; (b) it requires low process temperature, enabling module production on flexible and low-cost substrates; (c) the technological capability for large-area deposition exists; (d) its very thin film (= 1 μm) has low material requirements, due to the inherent high absorption coefficient compared with crystalline silicon; (e) it has a larger bandgap (which gives higher open-circuit voltage) (f) it exhibits low energy consumption during manufacture; and (g) there is a possibility of automation of the manufacturing process.

The main disadvantages are the lower efficiency and faster degradation due to higher internal resistance and therefore a smaller photon current. Drastic cost reduction in more efficient crystalline silicon has made this technology practically obsolete.

1.8.4 Cadmium telluride solar cell

Cadmium telluride is an excellent solar cell material with the highest theoretical conversion efficiency. Several preparation techniques such as vacuum evaporation, spraying, screen-printing and electro-deposition have been initially used to produce good solar cells. Owing to its optoelectronic and chemical properties, CdTe is an ideal absorber material for high efficiency and low-cost thin-film polycrystalline solar cells. CdTe is a direct bandgap material with an energy gap of 1.44 eV and an absorption coefficient of around 10^5/cm in the visible region, which means that a layer thickness of a few micrometers is sufficient to absorb ~90% of the incident photons.

Lab solar cells' efficiency is up to 16%, while commercial cells' efficiency of up to 6%–10% has been reported by a few manufacturers. The cells are relatively stable, although humidity affects them to some extent. Spraying and screen-printing are techniques with high economic potential. The toxicity of cadmium raises two main problems: the possibility of health hazards during production and after-life environmental pollution. In spite of these demerits, there has been significant progress in developing low-cost manufacturing processes for rapid commercialization of CdTe cells. One problem with CdTe is that p-type CdTe films tend to be highly resistive electrically, which leads to large internal resistance losses.

1.8.5 Copper indium di-selenide

$CuInSe_2$, having a bandgap of 1.53 eV, is considered an ideal material for photovoltaic application. The difficulties in controlling the Sulfur during deposition and the relatively rapid diffusion of metals and impurity species, even at low temperatures, slow down the development of this material. However, devices with an efficiency of 11.4% have been reported.

An increase in the bandgap and improved process conditions resulted in the fabrication of high-performance solar cells with efficiencies of 19.2% for small areas and 13.1% for large areas. Even though the efficiency and stability of the device are very promising, there are several factors that are less favorable for large-scale production of such devices. The increasing number of alloy components makes the multiple processes extremely complex, and thus intelligent processes are required for precise control of the composition during deposition. The use of expensive and rare metals such as In and Ga adds to the cost of manufacturing.

1.8.6 Copper indium gallium selenide solar cell

The beauty of this type of solar cell is that it is "polycrystalline" thin film, unlike most thin films of amorphous nature. Another advantage is that it is deposited on a large-area metal substrate, thus avoiding the use of delicate glass substrate. It has a bandgap 1.5–1.7 eV, ideally suited for solar cells. High efficiency (17–20%) with strong absorption has been observed. Even though it uses expensive rare metals, it is able to be partially commercialized. Higher cost compared to Si solar cells is the main deterrent for full commercialization. However, due to its many advantages, it is expected to achieve commercialization soon.

1.9 EMERGING TECHNOLOGIES

In addition to the above solar cells, which are either commercialized or on the verge of commercialization, there are other materials attracting large-scale attention from researchers with the potential for commercialization in the future. Some of them are explained below.

1.9.1 Gallium arsenide

Mostly used in space applications due to high cost, gallium arsenide (GaAs) and its variants gallium aluminum arsenide (GaAlAs) and gallium indium arsenide phosphide (GsInAsP) are the most efficient solar cell materials reported till today. These cells are generally combined in multiple junctions to achieve high efficiencies. These materials are highly suited to multiple bandgap cell designs because the band gaps are adjustable by changing the relative compositions of the components. Cell efficiencies of about 30%–34% are obtained for these structures, which are extremely high, though they are too expensive to be used for terrestrial applications.

1.9.2 Organic semiconductors

Organic materials are attractive for photovoltaics primarily because of the prospect of high throughput manufacture using processes such as reel-to-reel deposition. Additional attractive features are the possibilities for ultra-thin, flexible devices, which may be integrated into appliances or building materials and tuning of color through chemical structure. This field of research has made impressive progress since the late 1990s. Solar power conversion efficiencies of over 3% have been reported. A growing range of new photovoltaic materials have been studied, and increasing numbers of research groups and companies have declared an interest in "soft" solar cells.

Organic semiconductors can be classified into three categories, depending on their chemical properties, as insoluble, soluble, and liquid crystalline. They can be further classified as monomers, such as dyes, pigments, and polymers. Doping of organic semiconductors can be done by introducing foreign atoms or molecules, or by electrochemical oxidation/reduction processes. Organic solar cells have a stability problem that is common to conjugated polymers. However, these may not be very serious problems and may be overcome in the near future.

1.9.3 Dye-sensitized cells

The dye-sensitized cells are also considered thin-film cells. The principle of working is based on photosensitization of wide-bandgap semiconductors. A wide-gap semiconductor with a large surface area is covered with dye molecules. When the light is incident, the light is absorbed in the dye molecules, which are excited, and the electrons from their excited state are directly injected into the semiconductor, without the need for transport of photo generated carriers within the dye. The ground state of the molecules has to be filled again so that the process goes on. Since the technology does not require high-purity semiconductors, these types of cells are highly promising. In Germany, the company INAP GmbH has been working on the development of dye-sensitized TiO_2 cells, and efficiencies of 7% on 30 cm × 30 cm areas have been reached. There has been a lot of research activity and new photosensitizing chemicals are being developed. This is considered a potential and low-cost PV technology.

1.9.4 Nanotechnology solar cells

The basic concept is that tiny nano-rods prompt "spectrum modification". As we know, conventional solar cells absorb photons with energy equivalent to E_g. Photons energy less than E_g does not get absorbed at all. Photons with energy higher than E_g are only partially absorbed and most of them go to waste as heat. Scientists have found that nanostructures, e.g., quantum dots, luminescent dye molecules, and lanthanide-doped glasses, absorb photons at a certain wavelength and emit photons at a different wavelength, thus "squeezing" the wide solar spectrum (300–2500 nm) to a single small band spectrum. Such a quasi-monochromatic solar cell could "in principle" reach efficiencies over 80% (Wonderful concept!). Materials only with rare earth metals are used, e.g., Pr^{3+}, Yb^{3+}, Gd^{3+}, Nd^{3+}, Er^{3+}, Tb^{3+}, Ce^{3+}. Major advantages are: flexibility, lower costs, clean manufacturing, achievable efficiency of up to 65%, not requiring complicated equipment, low manufacturing cost, light weight, and transparent solar modules. They may be useful for military applications, remote areas, and windows. However, research is still in its infancy.

1.9.5 Perovskite solar cell

Any material with a crystal structure like calcium titanium oxide (CaTiO3) is called Perovskite. It is named after the Russian mineralogist L. A. Perovski. Perovskite belongs to a general chemical formula: ABX_3 – A and B cations and an X anion that bonds to both. A Perovskite solar cell is a Perovskite structured compound, hybrid organic-inorganic lead halide – as the light-harvesting active layer. It has emerged from the field of dye-sensitized solar cells. The most common materials are methylammonium lead trihalide $CH_3NH_3PbX_3$ (X=I, Br and/or Cl) with an ideal bandgap of 1.5-2.3 eV. It is reported to have excellent light absorption properties. It is possible to have low cost, high efficiency, thin, lightweight, and flexible solar modules (industry-scalable). As of today, a maximum of 27.3% conversion efficiency has been reported. The main reason for its non-use on a large scale is the doubt about its long-term stability, posing the biggest challenge to researchers.

1.9.6 Bio-nano solar cell

Normally, photosynthesis stops if a plant leaf is plucked. More than 40 years ago, scientists discovered that one of the proteins involved in photosynthesis (PS1) was able to function even after it was extracted from plants. PS1 impressively converts sunlight into electrical energy with nearly 100% efficiency compared to a maximum of 30% of silicon solar cells. Spinach is very rich in PS1 and hence is useful for bio-solar cells. Scientists at MIT, USA, extracted PS1 proteins from spinach and placed ~2 billion PS1 on a piece of glass in an artificial cell membrane. Then fixed a layer of proteins between layers of semiconductors and exposed to sunlight to produce electric current. This itself opened another door for a new type of solar cell, although it produced much less current (1 mA/cm^2) vs. Si (30 mA/cm^2). It has a lifespan of just weeks or months vs. Si (25 years).

1.9.7 Hot solar cells

Based on the principle of Thermophotovoltaics (Heat → Light → Electricity), sunlight is concentrated on an efficient absorber (e.g., Graphite) to achieve a temperature of 1000°C. The connected thermal emitter (photonic crystal), e.g., Tungsten or SiC converts heat into light which are mostly low energy light useful for lower bandgap solar cells. An optical filter transmits only desired light to "narrow band gap" PV solar cells, e.g., GaSb (0.72 eV), Ge (0.66 eV), and InAs (0.33 eV). An added advantage is that unwanted light is reflected back to the absorber (heat recycling). Low bandgap semiconductors, which are otherwise useless, become useful solar cells. In addition, if heat is stored, it can give light even at night. Alternatively, it can have dual use as CHP (Combined Heat and Power). Presently, such solar cells have low efficiency and high cost and are still under development.

1.10 SUMMARY OF DIFFERENT TYPES OF SOLAR CELLS

After having an insight into different types of solar cells, it is imperative to summarize them to give an overview and to select a practically applicable choice for field applications for specific applications. The summary presented in Table 1.2 also provides the platform for researchers to explore emerging fields for future commercialization. The present market status of different solar technologies is given ahead in Section 1.12.

1.11 SOLAR MODULE AND RATED POWER

A single silicon solar cell typically has a size of 6" × 6". Each crystalline silicon solar cell has V_{oc} 0.6–0.7 V and I_{sc} 35–40 mA/cm^2 at Standard Test Condition (STC). STC is explained below separately. The V_{oc} and I_{sc} of common single solar cells are given in Table 1.3.

Assembly of the solar module is done by simply connecting the top metal contact of one cell to the bottom metal contact of the other cell in series (Figure 1.19). Initially, modules with 36 cells were manufactured (Figure 1.20). Now we have modules in the market with 60, 72, and 120 solar cells. One can roughly estimate the total V_{oc} of full modules by just multiplying the voltages and I_{sc} by multiplying the area of total cells in modules by per cell I_{sc}.

While manufacturing, the desired number of cells connected in series are laminated with a mechanical scheme of Glass–Encapsulant–Solar cells–Encapsulant–Back sheet (Figure 1.21). The glass sheets are made of iron free and thermal tempered glass at a thickness of 2 to 4 mm, to achieve a high optical transmission. Encapsulant is a 0.5–0.7 mm thick layer of ethylene vinyl acetate (EVA). EVA is UV resistant to avoid "yellowing" or "browning" during long-term solar light exposure. The back sheet is a 0.5 mm thick white DuPont Tedlar polyvinyl fluoride (PVF) sheet. The lamination process is carried out at 150°C in a vacuum laminator. The EVA "cures" at that temperature and makes the lamination process non-reversible. The vacuum avoids air bubbles inside the laminate. Finally, the whole laminated

Table 1.2 A summary of different types of solar cells

S. N.	*Solar Cell Type*	*Bandgap (eV)*	*Efficiency (%)*	*Degradation (%/yr)*	*Mature*	*Commercial*
Crystalline Solar Cells						
1	Mono-crystalline	1.1	18–22	0.3–0.4	Yes	Yes (Dominant)
2	Polycrystalline	1.1	17–19	0.5–1.0	Yes	Yes
Thin-Film Solar Cells						
1	Amorphous silicon	1.7	06–09	2.0-3.0	No	Yes. Unstable
2	Cadmium telluride	1.44	09–13	1.5-2.0	Yes	Yes. But Cd Toxic
3	Copper indium di-selenide	1.53	10–11	No Data	No	No. High Cost
4	Copper indium gallium selenide	1.5–1.7	17–20	No Data	Yes	No. High Cost
Emerging Technologies						
1	Gallium arsenide	1.43	30–34	Very Stable	Yes	No. Very High Cost
2	Organic solar cell	No Data	05–10	Unstable	No	Under Research
3	Dye-sensitized solar cells	No Data	05–15	Unstable	No	Under Research
4	Perovskite solar cells	1.5–2.3	25–27	Unstable	No	No. Good Potential
5	Nano technology solar cells	No Data	60%	Unstable	No	No. Great Potential
6	Bio-Nano solar cells	No Data	No Data	Very Unstable	No	No. Under Research

Table 1.3 V_{oc} and I_{sc} of common single solar cells

S. N.	*Solar Cell Material*	*Typical Single Cell Voc*	*Typical Single Cell Isc*
1	Crystalline and Polycrystalline Silicon (x-Si)	~0.6 Volts	~35 mA/cm^2
2	Gallium Arsenide	~1.0 Volts	~27 mA/cm^2
3	Amorphous Si (a-Si)	~0.9 Volts	~15 mA/cm^2
4	Tandem a-Si	~1.8 Volts	~10 mA/cm^2
5	Copper Indium Di-selenide	~0.4 Volts	~35 mA/cm^2
6	Cadmium Sulfide, Cadmium Telluride	~0.7 Volts	~25 mA/cm^2

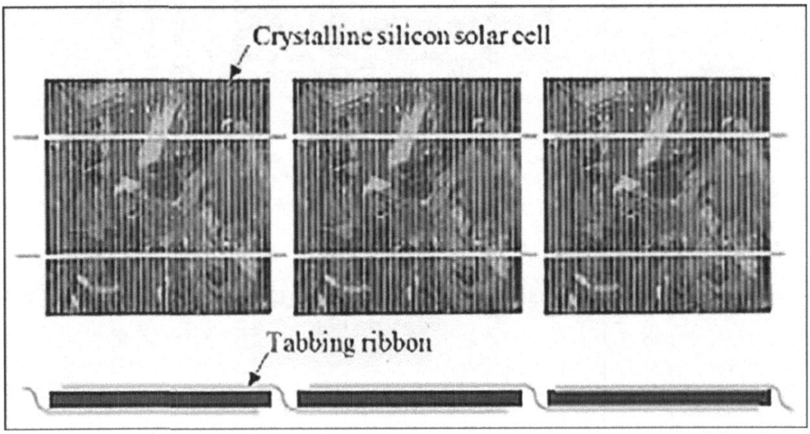

Figure 1.19 Connection of one solar cell to another in series. (Courtesy: Refer Annexure A)

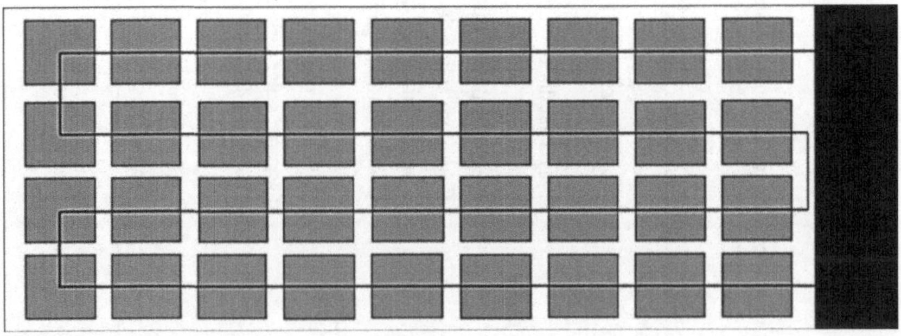

Figure 1.20 PV module with 36 cells interconnected to form a series string. (Courtesy: Refer Annexure A)

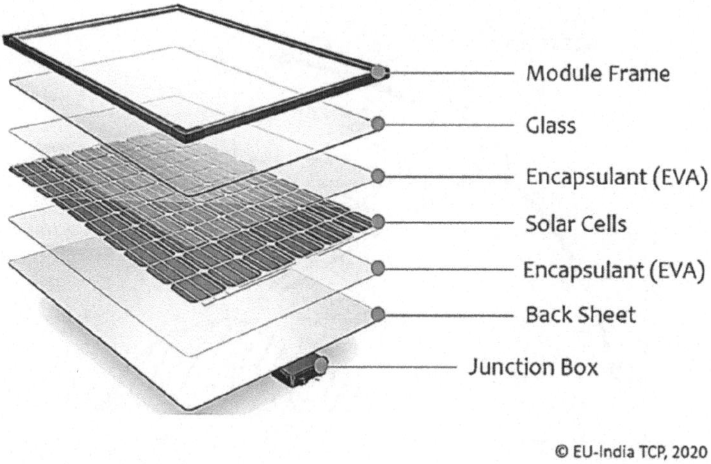

Figure 1.21 Different layers of a solar module. (Courtesy: Refer Annexure A)

set up is framed with an aluminum frame and hermetically sealed using silicon rubber suitable for long-term outdoor use.

The nameplate power of a solar module is the P_{mp} measured under Standard Test Condition (STC). STC is defined as the test condition of ambient temperature 25°C, irradiance 1000 W/m² (perpendicular) and Sun spectrum equivalent to Air Mass 1.5, or AM 1.5. AM 1.5 is defined as a spectral distribution according to solar irradiance passing at an elevation angle of 41.8° through the atmosphere. The maximum sunlight intensity occurs when the Sun is straight overhead (i.e., AM 1.0). Solar spectrum outside the Earth's atmosphere is the air mass zero condition (AM 0). All these positions of the Sun are shown in Figure 1.22.

There is a convention to represent the rating of a solar module in Wp, i.e., Watt peak. A five-hundred-watt module is called 500 Wp module. Solar modules have to keep working usefully for a minimum of 25 years under rough outdoor conditions. Hence it should also fulfill many other test conditions, which are called an international standard, well known as IEC standard (The International Electrotechnical Commission standard). The IEC is a world organization based in Geneva to develop international standards for electrical and electronic items. In addition, each country has its own standard such as IS for India, EN for European countries, BDS for Bangladesh, etc. The latest IEC standards for solar modules (2022) are:

- IEC 61215-2:2021 Terrestrial photovoltaic (PV) modules – Design qualification and type approval – Part 2: Test procedures
- IEC 61701:2020 Photovoltaic (PV) modules – Salt mist corrosion testing

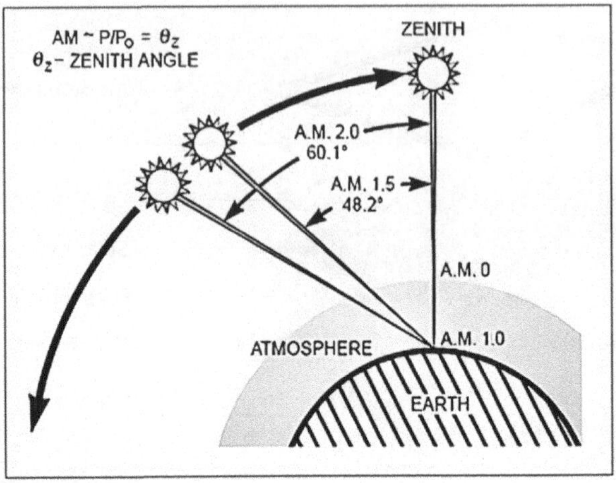

Figure 1.22 The concept of Air Mass. (Courtesy: Refer Annexure A)

- IEC 61730-1:2016 Photovoltaic (PV) module safety qualification
- IEC 62716:2013 Photovoltaic (PV) modules – Ammonia corrosion testing
- IEC 60068-2-68 Environmental testing – Part 2-68: Tests –Test L: Dust and sand
- IEC TS 62804-1-1:2020 Photovoltaic (PV) modules – Test methods for the detection of potential-induced degradation – Part 1-1: Crystalline silicon - Delamination
- IEC TS 62804-2:2022 Photovoltaic (PV) modules – Test methods for the detection of potential-induced degradation – Part 2: Thin-film

1.12 SPV TECHNOLOGY LANDSCAPE

Monocrystalline Si modules dominated the solar module market (market size itself was very small) from 1980 to 1995 [11]. Thereafter, polycrystalline Si modules percentage started growing due to cheaper manufacturing technology. There was only a little compromise in efficiency of polycrystalline compared to that of monocrystalline Si. Manufacturing procedure of monocrystalline was still costly. A major portion of market was captured by polycrystalline till 2016 untill the latest industry research enabled the cost reduction of manufacturing of monocrystalline Si due to latest technologies. Everybody was aware that monocrystalline is preferable anytime due to its lowest degradation rate, longer life, and better outdoor stability. Cost reduction further prompted the market to prefer it for commercial applications. By 2022, out of total PV production, monocrystalline Si is the most dominant in market, with a share of 120.6 GWp, while polycrystalline Si is only 23.3 GWp, and expectedly, thin film is only 7.7 GWp (Figure 1.23). Out of 7.7 GWp of thin-film solar modules, CdTe has a share of 6.1 GWp, CIGS 1.5 GWp and amorphous Si 0.2 GWp (Figure 1.24).

As regards installation of PV projects in the world, according to IRENA – International Renewable Energy Agency website (https://www.irena.org/) – the trend is shown in Figure 1.25, Figure 1.26 and Figure 1.27. The installed capacity of solar photovoltaic power projects has a share of 27.7% globally, which was just around 5% a decade ago. It shows the success story of the 100 GW to 1000 GW path from 2012 to 2022.

This data shows that the world is now nearing 1 TW installed capacity. This was unimaginable just a decade ago. Needless to say, that China is far far ahead in terms of solar PV installed capacity. After all, they have been the main trendsetter for the drastic reduction in the cost all over the world.

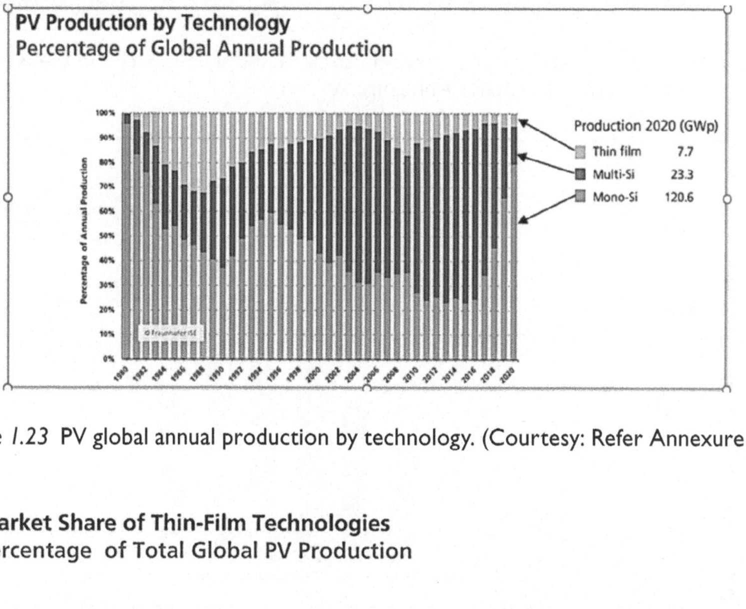

Figure 1.23 PV global annual production by technology. (Courtesy: Refer Annexure A)

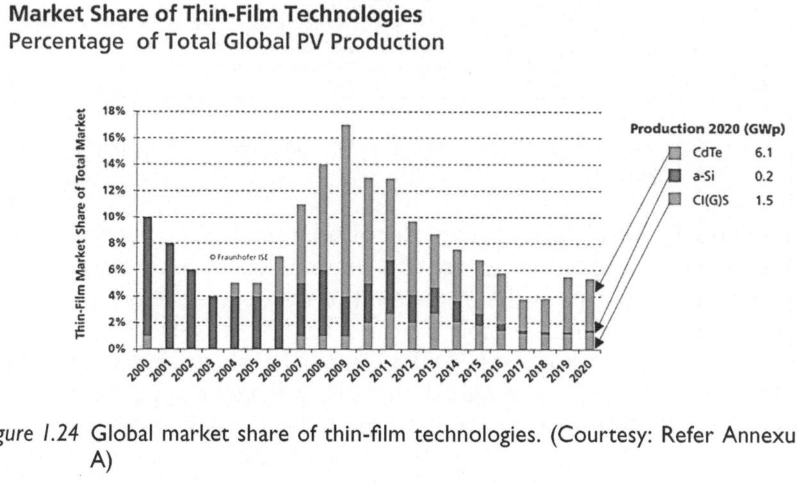

Figure 1.24 Global market share of thin-film technologies. (Courtesy: Refer Annexure A)

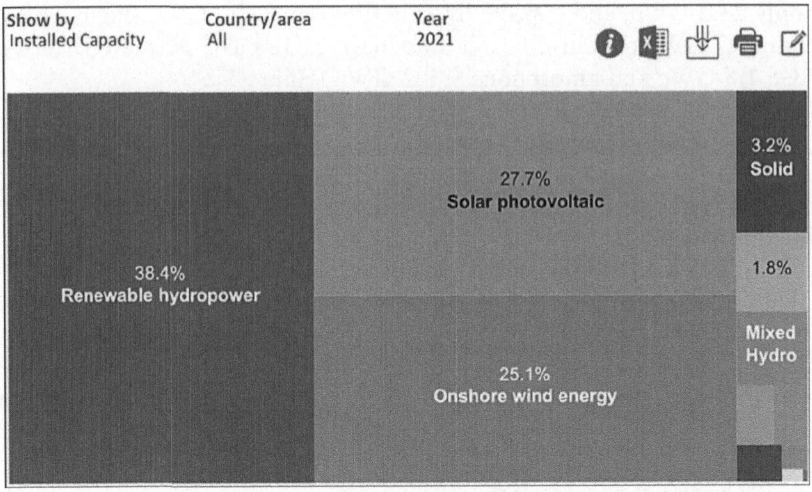

Figure 1.25 Share of solar power projects in total RE installed capacity in the world in 2021. (Courtesy: Refer Annexure A)

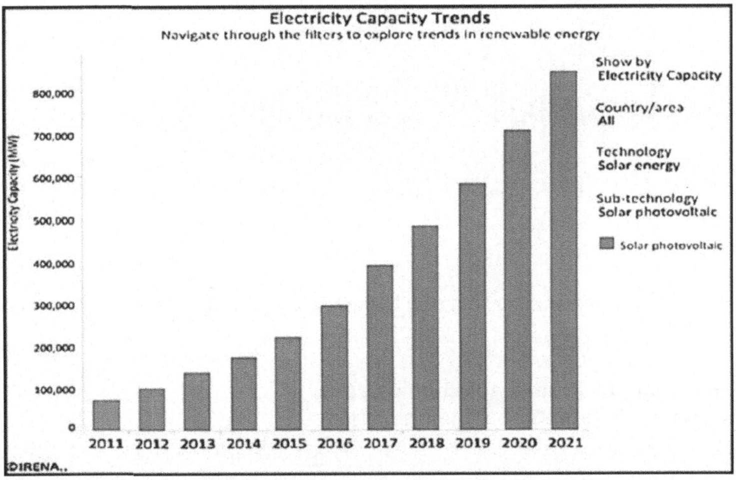

Figure 1.26 Year-wise PV installed capacity (MW) in the world. (Courtesy: Refer Annexure A)

In addition, China is the main exporter of solar modules to many countries in the world. This is the advantage of a common open world market.

The most interesting is the price trend of the Si solar cells [12]. This is mind boggling and unbelievable that a solar cell at USD 76 per watt in 1977 can be available for just USD 0.30 in 2020 as shown in Figure 1.28. If inflation of currencies is considered, the solar cell would appear to be available practically free for the past generation. Let the next generation enjoy

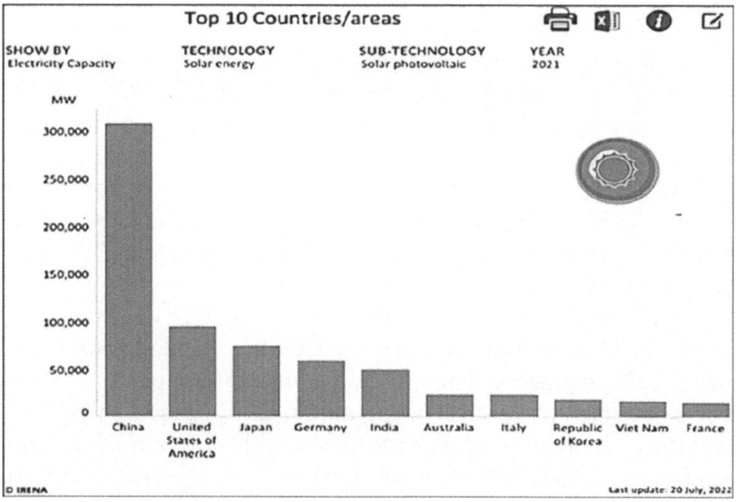

Figure 1.27 Country Ranking: PV Installed Capacity (MW) by 2021. (Courtesy: Refer Annexure A)

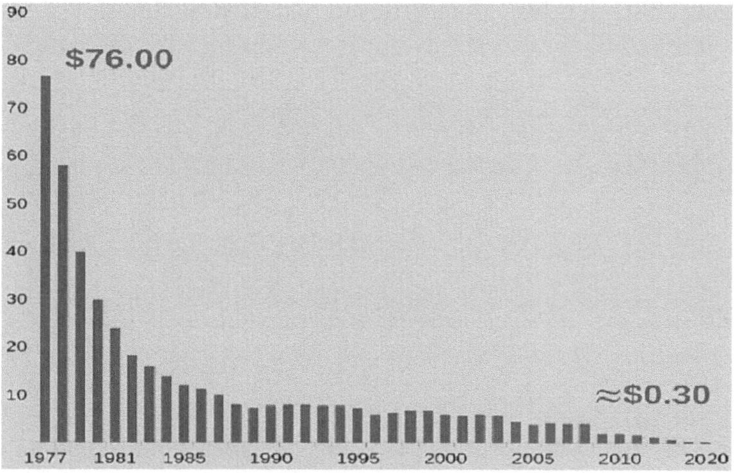

Figure 1.28 Price history of silicon PV cells. (Courtesy: Refer Annexure A)

it. That is why solar cells are now fully commoditized and commercialized. Power generation from solar energy is today cheaper than practically all sources of energy, whether conventional or nonconventional, renewable or non-renewable, local or imported – that too without any detrimental impact on the environment.

1.13 SOLAR PV FUTURE SCENARIOS

The world is moving confidently toward Net Zero Emissions (NZE) by 2050. Solar PV generation increased by a record 179 TWh (up 22%) in 2021 to exceed 1,000 TWh [13]. An average annual generation growth of 25% in the period 2022-2030 is needed to follow the NZE scenario. The growth rate is similar to the average annual expansion recorded in the past 5 years. This corresponds to a more than three-fold increase in annual capacity deployment until 2030 amounting to approximately 7,400 TWh in 2030 (Figure 1.29).

Out of the global solar PV capacity additions in 2021 the share of utility-scale plants is 52%, followed by the residential (28%) and commercial and industrial (19%) segments. The total global installed capacity of PV power projects is projected to 5,042 GW as per the NZE scenario by 2030.

As per the prevailing trend, the crystalline (poly- + mono-) Si PV modules were dominant, with over 95% market share. With the sharp narrowing of the gap between the costs of mono- and polycrystalline modules, the more efficient monocrystalline wafers technology is capturing almost all crystalline PV production. Even among the different types of Si monocrystalline modules, more efficient cell architecture such as Passivated Emitter Rear Contact (PERC) has been dominant since 2021, with almost 75% market

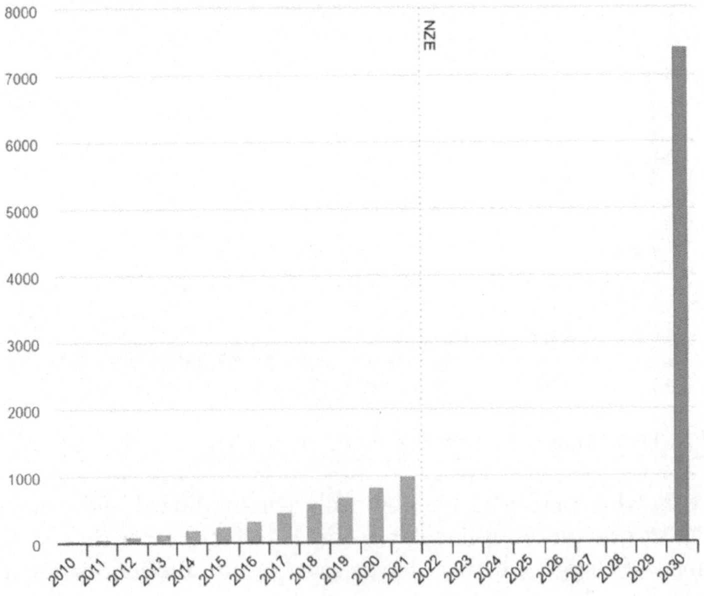

Figure 1.29 Solar PV power generation under the Net Zero Scenario, 2010-2030 (Courtesy: Refer Annexure A)

share. Now, even higher-efficiency cell designs (using technologies such as TOPCon, heterojunction, and back contact) saw expanded commercial production and captured about 20% of the market in 2021.

The above projected figures are based on the NZE scenario by 2050. Another report [14] is presenting the projections based on Low, Medium, and High scenarios. The challenges identified by the report are:

(a) Today, prices for silicon, wafer, cell, and modules are considerably higher than those at the beginning of 2021, and price relief is not expected in the near future.
(b) The world is in an inflationary phase; the negative impacts of the pandemic are not yet overcome.
(c) Russian war against Ukraine and its impact on the global market.

Despite all these challenges, solar energy is seeing impressive growth from 2022. The forecast for annual addition to the global solar PV market offers encouraging results (Figure 1.30). The Medium Scenario anticipates annual installed capacity additions to reach 328.5 GW, while the Low Scenario offers 243.5 GW and High Scenario 458.8 GW by 2026.

As regards the cumulative installed capacity of solar power projects, less than 10 years after the 100 GW mark was reached in 2012, the total operating on-grid solar power capacity has reached the 1 TW level in 2022. The solar power is expected to grow with rapid strides in the next decades. The Medium Scenario projects the cumulative installed capacities to reach 2,300 GW, while the Low Scenario offers 1,991 GW and the High Scenario 2,708 GW by 2026 (Figure 1.31).

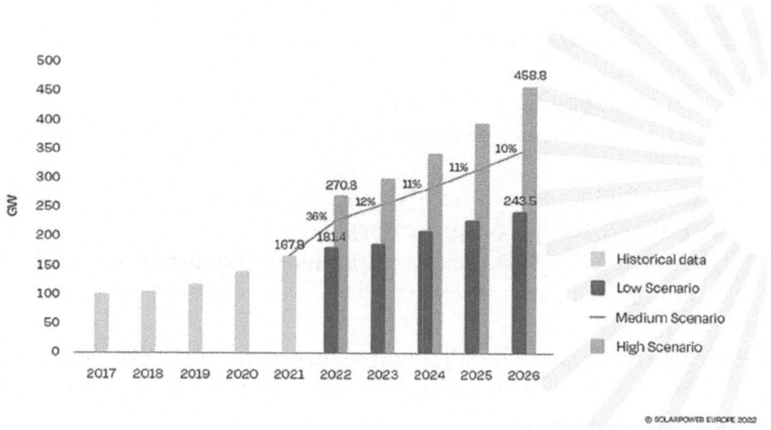

Figure 1.30 Projected annual solar PV addition 2022–2026 globally (Courtesy: Refer Annexure A)

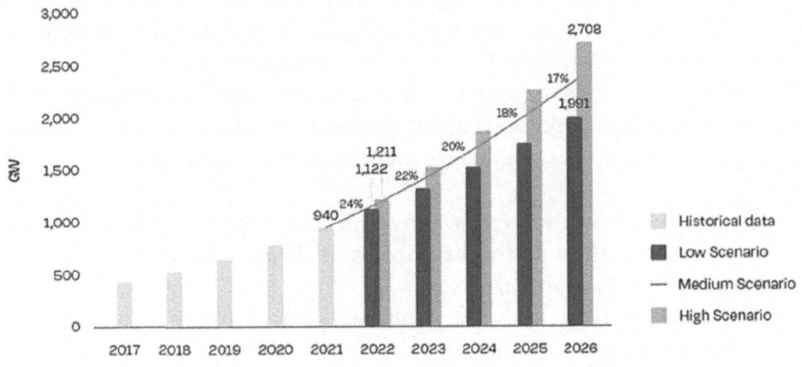

Figure 1.31 Global total solar PV market scenarios 2022–2026 (Courtesy: Refer Annexure A)

1.14 ADVANCES IN SOLAR MODULES

Once the efficiency and quality of solar cells were optimized, the solar industry started R&D to improve the efficiency of solar modules by fine-tuning the manufacturing processes. As a result, many variations of solar modules with great improvements are now available in the market. Some of them are:

- Glass-Glass SPV modules with both sides glass cover (weather proof, frameless, stronger, fire resistant)
- Bifacial glass-glass modules with metal fingers and busbar printed on both sides (efficiency gains of 11%, weather proof, frameless, stronger, fire resistant)
- Flexible monocrystalline silicon modules (no breakable glass, easy installation, can be fixed at curved surfaces – tilted surface, facade, sun shed, car parking)
- Solar modules integrated with IMM – Individual Module Monitoring ("real time" data, plant efficiency improvement: 3%; O&M cost saving: 50%)
- BIPV – Building Integrated PV modules (serves as building envelope material and also power generator)
- Half-Cut SPV modules (each cell produces 1/2 current with 1/4 resistance, higher fill factor, efficiency gain of 0.5%–1%, power gain of 3%–5%)
- Shingled Cells SPV Modules with seamless soldering like shingles on a roof (lower ohmic losses, reducing stress on cells, 10% gain in energy, improved aesthetics)

- PERC – Passivated Emitter Rear Contact SPV modules. A dielectric passivation layer is added to the rear of the cell to allow more sunlight to be captured. (1% efficiency gain)
- HIT – Heterojunction with Intrinsic Thin Layer SPV module (lower temperature coefficient, high efficiency)
- SPV Modules with $LiFePO_4$ battery (modular, better than central storage, shuts down over 50°C)
- SPV Modules with Optimizer with local DC/DC converter (constantly tracking MPPT of each module, monitor module performance, safe DC feature, accepts multiple orientations, tilts, and module types)
- SPV modules with micro-inverter (convert the DC to AC locally, no string/central inverter needed, modular, safer AC, no string sizing limitation)
- Wearable Solar Modules (solar textiles: tiny solar panels stitched into the fabric of clothing; application: window curtains, heating car seats)

We will witness many more developments in the future. It is an ongoing progress driven by a dynamic and competitive market. The above descriptions are just in brief. Details of "Advances in Solar Modules" are beyond the scope of this topic.

1.15 DEVELOPMENTS BEYOND THE LATEST

The PERC solar cell technology has become the dominant solar cell technology. In 2022, PERC solar cells held 75% of the market share. There is a continuous innovation effort from the solar manufacturing plants to improve solar module efficiency, lower inherent losses, and improve stability under outdoor conditions. The efficiency improvement potential of PERC now seems to have been exhausted and further optimization of the technology now seems elusive.

Manufacturers are now looking beyond it to:

- n-PERT (passivated emitter rear totally diffused)
- Heterojunction (HJT)
- Interdigitated back contact (IBC)
- n-TOPCon

Details of these newest technologies are beyond the scope of this topic; however, they are briefly touched upon below.

Unlike PERC, in the PERT structure, the rear surface is "totally diffused" with phosphorus (n-type). PERT technology is implemented on n-type Si solar cells with the advantages of lower temperature coefficient and lower light induced degradation than p-type Si wafers.

HJT technology was patented by Panasonic, which expired in 2010, and, thereafter, many manufacturers are trying in their own way to produce this kind of module due to its inherent higher qualities. HJT solar modules combine two different technologies into one cell: a crystalline silicon cell sandwiched between two layers of amorphous "thin film" silicon. Used together, these technologies allow more energy to be harvested compared to using either technology alone, resulting into high efficiency and low temperature coefficient.

IBC is an architectural approach to module design. The highest silicon wafer-based solar cell power conversion efficiencies reported to date have been achieved with the interdigitated back contact (IBC) architecture. IBC solar cells require interdigitated (or striped) doping on the rear surface and only have contacts on the rear. There are several advantages of the IBC architecture over the more commonly fabricated front and rear contact design: (a) the elimination of front grid shading, allowing for potentially higher short-circuit currents, (b) the elimination of front surface doping allows for a wider range of front surface texturing and light trapping schemes to be deployed on the front surface.

n-TOPCon or N-Type TOPCon (Tunnel Oxide Passivated Contact) is the new buzz in the solar industry. In a TOPCon cell, a tunnel oxide layer and an intrinsically polysilicon layer are added to the rear side. The front side is passivated by a dielectric stack of passivation and antireflection layer offering high efficiency and low temperature coefficient.

Only scattered information regarding these technologies is available on websites. Hopefully, very soon a well-documented report on these topics will be available for the benefit of all.

ANNEXURE A: SOURCES FOR FIGURES

Figure No.	*Caption*	*Source*
Figure 1.1	Figure 1.1. Solar spectrum in outer atmosphere and at sea level of the Earth.	Fondriest Environmental, Inc. https://www.fondriest.com/environmental-measurements/parameters/weather/solar-radiation/
Figure 1.2	Typical p-n junction solar cell.	https://www.neetprep.com/section/chapter/704/?sectionId=270
Figure 1.3	Molecular orbital diagram of hydrogen and helium.	https://www.chegg.com/homework-help/questions-and-answers/5-hydrogen-exists-diatomic-gas-helium-exists-free-atm-molecular-hez-complete-molecular-orb-q58231300

Figure No.	*Caption*	*Source*
Figure 1.4	Bonding diagram for the	
	(a) single unit of silicon 3D tetrahedron	https://commons.wikimedia.org/wiki/File:Silicon_Crystal_structure.svg
	(b) two-dimensional representation of a tetrahedron bond	http://hyperphysics.phy-astr.gsu.edu/hbase/Solids/sili.html
	(c) simplified sketch of the actual crystal structure.	https://www.pveducation.org/pvcdrom/pn-junctions/semiconductor-structure
Figure 1.5	Formation of conduction band, valence band, and bandgap in a solid semiconductor.	[4]
Figure 1.6.	Energy band diagram of insulator, semiconductor, and conductor based on bandgap energy.	[4]
Figure 1.7(a)	Extrinsic semiconductor with (a) phosphorus as donor atom	https://en.wikipedia.org/wiki/Donor_(semi conductors
Figure 1.7(b)	(b) boron as acceptor atom in the Si crystal lattice	https://en.wikipedia.org/wiki/Donor_(semiconductors
Figure 1.8	Creation of energy levels E_D (donor) and E_A (acceptor) in the bandgap due to impurity atoms.	[4]
Figure 1.9	A p-n junction at thermal equilibrium with no applied bias	https://www.google.com/url?sa=i&url=https%3A%2F%2Fedurev.in%2Fstudytube%2FP-N-Junction-Diode%2Fa335d78c-4053-4f63-bec4-11e5c4a4688e_t&psig=AOvVaw3JHjco6Wf5isjhbMSwcKvc&ust=1668270296884000&source=images&cd=vfe&ved=0CBEQjhxqFwoTCOCd8szFpvsCFQAAAAAdAAAAABAF
Figure 1.10	A p-n junction solar cell schematic.	https://www.imagesco.com/articles/photovoltaic/photovoltaic-pg4.html

Figure No.	*Caption*	*Source*
Figure 1.11	A single Si solar cell (a) 0.3 mm thick p-Si wafer, (b) entire back surface silver metal thin coating, and (c) front surface thin (0.05 mm) silver grid fingers and vertical bus bars of 0.2 mm flat strips of copper or aluminum.	[9]
Figure 1.12	Silicon solar cells with 2, 3, 5 busbars.	Open-source Google search
Figure 1.13	Devices to measure I-V characteristics and efficiency.	https://www.google.com/url?sa=i&url=https%3A%2F%2Fwww.electricaltechnology.org%2F2020%2F10%2Fcalculation-design-solar-photovoltaic-modules-array.html&psig=AOvVaw1UeurT-oSty63o4pscDHts&ust=1668321643943000&source=images&cd=vfe&ved=0CBEQjhxqFwoTCPiA99eEqPsCFQAAAAAdAAAAABAD
Figure 1.14	I-V characteristics of solar cells in (a) dark and under illumination and (b) its inverted representation.	[5]
Figure 1.15	Shockley–Queisser theoretical efficiency limit single p-n junction solar cells.	[7]
Figure 1.16	Current-voltage characteristics of a multi-crystalline silicon PV module at different radiation levels.	[5]
Figure 1.17	Current-voltage characteristics of a multi-crystalline silicon PV module at different temperatures.	[5]
Figure 1.18	Mono-crystalline, polycrystalline, and thin-film solar cells.	Open-source Google search
Figure 1.19	Connection of one solar cell to another in series.	Open-source Google search

Figure No.	*Caption*	*Source*
Figure 1.20	PV module with 36 cells interconnected to form a series string.	https://pv-manufacturing.org/solar-cell-manufacturing/pv-module-manufacturing/
Figure 1.21	Different layers of a solar module.	https://thesolarlabs.com/ros/photovoltaic-modules/
Figure 1.22	The concept of Air Mass.	https://www.azom.com/article.aspx?ArticleID=10817
Figure 1.23	PV global annual production by technology.	https://www.ise.fraunhofer.de/content/dam/ise/de/documents/publications/studies/Photovoltaics-Report.pdf
Figure 1.24	Global market share of thin-film technologies.	https://www.ise.fraunhofer.de/content/dam/ise/de/documents/publications/studies/Photovoltaics-Report.pdf
Figure 1.25	Share of solar power projects in total RE installed capacity in the world in 2021	https://www.irena.org/Statistics/View-Data-by-Topic/Capacity-and-Generation/Technologies
Figure 1.26	Year-wise PV Installed Capacity (MW) in the world.	https://www.irena.org/solar
Figure 1.27	Country Ranking: PV Installed Capacity (MW) by 2021	https://www.irena.org/Statistics/View-Data-by-Topic/Capacity-and-Generation/Country-Rankings
Figure 1.28	Price history of silicon PV cells.	https://www.researchgate.net/figure/Price-history-of-silicon-PV-cells_fig1_331445029
Figure 1.29	Solar PV power generation in the Net Zero Scenario, 2010-2030	IEA (2022), Solar PV, IEA, Paris https://www.iea.org/reports/solar-pv, License: CC BY 4.0
Figure 1.30	Projected annual solar PV addition 2022–2026 globally	SolarPower Europe (2022): Global Market Outlook for Solar Power 2022-2026, May 2022, ISBN: 9789464518610
Figure 1.31	Global total solar PV market scenarios 2022 - 2026	SolarPower Europe (2022): Global Market Outlook for Solar Power 2022-2026, May 2022, ISBN: 9789464518610

REFERENCES

1. Alan Chodos (Ed.), "This Month in Physics History", *American Physical Society, APS News*, Vol. 18, No. 4, April 2009. https://www.aps.org/publications/apsnews/200904/physicshistory.cfm#:~:text=In%201883%2C%20American%20inventor%20Charles,and%20thus%20not%20very%20practical
2. Elizabeth Chu and D. Lawrence Tarazano, "A Brief History of Solar Panels, U.S. Patent and Trademark Office", *Smithsonian Magazine*, pp. 1–6, 2022. https://www.smithsonianmag.com/sponsored/brief-history-solar-panels-180972006/
3. Lewis M. Fraas, "Chapter 1: History of Solar Cell Development", in *Low-Cost Solar Electric Power.* Springer International Publishing Switzerland, pp. 1–11, 2014
4. S. M. Sze and M. K. Lee, "Chapter 1: Energy Bands and Carrier Concentration in Thermal Equilibrium", in *Semiconductor Devices Physics and Technology.* John Wiley & Sons, Inc., pp. 15–42, 2012
5. Stefan C. W. Krauter, "Chapter 2: Photovoltaics", in *Solar Electric Power Generation – Photovoltaic Energy Systems, Pub.* Springer-Verlag, pp. 19–48, 2006
6. S. M. Sze and M. K. Lee, "Chapter 10: Photodetectors and Solar Cells", in *Semiconductor Devices Physics and Technology.* John Wiley & Sons, Inc., pp. 323–356, 2012
7. William Shockley and Hans J. Queisser, "Detailed Balance Limit of Efficiency of p-n Junction Solar Cells", *Journal of Applied Physics*, Vol. 32, No. 510, 1961
8. Sudhir Kumar, "Reasons for Failure of Solar PV Companies Globally: Lessons for Investors", *Energy Future*, Vol. 3, No. 3, pp. 52–56, April–June 2015, Pub. TERI, India
9. P. Jayarama Reddy, *Science & Technology of Photovoltaics*, BS Publications, 2004
10. K. L. Chopra, P. D. Paulson and V. Dutta, "Thin-Film Solar Cells: An Overview", *Progress in Photovoltaics: Research and Applications*, Vol. 12, pp. 69–92, 2004
11. S. Philipps, "Photovoltaics Report", Fraunhofer Institute for Solar Energy Systems and Werner Warmuth, PSE Projects GmbH, September 22, 2022. https://www.ise.fraunhofer.de/content/dam/ise/de/documents/publications/studies/Photovoltaics-Report.pdf
12. Mohamed Ibrahim A. Arafa and El-Sayed Soliman A. Said, "A Different Vision for Uninterruptible Load Using Hybrid Solar-grid Energy", *International Journal of Power Electronics and Drive System (IJPEDS)*, Vol. 10, No. 1, pp. 381–387, 2019
13. IEA, *Solar PV*, IEA, Paris, 2022. https://www.iea.org/reports/solar-pv, License: CC BY 4.0 (Solar PV Analysis)
14. SolarPower Europe, *Global Market Outlook for Solar Power 2022–2026*, SolarPower Europe May 2022, ISBN: 9789464518610.

Chapter 2

Significance of shadow and shading analysis in solar power projects

Sudhir Kumar

2.1 INTRODUCTION

A shadow-free area is a prerequisite for a successful solar power plant. This is the first and foremost criterion for a site engineer when visiting a location for a site survey to install a new solar power plant, and one has to carry out the site survey with a careful consideration of shading or shadows from nearby objects [2]. Naturally, no one wants solar panels to be in shadow from a high-rise structure or nearby trees for the entire day of solar radiation. Solar energy is the fuel for a solar power plant, and no power plant can run without fuel. If there are too many tall structures nearby, the site is deemed unfeasible. The best strategy, therefore, is to find a site that is completely shadow free. However, oftentimes, this is not possible, as there may be obstructing structures nearby (Figure 2.1). If there are many tall structures nearby, it is best to either remove the obstructing structures or avoid placing solar panels in areas where shadows are expected [9].

Shadows from solar light are not a constant phenomenon. They change every day and every hour throughout the year. In the Southern Hemisphere, shadows are at their maximum in December and at their minimum in June. One needs to find out the maximum length of the shadow on the shortest day in December. Once the shadow length is determined, a decision must be made whether to avoid the shaded area if sufficient land or roof area is available. In the event of non-availability of sufficient land or roof area, it is also important to find out how much energy loss is incurred annually. In the Northern Hemisphere, shadows are at their maximum in June and at their minimum in December. This process is called "shadow analysis", which involves the study of the sun's path and height angles corresponding to various azimuth angles. The impact of partial shading and its detrimental effect have been well documented by several studies [3, 5–7].

In a solar power plant, multiple panels (say 15–20) are connected in series to form a string to maintain the series voltage below 1000 volts and sometimes below 1500 volts. These strings are then connected in parallel to increase the current up to the limit required by the inverter, which can range from 30 to 3000 amperes depending on the size of the power plant.

DOI: 10.1201/9781003623168-3

Figure 2.1 Shadow of a tree on a rooftop solar power plant. (Source: Refer Annex. A)

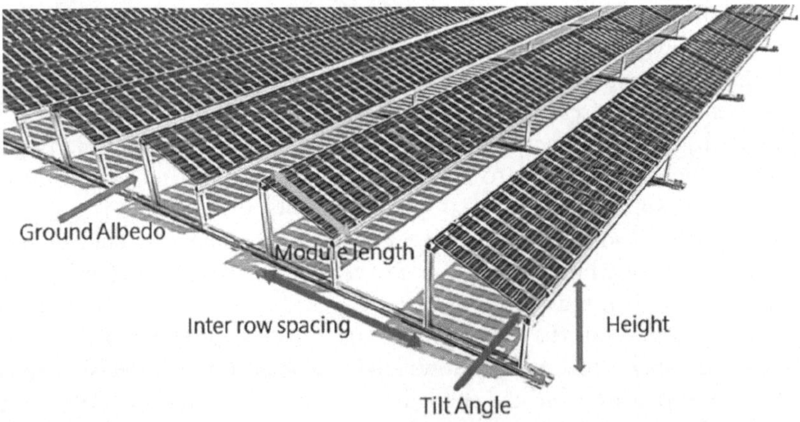

Figure 2.2 Inter-row spacing between solar strings in a solar power plant. (Source: Refer Annex. A)

Such a block of series-parallel combinations is called an array. Multiple arrays make a solar power plant. Each string is mounted on a metal structure, with panels placed side by side at a fixed tilt angle and height, depending on the size of the solar panel and the geographical location. Strings in parallel are placed one behind the other (Figure 2.2). The distance between two strings is determined by studying the shadow effect of one string on another (Solar Quarter). The inter-row spacing must be optimized for maximum solar output and efficient utilization of land. This process is called "shading analysis".

The following sections are devoted to studying "shadow analysis" using a simple manual approach and with an instrument called Pathfinder. Similarly, "shading analysis" for inter-row spacing using the manual method and software or apps is discussed.

2.2 GEOGRAPHICAL REASONS FOR SHADOW

Any one point on Earth can be identified by its latitude and longitude. The map of the Earth clearly shows the latitude and longitude lines for all locations (Figure 2.3). The vertical axis indicates latitude, with 0 degrees at the equator, spreading from 0° to 90° northward and 0 to 90 degrees southward. The horizontal axis indicates longitude, with 0 degrees at the Prime Meridian, spreading from 0° to 180° eastward and 0° to 180° westward. Thus, we now have four quadrants: North-East (N-E), South-East (S-E), North-West (N-W) and South-West (S-W). Latitude is always designated with suffixes N or S, with angles ranging from 0° to 90°. Longitude is always designated with suffixes E or W, with angles ranging from 0° to 180°. The examples below elucidate the positions of various cities.

Lat (N), Long (E): (Asia, Europe, North Africa) New Delhi 28.70° N, 77.10° E

Figure 2.3 World map showing longitude and latitude values. (Source: Refer Annex. A)

Lat (S), Long (E): (Australia, Indonesia, New Zealand) Brisbane 27.46° S, 153.02° E
Lat (N), Long (W): (Canada, USA, Mexico, W Indies) New York 40.71° N, 74.00° W
Lat (S), Long (W): (Brazil, Argentina, Chile, Bolivia) Buenos Aires 34.60° S, 58.38° W

Another convention is to denote the N, S, E, W by (+) or (–) signs. A negative latitude means South of the Equator, and a negative longitude means West of the Prime Meridian. That means N-E is + +, N-W is + –, S-W is – – and S-E is – +. New York will be thus represented as +40.71°, –74.00°.

Null Island, located in international waters in the Atlantic Ocean, is the point on the Earth's surface at zero degrees latitude and zero degrees longitude (0° N 0° E), i.e., where the Prime Meridian and the Equator intersect.

The shape of the Earth can be categorized as "oblate spheroid", where "oblate" refers to a slight elliptical appearance and "spheroid" means almost a sphere but not quite. This term describes the true shape of the Earth, which flattens at the poles and bulges at the equator. Its shape is more accurately defined as a geoid (https://www.vedantu.com/). The spheroid shape of the Earth also affects the angle at which sunlight strikes the ground. Shadow length depends upon this angle at a given point on the Earth at a given time, day, and season. Earth is further divided into four hemispheres: (a) the Northern Hemisphere i.e., the upper half of the Earth above the equator, (b) the Southern Hemisphere i.e., the lower half of the Earth below the equator (c) the Eastern hemisphere i.e., the right half of the Earth to the east of the Prime Meridian, and (d) the Western hemisphere i.e., the left half of the Earth to the west of the Prime Meridian (Figure 2.4).

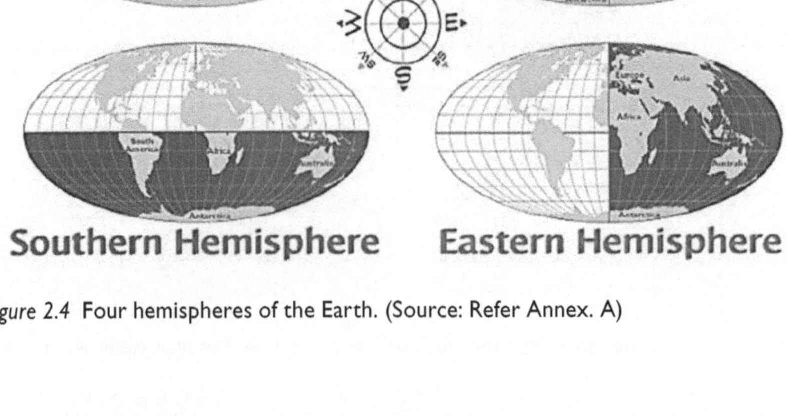

Figure 2.4 Four hemispheres of the Earth. (Source: Refer Annex. A)

The azimuth angle is of prime importance in solar power technology. However, some confusion surrounds the use of this term, which needs to be clarified. Solar scientists conveniently use the term "azimuth" for a given point on the Earth in two different ways for the same thing. The first is "geographical azimuth" and the second is "solar azimuth", which is also called "plane azimuth".

2.2.1 Geographical azimuth

The basic definition of azimuth is the angle of the direction from which sunlight is coming onto a plane. It is measured clockwise on the horizontal plane, from north to planar rotation (Figure 2.5). True south represents 180° azimuth and true north has an azimuth of 0°. Likewise, the true east is at 90° and the true west is at 270°. This distinction is important when we feed data in solar design software.

Most of the year, the sun's path is tilted toward the south in the Northern Hemisphere and toward the north in the Southern Hemisphere. At the equinoxes (21 March and 23 September), the sun moves directly from east to west. Thus, the azimuth angles are 90° at sunrise and 270° at sunset. The azimuth angle varies with latitude and time and day of the year.

2.2.2 Solar azimuth

Solar or plane azimuth considers 0° at true south in the Northern Hemisphere and at true north in the Southern Hemisphere. Solar design software such as PVsyst and PV*SOL uses solar azimuth, while SAM and PVWATT use geographical azimuth. Therefore, it is necessary to read software tutorials very carefully. For example, a screenshot of a PVsyst simulation for Pune (India), in the Northern Hemisphere, is shown in Figure 2.6, while that of Richmond (New Zealand), in the Southern Hemisphere, is shown in Figure 2.7. To get maximum power output, the solar panel should face true south in Pune (azimuth 0°) at a tilt angle of 22°. The left side of the panel is oriented toward the east, while the right side faces west. In Richmond, the azimuth is again 0° when the solar panel faces true north at the tilt angle

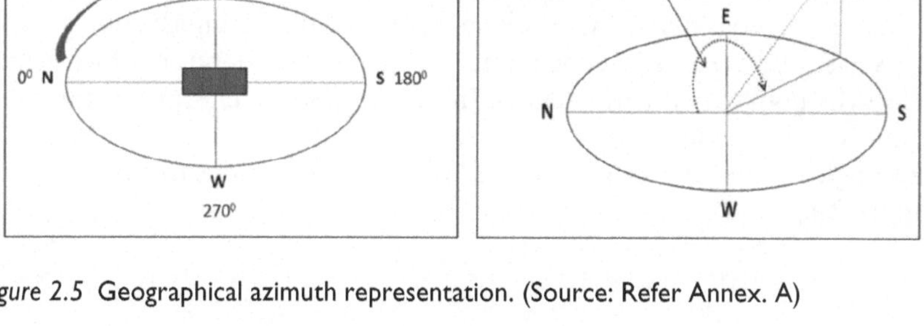

Figure 2.5 Geographical azimuth representation. (Source: Refer Annex. A)

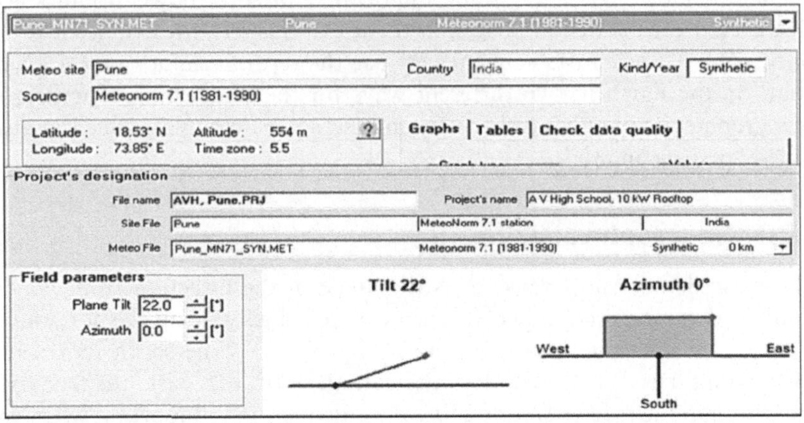

Figure 2.6 Screenshot of PVsyst simulation for Pune, India. (Source: Refer Annex. A)

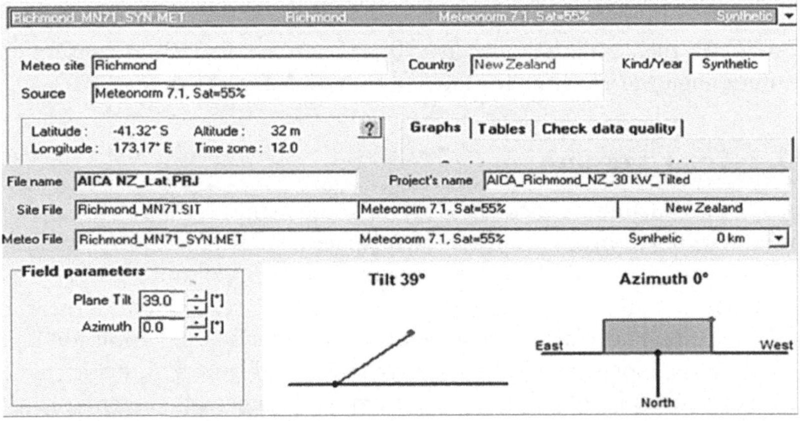

Figure 2.7 Screenshot of PVsyst simulation for Richmond, New Zealand. (Source: Refer Annex. A)

of 39°. Here the left side of the panel is oriented toward the west, while the right side faces east. It is worth noting here that according to the concept of "geographical azimuth", the solar panel in Pune would be placed at a 180° azimuth angle; however, in PVsyst, this is considered as 0° [11].

Unlike the geographical azimuth concept, the solar azimuth concept considers east and west as plus or minus 90° depending upon in which hemisphere the point of measurement is. In the Northern Hemisphere, the solar azimuth is the angle between the true south and the solar collector plane. This angle is taken as negative toward the east and positive toward the west (Figure 2.8).

	Northern hemisphere
North	180
East	-90
South	0
West	90

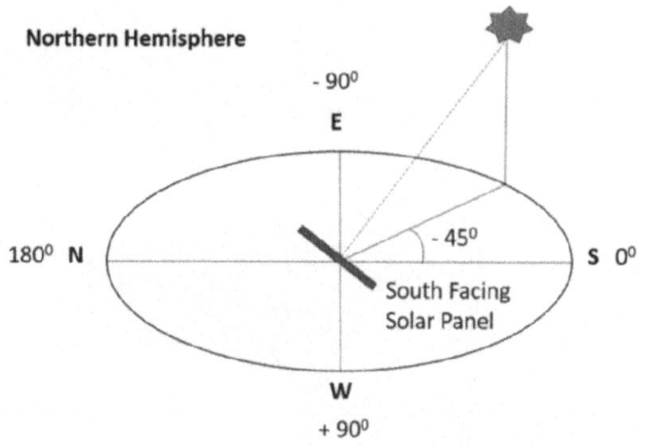

Figure 2.8 Solar azimuth measurements in the Northern Hemisphere. (Source: Refer Annex. A)

	Southern hemisphere
North	0
East	90
South	180
West	-90

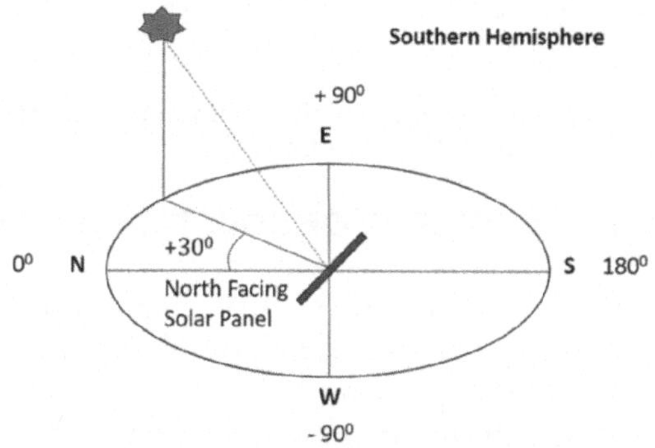

Figure 2.9 Solar azimuth measurements in the Southern Hemisphere. (Source: Refer Annex. A)

In the Southern Hemisphere, the solar azimuth is the angle between true north and the solar collector plane. This angle is taken as positive toward the east and negative toward the west (Figure 2.9).

A sun-path diagram is used to determine the location of the sun in the sky at any given moment during the day throughout the year. A 3-D view of a combined sun-path chart for 365 days and 12 hours for a single point on the Earth is shown in Figure 2.10, with hour lines and date lines included (Anthony et.al.). To represent it on paper, two methods are adopted: Rectangular Coordinates and Polar Coordinates. Each point on Earth with specific latitude and longitude has its own sun-path chart. However, for shading analysis, we do not need to use this complicated chart. There are many other simple approaches, which are described below.

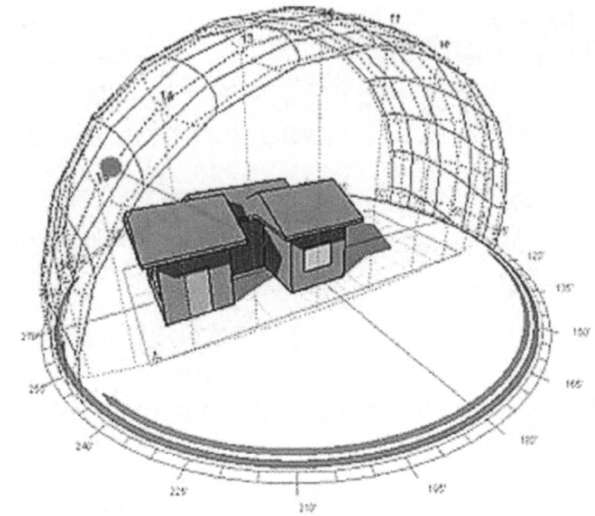

Figure 2.10 A 3D representation of the sun-path chart for a single point on the Earth. (Source: Refer Annex. A)

2.3 THE CONCEPT OF SOLSTICE AND EQUINOX

Earth moves around the sun in an elliptical path called Earth Orbit. Earth is slightly tilted on its axis as it travels around the sun (23.5°). That is why we have different seasons. Sunshine hours are different at different points on a given day and even vary every day and every season. There are two solstices: the June Solstice (20th or 21st) and the December Solstice (21st or 22nd). The exact date depends upon the year. Solstices are responsible for the shortest or the longest days or nights of the year. The two equinoxes occur during the March Equinox (on the 20th or 21st) and the September Equinox (on the 22nd or 23rd). At equinoxes, the sun is exactly above equator, which makes day and night of equal lengths [1].

In the Northern Hemisphere, the June Solstice is called the Summer Solstice (with the longest day and the shortest night). The December Solstice is called the Winter Solstice (with the shortest day and the longest night). The September Equinox is called the Autumn Equinox (with equal day and night), while the March Equinox is called the Vernal or Spring Equinox (also with equal day and night). The solstices and equinoxes in the Northern Hemisphere are shown in Figure 2.11 [13].

In the Southern Hemisphere, the June Solstice is called the Winter Solstice (with the shortest day and longest night). The December Solstice is called the Summer Solstice (with the longest day and shortest night). The September Equinox is called the Vernal or Spring Equinox (with equal day and night).

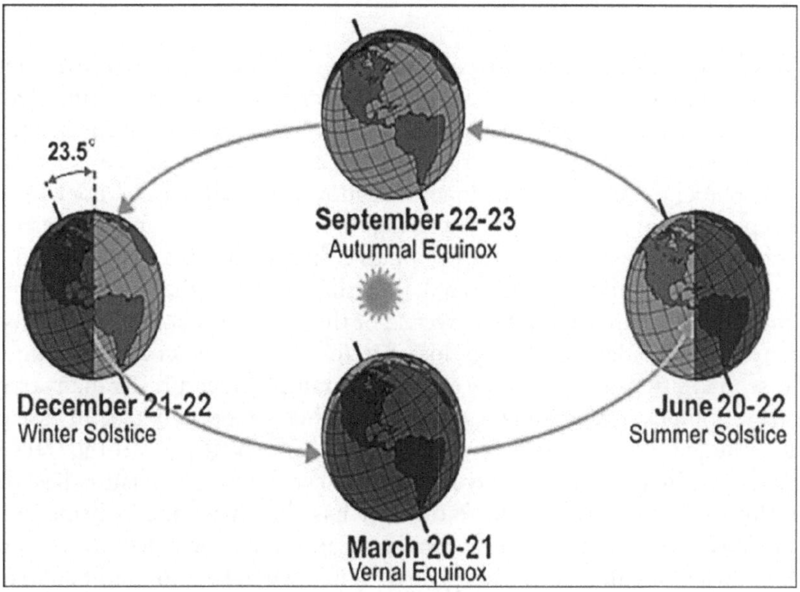

Figure 2.11 The solstices and equinoxes in the Northern Hemisphere. (Source: Refer Annex. A)

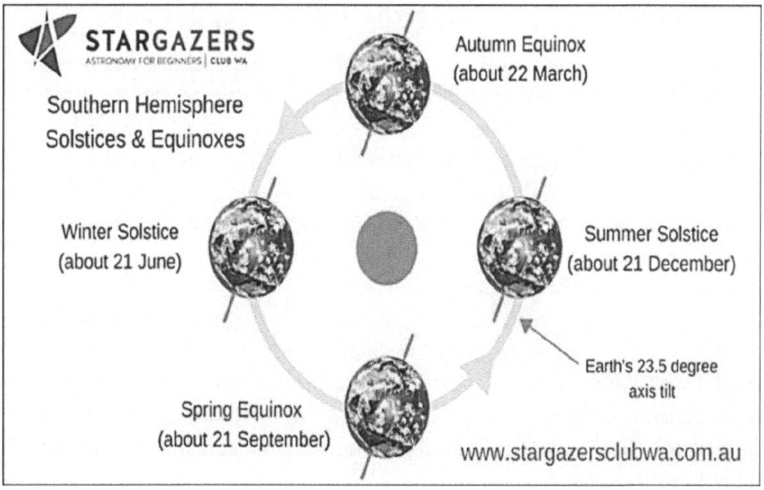

Figure 2.12 The solstices and equinoxes in the Southern Hemisphere. (Source: Refer Annex. A)

The March Equinox is called the Autumn Equinox (also with equal day and night). The solstices and equinoxes in the Southern Hemisphere are shown in Figure 2.12.

2.4 SHADOW ANALYSIS

While carrying out the shadow analysis, we may encounter two situations. The first situation is when there is a single structure or tree casting shadows on the proposed solar site. In this case, we can calculate the shadow area under a worst-case scenario, e.g., on 21 December in the Northern Hemisphere, the day with the longest shadows [1]. This can be done using a simple manual approach by measuring the shadow length at varying solar height angles. Then we have the option to either remove the shadow-casting structure (tree, building, pillar, etc.), if feasible, or avoid the shadowed area. If the shadowed area cannot be avoided or the structure cannot be removed, we need to calculate the energy loss due to the shadow as a compromise. The second situation is when we encounter multiple structures or many trees nearby, complicating manual calculations. This is especially true in case of a lone bungalow surrounded by tall trees, where cutting down the trees is prohibited. In such cases, a versatile non-electronic instrument called the Pathfinder is very useful. The Pathfinder has the advantage of being used every day throughout the year, providing an annual overview. Moreover, it also helps in calculating energy loss. Thus, we have two methodologies for shadow analysis: (a) Shadow analysis using a manual approach and (b) shadow analysis using the Solar Pathfinder.

2.4.1 Shadow analysis using the manual approach

Let us assume a 4-m-tall pillar on a roof at Pune, India, which is located in the Northern Hemisphere. We need to know the shadow length every hour of the day on the shortest day of the year, on 21 December, when the shadow is the longest. The first task is to get the hourly sun elevation (or sun's height angle θ). This information is available from the PVsyst software on a personal computer or the MySolarPanel Android app on a smartphone. We can easily calculate the shadow length using Tan θ and the height of the pillar (Table. 2.1).

The above data can also be obtained directly through online tools, without any need for calculations. One of the best websites for this purpose is https://www.suncalc.org/. One has to simply input the height of the shadow-casting object, along with the latitude, longitude, and date. The sun's altitude, azimuth, and shadow lengths for each hour of the day can directly be obtained.

The next task is to calculate the x-y coordinates of the shadow's tip every hour, which will help us in calculating the shadow coverage. We can use the shadow length values along with the zenith angle obtained from the PVsyst or MySolarPanel app (Figure 2.13).

Table 2.1 Calculation of shadow length at Pune on 21 December 2022.

Time on 21 December 2022	*Sun Height Angle (θ) at Pune (18.52° N, 73.85° E)*	*Shadow Length of 4 m Tall Structure*
08:00	11.4	19.8378
09:00	23.3	9.2879
10:00	33.9	5.9526
11:00	42.4	4.3806
12:00	47.3	3.6911
13:00	47.5	3.6653
14:00	42.9	4.3045
15:00	34.7	5.7767
16:00	24.3	8.8590
17:00	12.4	18.1930

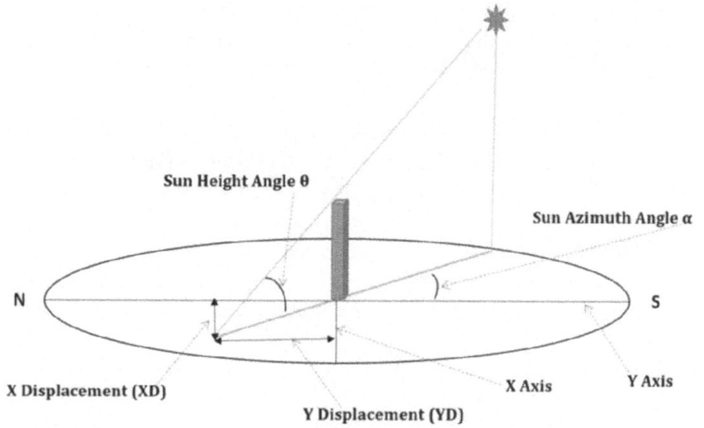

Figure 2.13 Shadow tip coordinates using shadow length at a given time. (Source: Refer Annex. A)

Such calculations provide Table 2.2, which shows the solar height angle (θ), the shadow height (4 m), the azimuth angle (α), and the coordinates (XD and YD).

Finally, we get the shadow diagram as depicted in Figure 2.14. The sun is southward and the total shadow area is toward the north of the pillar. Now the solar installer has many options:

(a) Leave a 10 m distance to completely avoid shadow. This luxury can be availed only if the roof area is much larger than the solar project area.
(b) Leave at least a 4 m distance as a good compromise, so that only the partial shadow areas from 8 am to 10 am and from 3 pm to 5 pm are compromised. Please note that the shortest shadow length is

Table 2.2 Calculation of shadow tip coordinates using shadow length every hour of the day.

4 m Tall Object Shadow Length: Pune 21 Dec 2022					
All length in meters (m)					
Time (h)	*Sun Height Angle (θ)*	*Shadow Length H*	*Sun Azimuth Angle (α)*	*Displacement X (XD)*	*Displacement Y (YD)*
08:00	11.4	19.8378	-60.3	-17.23	9.83
09:00	23.3	9.2879	-53.1	-7.43	5.58
10:00	33.9	5.9526	-43.1	-4.07	4.35
11:00	42.4	4.3806	-29.2	-2.14	3.82
12:00	47.3	3.6911	-11.0	-0.7	3.62
13:00	47.5	3.6653	9.3	0.59	3.62
14:00	42.9	4.3045	27.8	2.01	3.81
15:00	34.7	5.7767	42.1	3.87	4.29
16:00	24.3	8.8590	52.3	7.01	5.42
17:00	12.4	18.1930	59.8	15.72	9.15

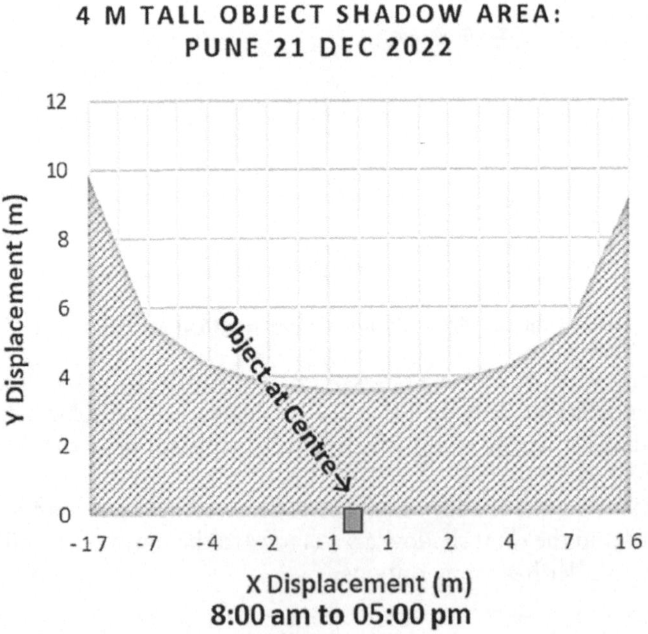

Figure 2.14 Shadow diagram of a 4 m tall pillar on a roof on 21 December, 2022. (Source: Refer Annex. A)

at 12 noon. These partially shadowed areas are on the worst day of year, but before and after these times, the shadow areas will keep on decreasing. It is worth mentioning here that field technicians in the Northern Hemisphere use a rule of thumb in case of a parapet wall on the roof, leaving a distance equal to the height of the parapet wall on the south side.

(c) Remove the pillar structure to keep the roof completely shadow free, provided, structurally, there is no objection or it is not too costly.

(d) Install the solar plant, including the shadow area, to mitigate the loss of power generation during the worst months. This can be the case when there is an extreme shortage of roof space and when maximum power is needed. The annual loss of energy can be calculated using the Pathfinder instrument.

We may encounter a situation when a power project site has a tall tree or structure in its vicinity. The first task then is to measure its height, which can be very easily done by using an inclinometer [14]. In modern days, even a smartphone can easily be used as an inclinometer. The abovementioned MySolarPanel has a built-in inclinometer, which can be used in the following steps, as shown in the Figure 2.15:

- Open the Inclinometer window in the app
- Keep the mobile flat at eye level while moving the other end up
- Align your eye level with both the upper end and the lower end of the mobile in line with the top of the tree
- Take a screenshot at this exact angle to get the inclination angle
- Find the height using the formula Tan $\theta \times$ Distance (your distance from the tree)
- Add your body height to get the final height of the tree

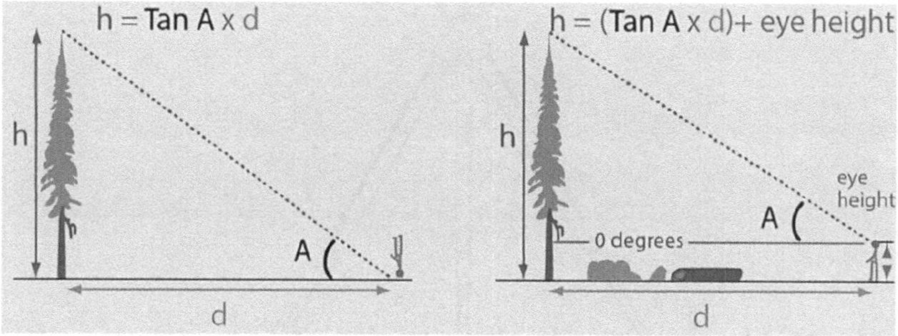

Figure 2.15 Measuring the height of a tree using an inclinometer. (Source: Refer Annex. A)

2.4.2 Shadow analysis with Solar Pathfinder

The Solar Pathfinder is a simple, non-electronic instrument [1] used for shadow analysis at any place on Earth. Tall objects nearby cast shadows on a plastic dome. Below this dome, a pre-printed sun-path diagram sheet is placed. The reflected shadow area is marked with a wax pencil. The scientifically engineered diagram provides data for the entire year. Interestingly, any one-day measurement gives data for all months. There is no need to get data only on the worst shadow day of the year. It can be set even on any uneven surface or sloping roofs using a tripod with rubber tipped telescopic legs. A compass and a bubble level are used to set the direction and level, respectively. The details can be found on the manufacturing company website https://www.solarpathfinder.com/. A simple scheme of a Solar Pathfinder in Figure 2.16 shows that it is made up of three sections:

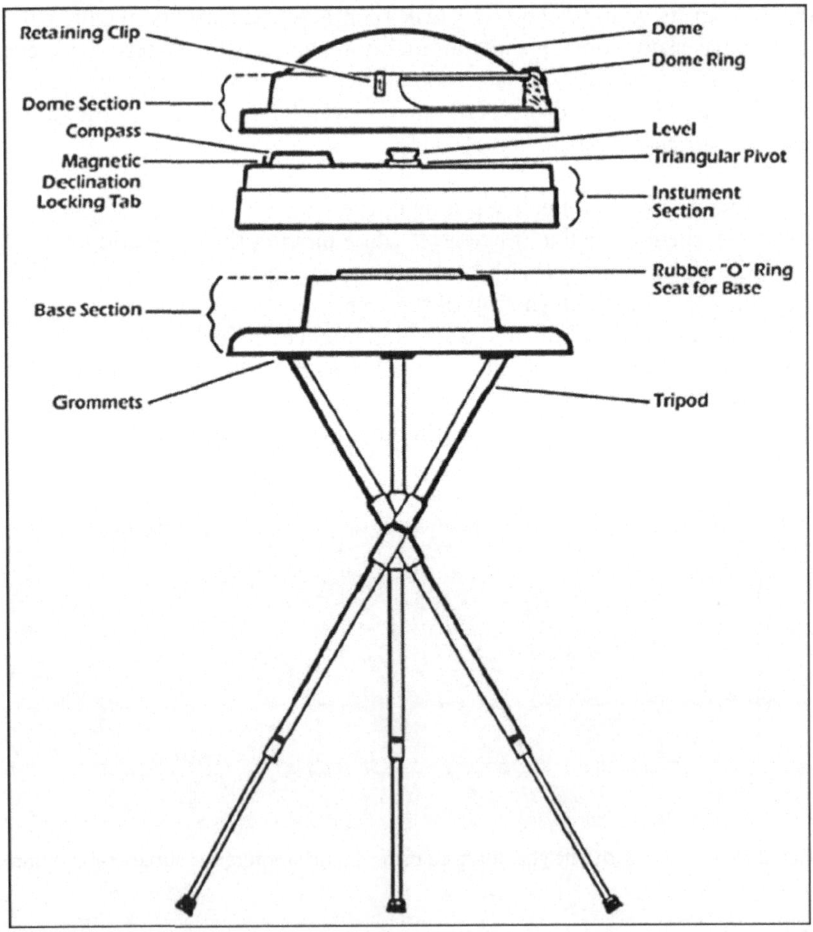

Figure 2.16 A simple scheme of a Solar Pathfinder. (Source: Refer Annex. A)

- A dome section: the part with the translucent lens/dome;
- An instrument section: the part with the compass and level;
- A base section: the part with the rubber grommets.

The Solar Pathfinder instrument is available in the market at the cost of USD 259.00 (2022). The cost includes pre-printed standard diagrams corresponding to different latitude ranges such as 0–25°, 37–43°, 54–65° N or S. Details are available on the Solar Pathfinder website (https://www.solarpathfinder.com/). The time line and month arcs are shown in Figure 2.17, which is self-explanatory.

Following are the steps to set up a Solar Pathfinder instrument:

- Step 1: Adjust the legs to get the base level
- Step 2: Put a paper sun-path diagram on the center pivot of the diagram platform
- Step 3: Adjust the Pathfinder for the proper magnetic declination by aligning the declination number with the small white dot on the rim of the base (e.g., Pune: −0.23°; Google search). The numbers to the left of "0" indicate "west of north" and are negative numbers; the numbers to the right of "0" indicate "east of north" and are positive numbers (Figure 2.18).
- Step 4: With the sun-path diagram locked to its proper declination, set the instrument section onto the base. Rotate the instrument section until the red pointer of the compass is pointing directly at the "N". This is magnetic south. When working near buildings with ferrous material (large metal beams), move the compass away from such material until the compass needle direction stabilizes.

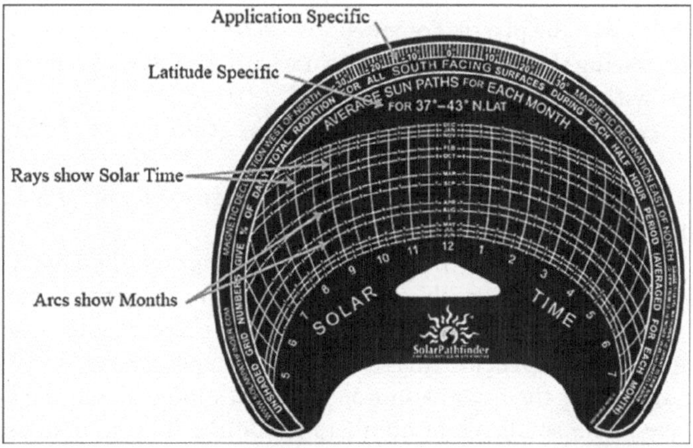

Figure 2.17 A sample of the Solar Pathfinder diagram. (Source: Refer Annex. A)

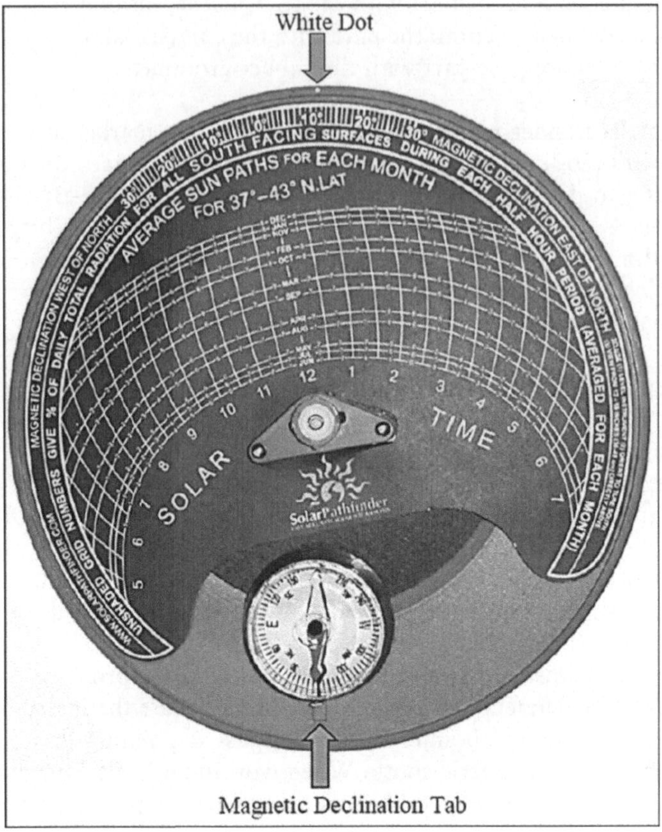

Figure 2.18 Setting the magnetic declination on Solar Pathfinder diagram. (Source: Refer Annex. A)

- Step 5: Place the plastic dome section on top of the instrument section. Start tracing the shadow area with a wax pencil. It is better to take pictures with a digital camera and analyze them with "Solar Pathfinder Assistant Base" software, available at the cost of $189 (2022).

The final output is shown in Figure 2.19 and the site tracing on the monthly sun-path diagram is shown in Figure 2.20

Let us now interpret the monthly sun-path diagram. The diagram contains 12 horizontal arcs, each representing the mean path of the sun for each month. Vertical lines represent solar time (different from actual time). The sun-path arc for December is the lowest winter sun path, while the sun-path arc for June is the highest summer sun path (opposite in the Southern Hemisphere). The small white numbers inside the half-hour divisions are the percentage of shadow-free area for each half-hour. The numbers along

Figure 2.19 Final output of Solar Pathfinder at a given site. (Source: Refer Annex. A)

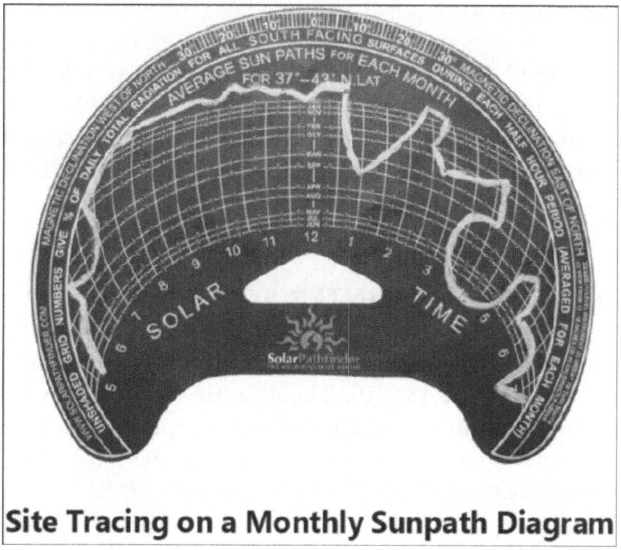

Figure 2.20 Site tracing on the monthly sun-path diagram. (Source: Refer Annex. A)

each sun-path arc add up to one hundred percent. For January, the figures from the shadow free arc are 2 + 3 + 4 + 5 + 6 + 7 + 7 +8 + 8 + 8 + 6 = 64%. This means that only 64% of shadow-free area is available.

Energy loss calculations can be done with the help of the "Solar Pathfinder Assistant Base" software. An example is shown in Figure 2.21 for an individual house with a 4-kW solar system with a fixed array on its roof. The house is situated at Franklin city in Johnson County, Indiana. The panels face a geographical azimuth of 180°at a tilt angle of 39.47° and a magnetic

Report Title	John Smith house 100 N. Main Street	Array Type	Fixed
Image File	37-43 back yard 1.JPG	Tilt Angle	39.47 degrees
Report Date	Saturday, December 23, 2006	Cost of Electricity	6 cents/kWhr
Declination	-3d 59m	DC Rate	4.00 kW
Latitude/Longitude	39.469 / -86.069	Derate Factor	0.77
Analysis Site	FRANKLIN, IN, Zipcode: 46131	Azimuth (180 = south)	180.00 degrees
Weather Station	INDIANAPOLIS, IN, Elevation: 246 m		
Station/Site Distance	21.56 miles		

Month	PVWatts Unshaded % Ideal Site Azimuth=180.0 Tilt=39.5	Actual Solar Rad w/o Shading Azimuth=180.0 Tilt=39.5 KWH/m /day	Actual Solar Rad w/ Shading Azimuth=180.0 Tilt=39.5 KWH/m /day	Actual AC Power (KWH) w/o shading Azimuth=180.0 Tilt=39.5	Actual AC Power (KWH) w/ shading Azimuth=180.0 Tilt=39.5	Actual Solar Savings
January	86.18%	3.17	2.74	316.0	277.27	$16.64
February	92.57%	4.14	3.83	371.0	349.98	$21.00
March	96.69%	4.44	4.30	417.0	410.36	$24.62
April	96.37%	5.16	4.97	454.0	448.0	$26.88
May	93.95%	5.74	5.39	509.0	490.79	$29.45
June	94.94%	5.85	5.56	480.0	470.0	$28.20
July	94.27%	5.93	5.59	499.0	483.69	$29.02
August	94.21%	5.68	5.35	486.0	469.74	$28.18
September	96.77%	5.20	5.03	439.0	432.3	$25.94
October	90.93%	4.70	4.27	427.0	397.0	$23.82
November	86.04%	3.07	2.64	279.0	245.62	$14.74
December	80.72%	2.37	1.91	232.0	190.34	$11.42
Totals	91.97% Unweighted Yearly Avg	-- Effect: 100.00% Sun Hrs: 4.62	-- Effect: 93.02% Sun Hrs: 4.30	4909.0	4665.09	$279.91

Figure 2.21 Energy loss calculation using Solar Pathfinder software output. (Source: Refer Annex. A)

declination of 3.98°. The percentage of unshaded areas is 91.97%, resulting in available radiation of only 93.02%. The actual AC energy output without shading is 4909 kWh per year, while with shading it is 4665 kWh per year. This means that the loss due to shading is 244 kWh, or 5% loss of energy per year.

2.5 SHADING ANALYSIS: INTER-ROW SPACING

Shading analysis means studying the shadow effects of one row of strings on another [4, 8, 10, 15, 18, 20]. The objective is to minimize inter-row spacing for efficient land utilization without compromising power output. Optimizing inter-row spacing can be achieved through different methods, such as (a) manual calculations, (b) PVsyst software, (c) SAM software, and (d) Android solar apps.

2.5.1 Shading analysis: Manual calculation

Let us consider the case of a manual study in Pune on 21 December (Winter Solstice). Unlike the shadow analysis described above, shading analysis, or inter-row spacing analysis, focuses on the shadow length at 8 am or 5 pm only. By using the tilt angle and the length of the panel, the vertical height of the panel is calculated [16]. The vertical height and the solar height angle give the shadow length at 8 am or 5 pm (Figure 2.22).

The next step is to get the Y displacement using the azimuth angle, which gives the distance between the outer edge of one row and the inner edge of

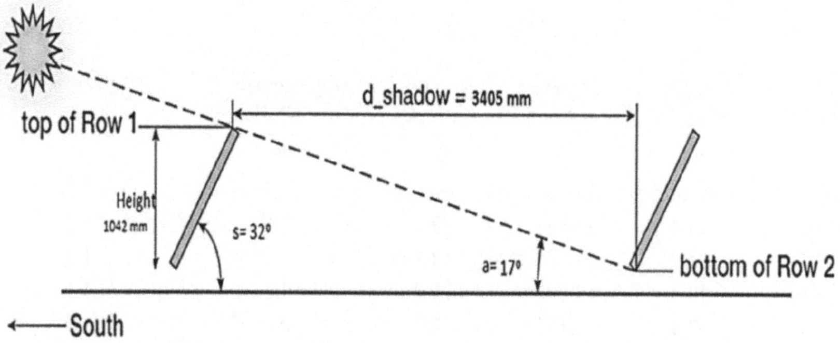

Figure 2.22 A representative scheme for calculating vertical height and horizontal coverage of solar panels. (Source: Refer Annex. A)

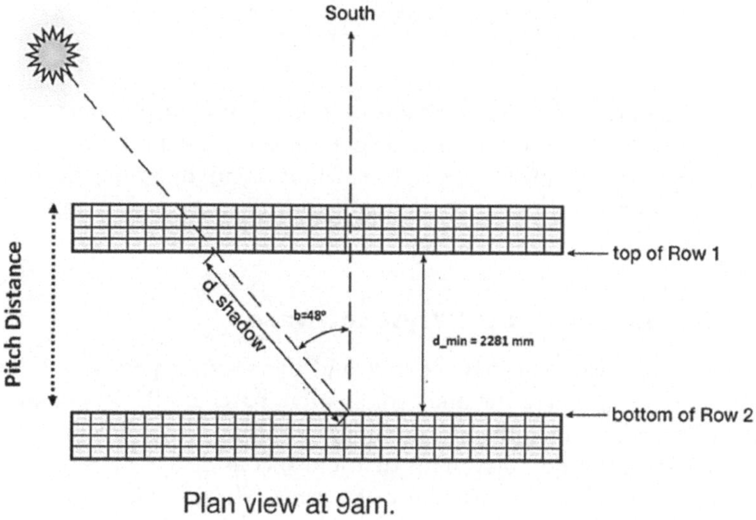

Figure 2.23 A representative scheme for calculating inter-row spacing of solar panels. (Source: Refer Annex. A)

the next row (Figure 2.23). Finally, the horizontal coverage of the panel is added to get the pitch distance (the distance between the lower edges or the upper edges of two rows of panels).

Based on the above scheme in Pune, let us consider a row of solar panels with a length 1.61 m and two panels placed vertically so that the total width of the panel row is 3.22 m. Pune is located at a latitude of 18.52° N and a longitude of 73.85° E. A tilt angle of 20° has been chosen as per PVsyst earlier to maximize energy generation. Accordingly, we get the vertical height of the panel as 1.10 m and the horizontal coverage as 3.02 m.

Table 2.3 Manual inter-row shading analysis for Pune, India.

Time (h)	*Sun Height Angle (θ)*	*Shadow Length H*	*Sun Azimuth Angle (α)*	*Displacement X (XD)*	*Displacement Y (YD)*
07:00	0.0		0.0	0	0.0
08:00	11.4	5.4554	-60.3	4.7	2.7
09:00	23.3	2.5542	-53.1	2.0	1.5
10:00	33.9	1.6370	-43.1	1.1	1.2
11:00	42.4	1.2047	-29.2	0.6	1.1
12:00	47.3	1.0151	-11.0	0.2	1.0
13:00	47.5	1.0080	9.3	0.2	1.0
14:00	42.9	1.1837	27.8	0.6	1.0
15:00	34.7	1.5886	42.1	1.1	1.2
16:00	24.3	2.4362	52.3	1.9	1.5
17:00	12.4	5.0031	59.8	4.3	2.5

Adopting the principles and formulas of Table 2.3, Y displacement is calculated corresponding to the solar height angle θ (11.4°) and the azimuth angle α (−60.3°) at 8 am. The Y displacement is 2.7 m. By adding the panel's horizontal coverage of 3.02 m, the total pitch distance is 5.72 m, or 6 m when rounded off.

2.5.2 Shading analysis: PVsyst software

The PVsyst software chooses the best and the shortest permissible inter-row spacing, by permutation and combination, based on the length and tilt angle of the solar panel, thereby minimizing energy loss to 1% or less. An added advantage of the software is that it also considers the ground coverage ratio (GCR) to enhance land use efficiency. The ground coverage ratio describes the ratio of the module area to the overall area of the power plant. In other words, it describes the proportion of the system area that is used to collect sunlight. The step-by-step procedure for using PVsyst to get inter-row spacing is as follows:

- Open PVsyst software – Tools – Tables and graphs of solar parameters – Shading optimization of sheds – Change the date to Winter Solstice i.e., 21 December – Enter location – Graphs
- Change the module dimensions and inactive band (thickness of module) – Fix tilt angle – Show optimization – Keep changing the module pitch and minimize losses (1%) till optimum inter-row spacing

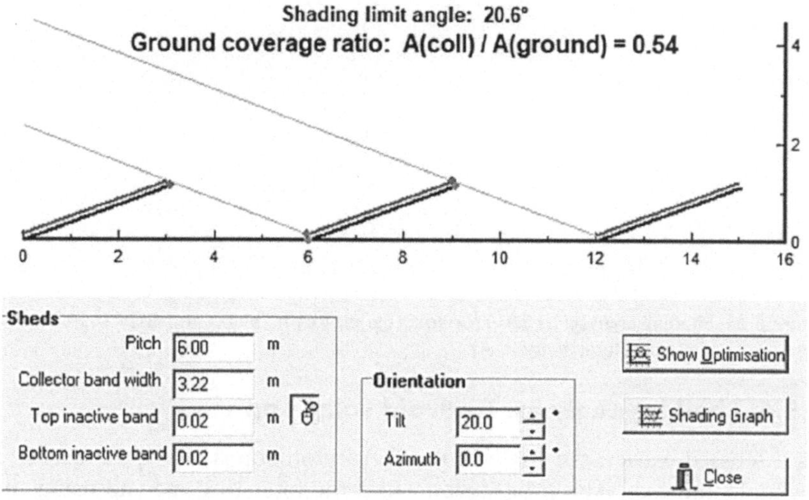

Figure 2.24 Shading analysis for inter-row spacing by PVsyst for solar panels of 1.61 m length with GCR 0.54 in Pune. (Source: Refer Annex. A)

- Maintain the Ground Coverage Ratio (GCR) by more than 0.45 – Change the pitch and find impact on GCR and optimize the distance between the two rows.

Keeping the parameters same as above, the PVsyst output for Pune will be as shown in Figure 2.24.

2.5.3 Shading analysis: SAM software

An open-source (free of cost) software developed by the USA's National Renewable Energy Laboratory (NREL) is the System Advisory Model (SAM). The software is available at https://sam.nrel.gov/download. A guideline book provides details about the various analytical tools available in this software [17]. The procedure is as follows:

Open SAM – Register with password or entry key – Start a new project – Photovoltaic – Detailed PV model – Choose module – Choose inverter – Fix system design – Shading and layout – Self Shading – Standard – Adjust module combination – Find the inter-row spacing.

Unlike PVsyst, SAM does not need a date for shading analysis. Keeping the parameters, the same as above, the SAM output for Pune is shown in Figure 2.25.

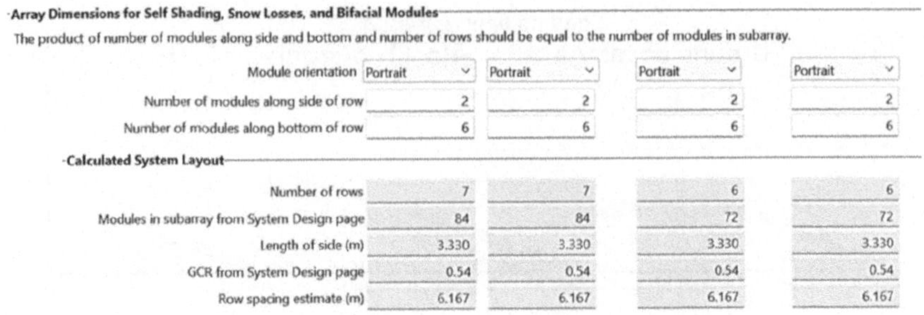

Figure 2.25 Shading analysis by SAM for solar panel: 1.61 m length and GCR 0.54 in Pune. (Source: Refer Annex. A)

2.5.4 Shading analysis: Android solar app

As discussed earlier, the MySolarPanel Android app is also quite useful. It is simple and fast. The procedure to find the inter-row spacing using this app is as follows:

Open the app – set location Pune – Set configurations by feeding the values of panel details, inverter efficiency, tilt angle, azimuth, etc. – Carry out simulation – Select option of shading model – Set tilt angle – Adjust row distance to set the energy loss at 1% or below. Similar to SAM, it does need a date for shading analysis. The final output is shown in Figure 2.26.

It can be seen from the above results that all of them give the inter-row spacing around 6 m. Design engineers will keep the inter-row spacing (pitch distance) of 6.5 as a safer margin.

2.5.5 FAR SHADING AND HORIZON DIAGRAM

All the discussions explained above regarding shading pertain to any shadow-casting structure, a tree or any object, in the vicinity of the solar power project. This phenomenon is termed as Near Shading. Imagine the roof of a house situated in a hilly area with a hill or hillock 1–2 km away to the south, and we want to have a solar rooftop plant. This house will naturally have late sunrise and early sunset for most of the year, especially during the winter season. This distant object will definitely reduce the sunshine hours at that place compared to planar location. This is termed Far Shading.

Far shading is best assessed using the Horizon Tool available in PVsyst, which allows us to define large distant structures such as mountains and other environmental features as well as large buildings. PVsyst treats horizon shading by completely blocking the beam component of irradiation when the sun is behind these horizon features. Horizon shading blocks direct irradiance for the entire PV system as opposed to the near shading,

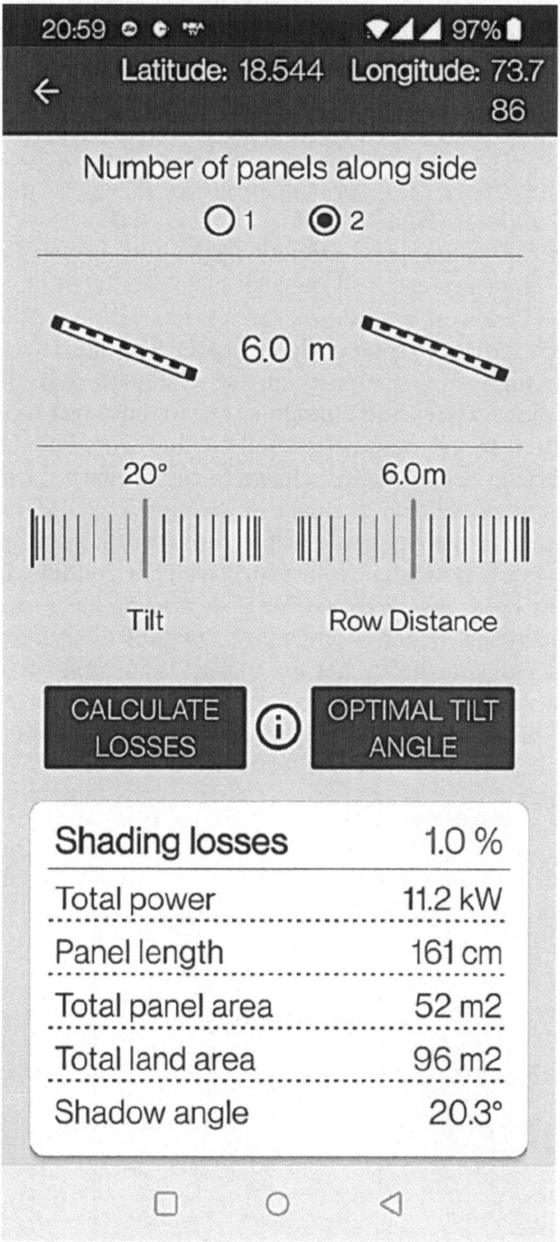

Figure 2.26 Shading analysis by MySolarPanel Android app for solar panel: 1.61 m length in Pune. The value of GCR 0.54 is an automatic output (52 m^2/96 m^2). (Source: Refer Annex. A)

where only a portion of the PV array is shaded at any given time. However, there will still be diffuse radiation even when the sun is behind a distant mountain. The horizon diagram is well explained in the PVsyst tutorial, which is briefly described below.

While carrying out the simulation exercise for "grid connected" systems under the category of "project design", click the "horizon" button. A window with some pattern diagram will appear for the given site with defined latitude and longitude (Figure 2.27).

On the X-axis, we have the azimuth angle, and on the Y-axis, the sun height angle. These represent all the possible positions of the sun throughout the year. The lowest arch shows the solar path on the Winter Solstice. The upper arch shows the solar path during the Summer Solstice. The thick lines on both sides show the parts of the solar path that are behind the photovoltaic plane. Here the PV modules receive no direct beam irradiance even when there are no obstructions on the horizon. The next task is to obtain the horizon height angles relative to the azimuth for the given site. This data was obtained from the open-source software PVGIS and was fed into the column on the right side of the sun-path diagram. As a result, a *dotted line* curve is created, also known as the horizon line, indicating far shading. It has been observed that horizon shading has a smaller relative impact in subtropical regions compared to regions farther from the equator, since sun paths in the former are almost vertical in the morning and evening. This means far shading will have little impact in New Delhi as compared to that in New York. For design simulation, PVsyst uses horizon data only if the horizon profile has features *higher than two degrees* above the horizontal.

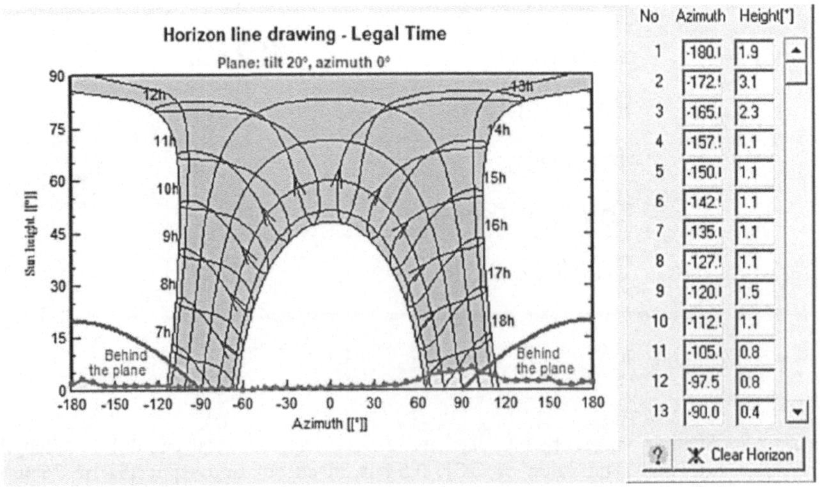

Figure 2.27 Sun-path diagram with horizon line for a site at Pune, India Latitude 18.53° N Longitude 73.85° E and panel tilt angle 20°. (Source: Refer Annex. A)

2.6 SHADING ANALYSIS TOOLS REVIEW

There are many online tools and desktop software available for near shading analysis. Some of them are free of cost and some are paid versions. For example, using the Solar Pro software version 4.1 provides a detailed simulation of analysis of shading and therefore clear insights into the site under various shading conditions [4]. Most of them need measurements of shadow-casting objects, requiring a site visit. Only the Aurora Solar design tool can calculate the dimensions of obstacles without a site visit as it uses Light Detection And Ranging (LIDAR) [9]. Some popular tools are explained below.

2.6.1 ScanTheSun Android App

This is available for free download on Google Play Store. The app has three main sections: (a) Define your geolocation, (b) Contour the horizon shape, and (c) Adjust solar panel. Each section has a Help tab at the bottom right corner, which provides tutorials and step-by-step procedures. One can get data from the app and validate it with site data.

2.6.2 Online SketchUp tool

SketchUp is a simple user-friendly online tool for 3D designing, available for free at https://www.sketchup.com/plans-and-pricing/sketchup-free. A step-by-step procedure for the first-time users can be found at https://www.linkedin.com/pulse/free-easy-shading-calculation-sketchup-eduardo-rodriguez-e-i-/. A paid version is available for advanced design, ranging from USD 119–699 per year (2022), depending upon the features one chooses. Models of solar panels and shadow-casting objects can be created and the shadow length on a particular day and time of the year can be determined.

2.6.3 Skelion tool

Skelion is a paid add-on for SketchUp Pro on desktop. It costs around $200/year (2022) and allows users to do really fast and easy shadow analysis. It also allows users to export data to other modeling software (PVsyst or SAM) for a more in-depth study. It is useful for designing residential installations or ground-mounted power plants with just a few clicks using SketchUp and Google Earth.

2.6.4 SAM (system advisor model)

SAM is free software from National Renewable Energy Laboratory (NREL), USA. After downloading the latest version, enter your email address and click Register. You should receive an email with a registration key. Copy and paste the key from the email and click Confirm. SAM includes a "shading losses" option that allows you to design a 3D layout in order to simulate shading. The 3D design tool within SAM is less effective compared to SketchUp, and it can be really hard to model the obstacles. Ideally, it is better to obtain the shading losses by using other methods and input them into SAM for analysis.

2.6.5 PVsyst software

One of the most recognized downloadable software programs, PVsyst is useful for designing solar power plants of any size. It includes a powerful shading calculator similar to SAM, but a bit complicated and not user friendly. It also allows you to import 3D models from SketchUp. It has only paid versions of software at the rate of around USD 630 per year (2022). The tutorial of PVsyst says that horizon shading is fairly accurate and easy. But for near shading, especially for large power projects, the analysis is subjected to too many conditions and its accuracy has not yet been validated by any third party.

2.6.6 HelioScope online tool

The software from Folsom Labs is a paid online tool that allows you to create quick 3D design and simulations of potential shading losses. It is one of the most preferred PC design software programs for creating proposals, SLD, and energy simulations. Although it requires manual input of dimensions (height of objects), the shading calculation results are really easy to obtain in a visually simple way. It has two price options: USD 950 per year or USD 95 per month (2022). Its accuracy has been validated by an independent study from the NREL, with the conclusion that the variance is only ± 3% compared to the ground data.

2.6.7 PVCad software

The software from the organization PVComplete is a solar PV design tool built into AutoCAD. It can be used as an AutoCAD plug-in or as a standalone application. It is optimized for quick 2D design and also provides energy simulation reports. Its free online tool is also available at PVScketch.com. The paid version is available at USD 150/month or USD 1,500/year (2022).

2.6.8 Aurora solar

One does not need to visit the site and get the actual dimensions of shadow-casting objects while using this software. It is supposed to be one of the most high-end online tools validated by the NREL. When it comes to shadow analysis, it provides LIDAR data that allows to remotely get the dimensions of objects like trees without an on-site survey. For more precise calculations one can manually input the dimensions of obstacles. If one can afford it, this would be the best tool for shadow analysis. It is available at USD 159–259 per month (2022).

2.6.9 SunEye on-site tool

This is a good alternative to the Solar Pathfinder. The instrument with software, available from Solmetric, provides high-end on-site shadow analysis. It simplifies the process and increases the quality and speed of data collection. The device includes a wide-angle lens camera that takes a picture of the surroundings and detects the obstructions in the sky. Then it uses the built-in sensors to determine the shading profile in that place. The software, including the tool, costs USD 2,415 (2022). This hand-held electronic tool measures the available solar energy by day, month, and year at the press of a button by determining the shading patterns of a particular site. The SunEye incorporates a calibrated fisheye camera, an electronic compass, a tilt sensor, and GPS to give immediate measurements in the field. Its one-handed operation, rugged enclosure, and outdoor readable display make it reliable for field use. The SunEye-210 now includes a lifetime license for the PV Designer software, enabling you to import shade measurements, select modules and inverters, layout modules, simulate kWh production, and generate PDF reports.

2.7 CONCLUSIONS

In case of a single obstruction element such as a pillar, an industrial chimney, or a large tree nearby, a manual approach of shadow analysis is always preferable due to its reliability, accuracy, and cost effectiveness.

If there are many trees or tall structures surrounding the site, the Solar Pathfinder is the most reliable method for shadow analysis. At an affordable cost of only around USD 300, it is the most useful tool with data at high confidence level, especially for rooftop projects. The SunEye on-site tool, although costlier than the Solar Pathfinder, is also an equally good on-site tool.

As regards the inter-row spacing of solar modules or shading analysis, a manual approach is once again the best and cheapest option. Another very low-cost option is the MySolarPanel Android app, which is equally good. If you are well conversant with solar design software, the open-source SAM and the paid PVsyst can be confidently used. Based on the author's experience, the results from PVsyst are found to be more consistent with the results from other tools.

Among the other technologies, HelioScope is one of the best online tools since it allows you to create quick 3D design and simulations of potential shading losses. Although it requires manual input of dimensions (height of objects), the shading calculation results are really easy to obtain in a visually simple way.

For large and very large solar projects, it is always advisable to use Aurora Solar software mainly due to its ease and off-field accuracy. Probably it is the only tool which provides LIDAR data that allows to remotely get the 3D dimensions of objects like trees without an on-site survey, with visually presentable data and plots.

ANNEXURE A: SOURCES OF FIGURES

Figure No.	*Caption*	*Source*
Figure 2.1	Shadow of a tree on a rooftop solar power plant	https://www.linkedin.com/pulse/shadow-analysis-tools-software-eduardo-rodriguez-e-i-t-/
Figure 2.2	Inter-row spacing between solar strings in a solar power plant	https://solarquarter.com/2020/05/30/designing-bifacial-pv-power-plants-getting-the-details-right-part-1/
Figure 2.3	World map showing longitude and latitude values	https://www.printablee.com/post_printable-blank-world-maps-with-grid_397827/
Figure 2.4	Four hemispheres of the Earth	https://www.pinterest.com/pin/163044448980515773/
Figure 2.5	Geographical azimuth representation	By author
Figure 2.6	Screenshot of PVsyst simulation for Pune, India	[12]
Figure 2.7	Screenshot of PVsyst simulation for Richmond, New Zealand	[12]
Figure 2.8	Solar azimuth measurements in the Northern Hemisphere	By author
Figure 2.9	Solar azimuth measurements in the Southern Hemisphere	By author

Figure No.	*Caption*	*Source*
Figure 2.10	A 3D representation of the sun-path chart for a single point on the Earth	Anthony et. al.
Figure 2.11	The solstices and equinoxes in the Northern Hemisphere	https://www.weather.gov/cle/seasons
Figure 2.12	The solstices and equinoxes in the Southern Hemisphere.	www.stargazersclubwa.com.au
Figure 2.13	Shadow tip coordinates using shadow length at a given time.	By author
Figure 2.14	Shadow diagram of a 4 m tall pillar on a roof on 21 December, 2022.	By author
Figure 2.15	Measuring the height of a tree using an inclinometer.	https://bigtrees.forestry.ubc.ca/measuring-trees/height-measurements/
Figure 2.16	A simple scheme of a Solar Pathfinder.	https://www.solarpathfinder.com/
Figure 2.17	A sample of the Solar Pathfinder diagram.	https://www.solarpathfinder.com/
Figure 2.18	Setting the magnetic declination on Solar Pathfinder diagram.	https://www.solarpathfinder.com/
Figure 2.19	Final output of Solar Pathfinder at a given site.	https://www.solarpathfinder.com/
Figure 2.20	Site tracing on the monthly sun-path diagram.	https://www.solarpathfinder.com/
Figure 2.21	Energy loss calculation using Solar Pathfinder software output.	https://www.solarpathfinder.com/
Figure 2.22	A representative scheme for calculating vertical height and horizontal coverage of solar panel.	http://esunsolar.in/how-to-calculate-inter-row-spacing-on-ground-or-flat-roof-top-solar-plant/
Figure 2.23	A representative scheme for calculating inter-row spacing of solar panels.	http://esunsolar.in/how-to-calculate-inter-row-spacing-on-ground-or-flat-roof-top-solar-plant/
Figure 2.24	Shading analysis for inter-row spacing by PVsyst for solar panel of 1.61 m length with GCR 0.54 in Pune.	[12]

Figure No.	*Caption*	*Source*
Figure 2.25	Shading analysis by SAM for solar panel: 1.61 m length and GCR 0.54 in Pune.	SAM manual
Figure 2.26	Shading analysis by MySolarPanel Android app for solar panel: 1.61 m length in Pune. The value of GCR 0.54 is an automatic output (52 m^2/96 m^2).	MySolarPanel app
Figure 2.27	Sun-path diagram with horizon line for a site at Pune, India Latitude 18.53° N Longitude 73.85° E and panel tilt angle 20°.	[12]

SUGGESTED READINGS

1. Roger Messenger and Jerry Ventre, "Chapter 2: Shading", in *Photovoltaic Systems Engineering*, CRC Press LLC, pp. 38–44, 2004
2. Michael Boxwell, "Surveying Your Site: Shading", in *Solar Electricity Handbook*, Greenstream Publishing, 2012
3. Andhika Giyantara, Wisyahyadi, Rifqi Bagja Rizqullah, and Yun Tonce Kusuma Priyanto, "Analysis of Partial shading Effect on Solar Panel Power Output", *Journal of Physics: Conference Series*, Vol. 1726, p. 012022, 2021
4. Vinodh Annathurai, Chin Kim Gan, Kyairul Azmi Baharin, and Mohd Ruddin Ab. Ghani, "Shading Analysis for the Siting of Solar PV Power Plant", *ARPN Journal of Engineering and Applied Sciences*, Vol. 11, No. 8, Apr. 2016
5. Savita Kumari, "Shading Impact on Solar PV Module", *AIJRSTEM*, Vol. 14, No. 374, 2014 (http://www.iasir.net)
6. E. E. Ekpenyong and F. I. Anyasi, "Effect of Shading on Photovoltaic Cell", *IOSR Journal of Electrical and Electronics Engineering*, Vol. 8, No. 2 (November–December 2013), pp. 01–06 (www.iosrjournals.org)
7. Kruthi Jayaram, "Analysis of Simulation of PV Array with Partial Shading Effect", *Journal of University of Shanghai for Science and Technology*, Vol. 24, No. 5, May 2022, pp. 106–206
8. Anthony N. Ede, Oluwarotimi M. Olofinnade, Edwin Enyi-Abonta and Gideon O. Bamigboye, "Implications of Construction Materials on Energy Efficiency of Buildings in Tropical Regions", *International Journal of Applied Engineering Research*, Vol. 12, No. 18, pp. 7873–7883, 2017
9. P. E. Eduardo Rodriguez, "Shadow Analysis. Tools and Software", *LinkedIn Online Publications*, April 9, 2020 (https://www.linkedin.com/pulse/shadow-analysis-tools-software-eduardo-rodriguez-e-i-t-/)

10. Solar Quarter Editorial Desk -30th May 2020, "Designing Bifacial PV Power Plants – Getting the Details Right" (https://solarquarter.com/2020/05/30/designing-bifacial-pv-power-plants-getting-the-details-right-part-1/)
11. PVsyst pdf tutorials (https://www.pvsyst.com/pdf-tutorials-grid-connected/)
12. PVsyst user's manual (https://d3pcsg2wjq9izr.cloudfront.net/files/73830/download/660275/100.pdf)
13. National Weather Service (NWS), "The Seasons, the Equinox, and the Solstices" (https://www.weather.gov/cle/seasons)
14. UBC (The University of British Columbia), BC BigTree Website, Faculty of Forestry (https://bigtrees.forestry.ubc.ca/measuring-trees/height-measurements/)
15. Solar Pathfinder, "A Professional Tool for Professional Installers" (https://www.solarpathfinder.com/)
16. eSun Solar, "How to Calculate Inter-Row Spacing on Ground or Flat Roof Top Solar Plant Installation", March 10, 2020 (https://esunsolar.in/how-to-calculate-inter-row-spacing-on-ground-or-flat-roof-top-solar-plant/)
17. P. Gilman, A. Dobos, N. DiOrio, J. Freeman, S. Janzou, D. Ryberg, "SAM Photovoltaic Model Technical Reference Update", p. (2018). NREL/TP-6A20-67399
18. Simon Wakter and Fredrik Wikerman, "A Novel Shade Analysis Technique for Solar Photovoltaic Systems", KTH School of Industrial Engineering and Management Energy Technology EGI-2014, SE-100 44, Stockholm (https://www.diva-portal.org/smash/get/diva2:735243/FULLTEXT01.pdf)
19. Suverna Sengar, "Performance Analysis of Different Shading Effects on Solar Photovoltaic Cells", JETIR, Vol. 6, No. 4, April 2019 (www.jetir.org) (ISSN-2349-5162)
20. N. Alam, V. Coors, S. Zlatanova and P. J. M. Oosterom, "Shadow Effect on Photovoltaic Potentiality Analysis Using 3D City Models", International Archives of the Photogrammetry", Remote Sensing and Spatial Information Sciences, Vol. XXXIX-B8, 2012, XXII ISPRS Congress, 25 August–01 September 2012, Melbourne, Australia

Section II

Solar photovoltaic modeling and performance analysis

Chapter 3

Prediction of photovoltaic generation using deep learning neural networks with hyperparameter optimization

Reinier Herrera Casanova and Arturo Conde Enríquez

3.1. INTRODUCTION

The production of electrical energy is an issue of vital importance for humankind. In recent years, there has been a sustained growth in worldwide electricity consumption, which is directly related to the degree of economic and social development of each country. In many countries, the main source of electricity generation is based on the use of fossil fuels. It is well known that this form of electricity production pollutes the environment, emits greenhouse gases, and leads to climate change and pollution of the planet. Moreover, fossil fuels are considered to be a non-renewable energy source, and it is estimated that their reserves will be depleted within a few years [1]. In this context, electricity generation through renewable energy sources (RES), such as wind, solar, geothermal, hydroelectric, and others, has emerged as a promising solution to solve the problems related to environmental pollution and the growing demand for electricity [2, 3].

One of the most widely used RES in recent years is solar energy, due to its high availability and high potential worldwide. Solar energy can be broken down into two main forms of energy: solar thermal and solar photovoltaic (PV). Solar photovoltaic energy has many advantages over conventional energy sources, including the following: it is a clean and safe energy source, it does not pollute the environment, it is available practically anywhere on the planet, it is easy to install and operate, it is a very flexible energy source, and it has a high level of adaptability. In addition, the costs of this technology have been significantly reduced in recent years. For these reasons, many countries have become interested in this type of energy, as it represents a viable and safe solution for the energy future of humankind [4].

The generation of electricity based on solar energy has been growing at an accelerated rate every year. The installed capacity of PV generation increased from 9 GW in 2007 to 512 GW in 2018 [5]. A study issued on the status of renewable energy in 2019 made it known that more than 181 GW of renewable energy had been installed worldwide during that period, and approximately 55% of that figure was accounted for by solar PV [6]. Moreover, during the year 2021, global solar power generation experienced

DOI: 10.1201/9781003623168-5

a 23% growth. In fact, it is considered to be the fastest-growing power generation source for 17 consecutive years. During 2021, this energy source produced approximately 3.7% of the world's electricity, which is significant compared to 2015, when only 1.1% was produced. By 2030, solar power generation is estimated to increase seven-fold, which is equivalent to approximately 19% of global electricity production [7]. According to a report by the Global Future Report 2013 REN21, by 2050 the global PV generation capacity may reach 8000 GW [8].

Solar PV can be integrated into electricity generation plants of various capacities, ranging from small-scale plants that can be operated in standalone or grid-connected operations, as well as larger PV installations that are usually operated with a direct connection to the grid. Despite all the advantages of solar PV, its integration into the electricity system is a major challenge. This energy source is greatly affected by climatic factors such as temperature, cloud cover, wind, and rain. It is also affected by the periodicity of days and nights, geographical location, and seasonal variations that occur during the year. All of this leads to the fact that the energy generated in PV installations presents a stochastic, intermittent, and variable character, which causes serious problems for its integration into power grids [9, 10].

The main problems associated with the connection of PV systems in power grids include frequency stability problems, which are associated with intermittent variations of PV generation during short time intervals and which, in turn, cause imbalances between load and generation. Voltage quality problems include voltage magnitude problems (under voltage and over voltage), voltage fluctuations, and voltage imbalances. Reverse power flows can also occur. This can be a problem when the network is designed for unidirectional power flow propagating from the higher voltage node to the lower voltage node. In addition, distortion in voltage and current waveforms, voltage flickers, and other power quality problems can occur. Moreover, the intermittency of this power source directly impacts grid stability, operation reliability, planning, and economic dispatch [11].

To address the above problems, grid operators highlight the need to decrease the degree of uncertainty of RES before it continues to increase its penetration in the system. In this context, accurate forecasting of the power generated in PV plants can be considered an extremely important aspect. Accurate forecasting allows efficient management of the operation of power grids, improves system stability, and leads to proper power generation planning. In addition, operating costs are reduced and a more economical and efficient operation of the system is achieved. It also helps extend the lifetime of battery energy storage systems, optimizing their charge and discharge cycles based on the prediction results and increasing the level of PV penetration in power grids. However, obtaining an accurate prediction of future PV energy production is difficult because the behavior of this energy source is governed by complex and unpredictable meteorological circumstances [12].

In recent years, researchers and academics from all over the world have proposed different PV power prediction models. These models can be classified and grouped into different categories. First, according to the parameter to be predicted, PV prediction can be classified into direct or indirect predictions. In the direct strategy, PV power is predicted from historical generation data and/or meteorological conditions. In the indirect strategy, the solar irradiance is initially determined to subsequently calculate the PV power using a model of the PV installation [13]. Regarding the prediction horizon, there is no unified criterion in the literature, as several authors have presented different time intervals for the prediction horizons. A frequently used classification for the prediction horizon is shown in Figure 3.1.

Very short-term forecasting is useful for voltage and power regulation, grid stability, and real-time power dispatch. Short-term forecasting impacts grid security, electricity market participation, and power reserve management. Medium and long-term forecasts are characterized by high randomness and high prediction errors, so their application is mainly focused on issues related to grid planning. In fact, medium-term forecasting is beneficial for unit commitment decisions, planning, and power system maintenance scheduling. Long-term forecasting is of interest for planning power generation, transmission and distribution, and structuring and operating electricity markets [14].

Depending on the spatial scale of the prediction, it can be classified into regional prediction or single-plant prediction. On the other hand, depending on the type of prediction, models can be classified into point prediction (deterministic) or probabilistic prediction. The deterministic prediction obtains a unique value at each instant of time. Probabilistic predictions report the confidence at an instant of time in which the value falls within a certain range defined by the model. In both cases, the prediction can be one step ahead or multiple steps ahead [15]. Various types of data can be used for PV prediction, e.g., time series, sky images, and spatial data. They can be univariate data (e.g., PV power time series) or multivariate (e.g., when combining PV power time series with meteorological data). In order

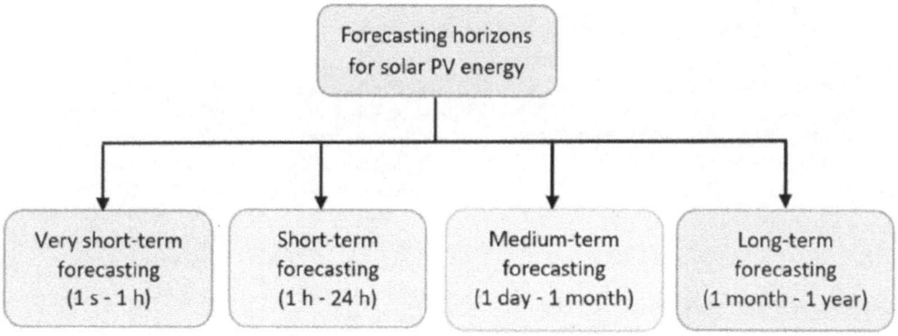

Figure 3.1 Classification of PV energy prediction horizons.

to obtain satisfactory performance of forecast models, preprocessing of the data is required. Preprocessing may include plotting the data, correcting outliers, changing the temporal resolution of the data, decomposing and clustering data, correlation analysis, selection of the main input features, and data normalization or standardization, among other aspects of importance [16].

Another possible classification is related to the method used in the prediction, in this case the following categories are defined: statistical methods, physical methods, artificial intelligence or machine learning methods (which also include deep neural networks), and hybrid methods [15]. A more detailed description of these classifications is shown in Figure 3.2.

Physical forecasting methods establish a mathematical model of the PV plant, considering several characteristics and parameters of the installation. They then use meteorological predictions of some variables (e.g., solar irradiance and ambient temperature) to combine them with the established model and obtain the PV power output. For example, the PV system output power, as a function of temperature and solar irradiance, can be obtained by Equation 3.1 [17].

$$P_{PV}(t) = N_{PV} \cdot P_{PV_{STC}} \cdot \frac{G(t)}{G_{STC}} \cdot \left(1 + \frac{\alpha_P}{100} \cdot \left(T_{cell}(t) - T_{STC}\right)\right) \cdot f_{PV} \tag{3.1}$$

where $P_{PV}(t)$ represents the PV system output power at time t, N_{PV} is the number of panels, $P_{PV_{STC}}$ represents the PV power under standard test conditions ($G_{STC} = 1000 \frac{W}{m^2}, T_{STC} = 25°C, AM = 1.5$), and $G(t)$ represents the

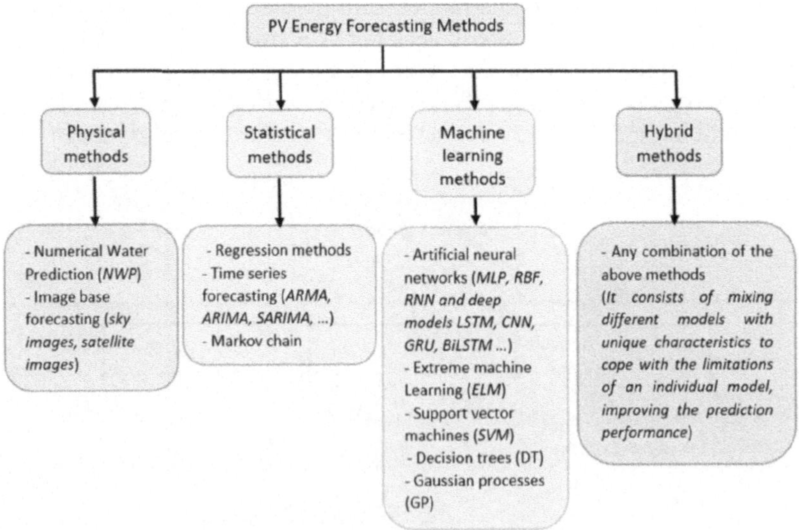

Figure 3.2 Classification of PV energy forecasting methods.

solar irradiance at time instant t. In addition, $T_{cell}(t)$ is the PV cell temperature at time t, α_P refers to the temperature coefficient for maximum power, and f_{PV} is a correction factor that takes into account the effect of shading losses and the degradation level of the PV panel.

The cell temperature is related to the ambient temperature and solar irradiance by Equation 3.2).

$$T_{cell}(t) = T_{amb}(t) + G(t) \cdot \left(\frac{NOCT - 25}{1000} \right) \tag{3.2}$$

where $T_{amb}(t)$ represents the ambient temperature at time t and $NOCT$ represents the nominal operating temperature of the PV cell.

This model does not require historical data; its performance is mainly based on detailed plant information and weather forecasts for that geographic location. However, it presents low robustness and limited applicability because it requires PV plant data that, in many situations, are not known and must be estimated. Moreover, its accuracy also depends on the accuracy of weather forecasts, which can sometimes have a high degree of uncertainty, especially in adverse weather conditions such as rainy or cloudy days [18]. Statistical methods consist of establishing a model between input and output data based on the correlation that may exist between historical data, estimation of some parameters, and curve fitting. Compared to physical methods, it can be said that statistical methods are simpler and more applicable in different locations. Their application requires a large number of historical measurements, and their accuracy is closely related to the quality of the data collected. It also requires a high number of numerical calculations, and sometimes, its execution time is high, which limits its application, especially for very short-term prediction horizons [19].

Machine learning (ML) based methods have become one of the most widely used methods in time series forecasting. These models have the ability to accurately extract high-dimensional complex nonlinear relationships between input and output data. In addition, they can achieve higher accuracy and prediction quality compared to other methods. However, shallow ML models, when faced with prediction problems with large data volumes and large numbers of time series of different variables, may have limited learning capability. Moreover, they may be affected by problems of overfitting, falling into local minima, and problems of vanishing or exploding gradients during the training process [20].

To solve the above problems, deep learning (DL) models are proposed. These models can extract nonlinear and complex behavioral patterns from time series, achieve higher generalization capability, and efficiently analyze large volumes of data. Compared with ML methods, DL models exhibit better performance, higher stability, and robustness in time series prediction [15]. Hybrid methods combine different individual models in order to complement strategies, improve accuracy, optimize resources, and diversify their usefulness. In recent literature, it is common to find the union of ML or DL methods with heuristic methods (genetic algorithms (GA), particle swarm optimization (PSO), differential evolution (DE), ant colony optimization (ACO), tabu search algorithm (TSA), and Bayesian optimization (BO), etc.). These heuristic methods can be used to optimize the set of hyperparameters of the ML or DL model, to obtain optimal initial weights, and to generally improve the performance of forecasting models [16].

3.1.1 State of the art of PV forecasting

In recent years, the topic of PV forecasting has attracted the attention of many researchers around the world, and a large number of papers have been published on this subject.

In [21], a solar irradiance forecasting model was proposed based on satellite data, meteorological predictions, and solar irradiance measurements at various locations. The proposed model outperformed another set of models that were trained with data from a specific location. On the other hand, [22] presented a PV power prediction approach that includes solar irradiance information obtained from satellite data, numerical weather prediction (NWP), and temperature measurements and power generated at a solar plant. The model performed favorably in PV prediction, with a forecast horizon up to 36 hours ahead. A prediction model based on satellite data and NWP forecasts was developed in [23]. Using these sources of information and a Kalman filter, the cloudiness index for a city was predicted with intra-hour resolution. The results obtained can be very useful for solar energy forecasting.

The author of [24] performed the short-term prediction of solar radiation and energy production of a PV plant using an autoregressive integrated moving average (ARIMA) model. The obtained results were compared with an intelligent persistence model, and the proposed model showed superior performance. In [25], a statistical model for short-term PV power prediction (up to 6 hours ahead) was proposed. The proposal was based on a vector autoregressive (VAR) model and a least absolute reduction and selection operator (LASSO) for variable selection. It was shown that using spatio-temporal data from several distributed plants can improve the quality of predictions. In [26], the prediction of the PV power of the next hour was performed using support vector regression (SVR). The conducted studies showed that the input feature selection process could significantly impact the prediction. Compared to other regression models (linear regression, quadratic regression and LASSO regression), the SVR model showed superior performance.

The authors of [27] presented a comparison of several prediction methods, such as artificial neural networks (ANN), decision trees (DT), extreme gradient boosting (XGB), and random forests (RF). In this case, the tests performed showed that the best-performing methods were ANN and DT. In

[28], a hybrid method, combining iterative filtering (IF) and extreme learning machine (ELM), was used for very short-term PV power forecasting. The performance of the proposed method was evaluated in three PV plants of different locations, and the obtained results were satisfactory. On the other hand, [29] implemented a hybrid model based on the combination of four feed-forward neural networks (FFNN) with wavelet transform (WT) to forecast PV power ten minutes in advance. The performance of the proposed model was found to be superior compared to a persistent model and several ML models reported in the literature. In [30], short-term PV power prediction (30 minutes to 6 hours ahead) was performed using a hybrid method combining a support vector machine (SVM) and an improved ant colony optimization (IACO) algorithm. The results obtained were compared with other simple models, and it was shown that the optimization of the SVM model parameters using the IACO algorithm was very beneficial.

In [31], very short-term PV power prediction was addressed by combining an instance-based feature selection method, known as RReliefF (RRF), and a multilayer perceptron (MLP) ANN. The proposed model was found to be more accurate than three widely used ML algorithms (ANN, SVR and RF). A neural network model was also used in [32] to predict PV energy with a very short time horizon; in this case, the nonlinear autoregressive neural network with exogenous inputs (NARX) optimized by (GA) was used. Comparison of the NARX-GA model with other models in the literature showed satisfactory results.

In [33], a long short-term memory (LSTM) model was combined with location-specific synthetic weather predictions to predict short-term PV generation. The model predictions were compared with a recurrent neural network (RNN) model, a generalized regression neural network (GRNN), and an extreme learning machine (ELM). In all cases, the predictions of the proposed model showed higher accuracy and stability. In [34], an LSTM model was also used to predict PV power one hour in advance in three different plants. A comparison of the proposed model with some ML models proved superior in terms of robustness and accuracy. In [35], an LSTM model was employed to forecast the next day's PV power. This study highlighted the importance of data preprocessing to improve the quality of predictions. The LSTM model was compared with an MLP neural network model on different types of days (stable and variable) and at two different times of the year (winter and summer). In all cases, the LSTM model showed better performance and accuracy in its predictions.

In [36], a hybrid NARX-LSTM prediction model was proposed. In this case, the most relevant hyperparameters of the LSTM model were optimized using the tabu search algorithm (TSA). The prediction horizon analyzed ranged from 5 minutes to 24 hours ahead. To demonstrate the effectiveness of the proposed model in PV prediction, two real databases, with different temporal resolutions and geographical locations, were used. In [37], a four-core convolutional neural network (CNN) model was presented for PV forecasting with a horizon of 5 minutes to 1 hour-ahead.

The model was compared with a CNN model and a CNN-LSTM model, and its results were superior. In addition, it was shown that the quality of predictions worsens with increasing prediction horizon and decreasing temporal resolution of the data. In [38], a gated recurrent unit (GRU) based DL model was proposed to predict the solar irradiance of the next hour. The proposed model was compared with an LSTM model. Both models showed similar accuracy, but the GRU model required less training time. It was shown that including exogenous weather variables and cloud cover data as inputs to both models significantly improved forecast accuracy.

Reference [39] presented a very short-term PV prediction model based on a bidirectional long short-term memory (BiLSTM) neural network and a GA, for optimization of the fundamental hyperparameters of the BiLSTM model. The proposed model was compared with a group of ML and DL models. The obtained results showed that the BiLSTM model is a powerful tool in time series prediction. A BiLSTM model based on an attention mechanism was also proposed in [40], to predict solar irradiance for a short-term horizon. The model was able to forecast variations in solar irradiance at short intervals efficiently. In [41], a hybrid deep learning model (BiLSTM-CNN) was implemented for PV energy prediction. In this case, the BiLSTM model was responsible for extracting the temporal features of PV energy, while CNN was responsible for extracting the spatial features. The predictions of the proposed model showed higher accuracy, compared to other simple and hybrid deep learning models. In [42], different types of statistical models and various types of neural networks (LSTM, BiLSTM, FFNN, MLP) were compared in PV prediction for a short-term time horizon. The results obtained showed that ANN-based models were more accurate than statistical methods. Of all the models analyzed, BiLSTM showed the best performance for all prediction horizons and for different climatic conditions.

In [43], a hybrid model based on variational mode decomposition (VMD), improved sparrow search algorithm (ISSA), and GRU was proposed for short-term prediction of PV power. The VMD method was applied to decompose the original PV series into several subseries to reduce the non-stationarity of the original data. The optimization algorithm was used to fit some hyperparameters of the GRU model. Compared with other ML and DL algorithms, the proposed model ensured efficient prediction of PV power under different weather conditions.

Also, in [44] a hybrid model for the prediction of the hour-ahead power output in three different types of PV modules was presented. The model was based on the combination of LSTM and the salp swarm algorithm (SSA), for the optimization of five relevant hyperparameters. The proposed model was compared with GA-LSTM, PSO-LSTM, and LSTM and showed higher robustness and effectiveness in the predictions made.

From the literature review, it can be seen that a large number of recent publications have focused on artificial intelligence methods for PV power prediction in very-short and short-term time horizons. Within the AI field, machine learning and deep learning methods have gained in popularity.

In the case of DL-based models (LSTM, GRU, CNN, BiLSTM, among others), they have a great capacity to extract complex behavior patterns in time series, handle large volumes of data, and predict the future behavior of the time series with high accuracy. In many cases, the union of several deep learning models or the optimization of their hyperparameters through optimization methods leads to results of higher quality and robustness. However, there are many models in the literature that, despite their high accuracy, are very computationally expensive. In addition, they handle large amounts of input data that are sometimes difficult to obtain or are simply not available, making their practical application difficult and limited. On the other hand, in some cases, the data preprocessing and model input selection stage, which is a key factor in ensuring the satisfactory performance of forecasting models, is not analyzed in the required depth.

In this work, a BiLSTM deep learning model is proposed to predict the power generated in different PV installations for an hour-ahead and day-ahead forecast horizon. The main contributions of this work are as follows:

- An efficient algorithm for data preprocessing is proposed in order to improve the quality of the data and obtain a better performance of the prediction model.
- A Bayesian optimization algorithm (BOA) is implemented to obtain the optimal adjustment of the hyperparameters of the model, which contributes to improving its efficiency and generalization capacity.
- An efficient and accurate working methodology is proposed, applicable to any photovoltaic installation with the necessary historical data.
- A comparative analysis of the results obtained by the BiLSTM model with respect to a NARX model and an MLP model is presented. In this way, the efficiency and quality of the proposed model in predicting the PV power of the hour-ahead and the day-ahead is demonstrated.

For future research, it is intended to analyze some hybrid DL models and use other larger data sets and from other PV installations. In this way, the performance of the proposed model in other operation scenarios could be evaluated.

3.2. FORMULATION OF THE MODELS

This section describes the basic formulation of some artificial neural networks used in PV energy prediction. The general characteristics of RNNs are outlined and then the performance of the LSTM type RNN model is discussed. Subsequently, a more advanced and deeper variant of the recurrent neural network, the BiLSTM network, is analyzed. The Bayesian optimization algorithm, which is used to adjust the most relevant hyperparameters of the prediction models, is also described.

3.2.1 Recurrent neural networks

Recurrent neural networks [45–47] are a family of artificial neural networks that specialize in processing sequential data. Therefore, they are useful for problems where time dependence exists, e.g., in time series. The following is a mathematical description of the RNN model and its fundamental characteristics.

An RNN can be viewed as a dynamic system, as follows:

$$y^{(t)}, s^{(t)} = g\left(x^{(t)}, s^{(t-1)}\right) \tag{3.3}$$

where $x^{(t)}$ is the input vector of the sequence at instant t, $g\left(x^{(t)}, s^{(t-1)}\right)$ is the transition function that maps from the previous instant to the current instant, $y^{(t)}$ is the output vector at instant t and $s^{(t)}$ is the internal state of the model. Figure 3.3 shows the recursive development of an RNN.

Depending on the RNN architecture, the transition function may vary. In general, the transition function can be represented by a three-layer graph consisting of an input layer, a hidden layer, and an output layer. The nodes found in the input layer represent the values or features of the input vector $x^{(t)}$ [47]. The nodes in the other layers take as input the values of the previous layer and deliver an output value through a linear combination, as follows:

$$a_{q,d} = f\left(w_0 + \sum_{j=1}^{m} w_{qj} a_{j,d-1}\right) \tag{3.4}$$

where $a_{q,d}$ is the output of node q in layer d, w_0 is the bias, and w_{qj} is the weight or connection between node j in the previous layer and node q in layer d. The term $a_{j,d-1}$ is the previous output of layer d, and f is a function that may be nonlinear, called the activation function.

The activation function of an RNN is responsible for giving it the ability to find nonlinear relationships between input values. Among the most

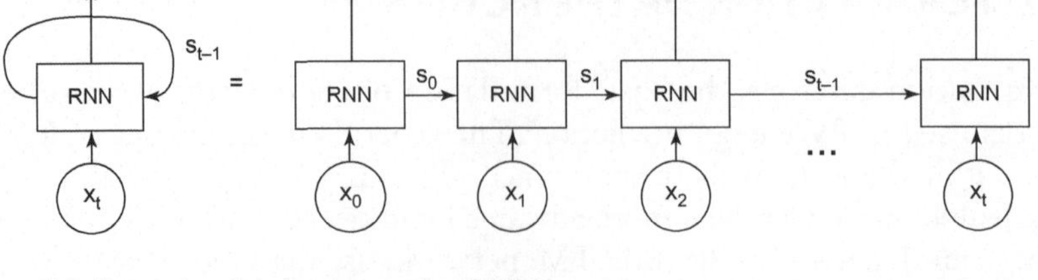

Figure 3.3 Recursive development of an RNN.

commonly used activation functions are the sigmoid (σ), the hyperbolic tangent (*tanh*), and the rectified unified unit (*ReLU*) [45].

The sigmoid function is defined as:

$$\sigma(z) = \frac{1}{1+e^{-z}} \tag{3.5}$$

The hyperbolic tangent function is defined as:

$$tanh(z) = \frac{e^{z} - e^{-z}}{e^{z} + e^{-z}} \tag{3.6}$$

The rectified unified unit function is defined as:

$$\mathrm{ReLU}(\mathrm{z}) = \mathrm{z}^{+} = \max(0, \mathrm{z}) \tag{3.7}$$

The functions σ and *tanh* are characterized by a smooth transition between [0, 1] and [1] respectively. On the other hand, the *ReLU* function is not bounded and its transition is not smooth, so there is no derivative when z is equal to zero.

After defining the architecture and the activation function, it is important to find the parameters of the w_{qj} model that best fit the data. There are several techniques to perform this optimization, and in general what is sought are the parameters w_{qj} that minimize a loss function L. In prediction problems, the mean square error (MSE) can be used as the loss function, as shown in Equation 3.8 [46].

$$L = \frac{1}{n}\sum_{i=1}^{n}(y_i - \hat{y}_i)^2 \tag{3.8}$$

where y_i are the actual values of the variable, $\hat{y}_i$ are the output values of the RNN and n is the number of measurements.

The major difficulty encountered in the use of RNNs is their limited learning capability when faced with large time series. In this situation, vanishing or gradient explosion problems may occur, leading to a model with poor training, poor generalization capability, and high prediction errors [47].

3.2.2 Recurrent neural networks LSTM

To solve the problems concerning RNNs, Sepp Hochreiter and Jürgen Schmidhuber proposed the LSTM recurrent neural network model in 1997 [48]. Since then, LSTM networks have proven their effectiveness in multiple tasks and applications, including large time series prediction. In traditional neural networks, the main element is the artificial neuron, while the basic element of LSTM networks is the memory cell. The model can contain a certain number of memory cells depending on the problem to be solved.

Each memory cell presents three main gates in its architecture, known as: input gate, output gate, and forgetting gate [49]. Figure 3.4 shows the structure of a memory cell of the LSTM model.

The memory cell remembers values over arbitrary time intervals, and the three gates control the flow of information into and out of the memory cell. The information entering and leaving the memory cell is controlled by the input (I_t) and output (O_t) gates respectively. The forgetting gate (F_t) controls the information in the internal state of the memory cell and determines whether features should be forgotten or retained in the learning process. The operation of the LSTM model is described by Equations 3.9–3.14) [36]:

$$\tilde{C}_t = \tanh\left(x_t \cdot W_{xc} + H_{t-1} \cdot W_{hc} + b_c\right) \tag{3.9}$$

$$I_t = \sigma\left(x_t \cdot W_{xi} + H_{t-1} \cdot W_{hi} + b_i\right) \tag{3.10}$$

$$F_t = \sigma\left(x_t \cdot W_{xf} + H_{t-1} \cdot W_{hf} + b_f\right) \tag{3.11}$$

$$O_t = \sigma\left(x_t \cdot W_{xo} + H_{t-1} \cdot W_{ho} + b_o\right) \tag{3.12}$$

$$C_t = \tilde{C}_t \cdot I_t + F_t \cdot C_{t-1} \tag{3.13}$$

$$H_t = \tanh\left(C_t\right) \cdot O_t \tag{3.14}$$

where W_{xc}, W_{xi}, W_{xf}, W_{xo},W_{hc}, W_{hi}, W_{hf}, W_{ho} represent the model weights; the previous and current time instant are represented by t–1 and t, respectively. The biases are represented by b_c, b_i, b_f, b_o. The hidden layer

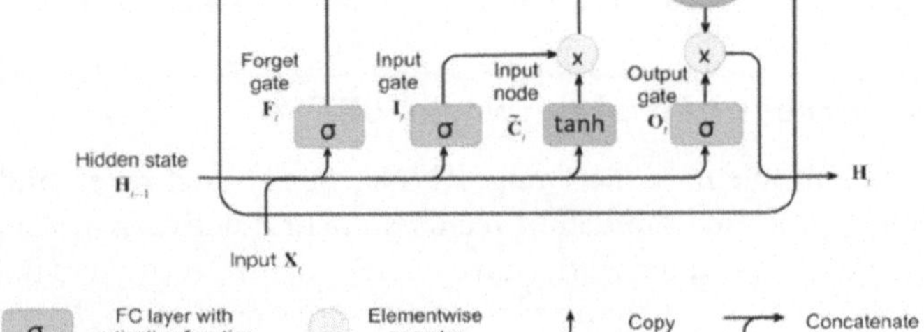

Figure 3.4 Structure of a memory cell of the LSTM model.

at the previous instant is represented by H_{t-1}. The element $\tilde{C}_t$ represents the candidate memory cell, C_t is the memory cell of the current time step, C_{t-1} represents the memory cell at the previous time step, H_t is the hidden state, and x_t is the current time entry., and the symbol (*) represents an element-by-element multiplication, also known as the Hadamard product.

3.2.3 Recurrent neural networks BiLSTM

In the unidirectional LSTM network, the information is transmitted in one direction and the processing of the data sequence is performed only forward. The BiLSTM model can perform data processing in two directions – forward and backward. This allows it to more effectively extract deep features and comprehensively learn time-correlated information. In this way, it can achieve higher accuracy in predicting complex time series, such as PV energy [39].

In the BiLSTM model, there are LSTM layers that perform forward and backward propagation of information. In the horizontal direction, the model operation is based on performing the calculation of the hidden vector in the forward direction $\vec{h}_t$ and in the backward direction $\overleftarrow{h}_t$ for each time step t, simultaneously. In the vertical direction, a unidirectional flow is realized that propagates from the input layer to the hidden layer and then to the output layer [50, 51]. The basic structure of the BiLSTM model is shown in Figure 3.5.

The operation of the BiLSTM model is described by:

$$\vec{h}_t = LSTM\left(x_t, \vec{h}_{t-1}\right) \tag{3.15}$$

$$\overleftarrow{h}_t = LSTM\left(x_t, \overleftarrow{h}_{t+1}\right) \tag{3.16}$$

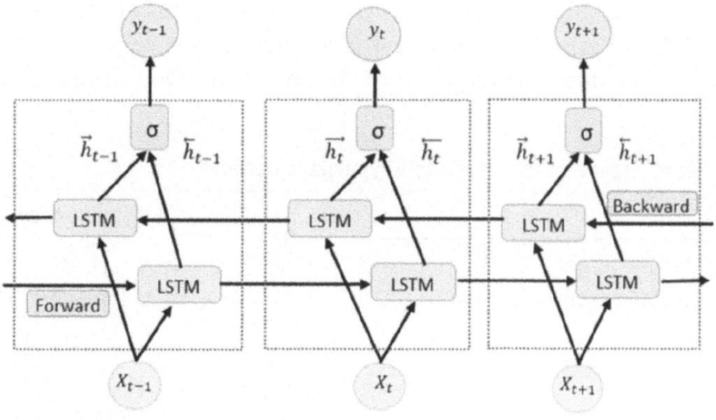

Figure 3.5 Basic structure of the BiLSTM model.

$$y_t = W_{\vec{h}_y} \vec{h}_t + W_{\overleftarrow{h}_y} \overleftarrow{h}_t + b_y \qquad (3.17)$$

where x_t and y_t correspond to the input and output at time t, $W_{\vec{h}_y}$ and $W_{\overleftarrow{h}_y}$ represent the weights of the LSTM model in the forward and backward direction, respectively. The term b_y represents the output layer bias.

3.2.4 Bayesian optimization algorithm for hyperparameter adjustment

Proper tuning of the hyperparameters of a prediction model ensures good performance and higher quality of predictions. Bayesian optimization is a method that can be efficiently applied to fit the hyperparameters of a prediction model. Compared to other automatic search methods, such as grid search and random search, Bayesian optimization is more efficient. Its operation is based on finding a point to minimize a real valued function called the objective function. To perform a more efficient search, Bayesian optimization can rely on the previous data set and use this information to update the future values of the optimization function [52, 53].

In this work, the Bayesian optimization algorithm (BOA) is used to tune the following hyperparameters: number of hidden layers (N_{hl}), number of units in the hidden layer (N_{hu}), initial learning rate (I_{lr}), regularization factor L_2 (R_{L2}), and mini-batch size (S_{mb}). Table 3.1 shows the possible search spaces for each hyperparameter considered.

The proper adjustment of these hyperparameters is of great importance to ensure a satisfactory performance of the prediction model. For example, the number of hidden layers of the model is an important parameter, since using a larger number of hidden layers may improve the accuracy of the predictions, but increases the computational cost of the model. The number of hidden units also has a significant influence on the performance of the predictive model. A very small value of this parameter can lead to poor training and poor generalization capability of the model. On the other hand, too large a value can lead to overfitting problems and very long training times. With respect to the initial learning rate, a very small value may cause extremely slow convergence of the model or even failure to converge

Table 3.1 Search space of the proposed hyperparameters

Hyperparameter	*Search space*
Number of hidden layers	[1–2]
Number of units in each hidden layer	[50–200]
Initial learning rate	[0.001–0.010]
Regularization factor L_2	[10^{-4}–10^{-2}]
Mini-batch size	[32–256]

for a given number of iterations. Conversely, a very high initial learning rate may cause a very fast convergence of the model, but the solution may be of low quality. The regularization factor L_2 is used to avoid overfitting of the model, and its value should be selected appropriately to obtain higher robustness and stability in the predictions. The size of the mini-batch is related to the number of examples that are introduced for training the network in each iteration. A small value causes the network to have a small amount of data in its memory; under these conditions, the training is faster and the model response can be more unstable. If the mini-batch size has a large value, the opposite effect occurs [53].

In this case the Bayesian optimization problem is established through a functional association between the hyperparameters of the prediction model and the loss function:

$$h^* = argmin_{h \in H}(Opt_{obj}) \tag{3.18}$$

$$Opt_{obj} : H\left(N_{hl}, N_{hu}, I_{lr}, R_{L2}, S_{mb}\right) \tag{3.19}$$

where h^* represents the optimized hyperparameter combination, H represents the hyperparameter search space, h denotes the set of hyperparameter combinations in H and Opt_{obj} represents the objective function.

Initially, BOA fits a surrogate probability model based on Bayes' rule. Then, the surrogate probability model is updated through a Gaussian process (GP) model in order to obtain a posterior distribution. The next step is to use the posterior distribution to obtain an acquisition function that is used to determine the next point to be evaluated. In this case the acquisition function, called expected improvement (EI), is used. The BOA performs exploration in regions of the search space where the surrogate probability model expects to find the global optimal solution to the problem. The above steps are continuously repeated until a certain number of user-defined iterations is reached. Finally, a combination of hyperparameters that guarantees a minimum prediction error of the model is obtained [54].

3.3. DATA ANALYSIS AND PREPARATION

This section initially describes the data used in the study. These data correspond to two photovoltaic plants of different power that are located in different places. Then, the different parts that make up the proposed algorithm for data preprocessing are explained. This algorithm includes outlier removal, filtering of the night values, selection of the input features of the forecast model by Pearson correlation analysis, change of temporal resolution, and normalization of the data.

3.3.1 Description of the data

The historical data used in the study were obtained from two PV plants of different installed capacity, which are located at different latitudes. The first case is a PV system with a fixed tilt angle of 15° and facing south. This plant has a nominal installed capacity of 3 MW and covers an area of about 61,200 m^2. The plant is equipped with 12,000 monocrystalline silicon photovoltaic modules with a nominal power of 250 W, which feed three inverters. The power data is measured at the final power delivery point of the PV system. Measurements of generated power, solar irradiance, ambient temperature, and PV panel temperature are available at this plant. The measurements were recorded with a time resolution of 5 minutes from January 1 to December 31, 2016. Table 3.2 shows the description of the main characteristics of each variable present in the database.

Figure 3.6 shows the behavior of each time series over a 14-day time period from December 18 to December 31, 2016. It is evident that the PV power generated has very similar behavior to the solar irradiance. Both variables present a zero value during the night time, and start to increase their value at dawn, reaching their maximum value in the midday hours. Then they begin to decrease as the afternoon progresses and return to zero at dusk. In the case of the ambient temperature and the temperature of the photovoltaic panels, a pattern of behavior is also observed that shows how temperatures increase during the day, when the solar irradiance is higher, and decrease during the night, because the solar irradiance becomes zero.

The second case corresponds to the eco-Kinetics PV system, which is located at the Desert Knowledge Australia Solar Center (DKASC), Alice Springs, Australia [55] and was installed in 2010. It is a two-axis solar tracking system. The system has a nominal installed power of 26.52 kW and occupies an area of approximately 199.16 m^2. There are 156 monocrystalline silicon photovoltaic panels, with a nominal power of 170 W, installed in this plant. The system is equipped with two Fronius Symo 15 M X2 inverters. The data collected includes the active power generated, global

Table 3.2 Description of the variables present in the database I

Variables present in the database	*Unit of measure*	*Minimum value*	*Maximun value*	*Mean*
PV power generated (*Ppv*)	[MW]	0	2.73	0.76
Global horizontal irradiance (*GHI*)	[W/m²]	0	1296	368
Ambient temperature (*Ta*)	[°C]	13.90	37.50	27.51
Temperature of PV panels (*Tp*)	[°C]	6.80	64.90	36.97

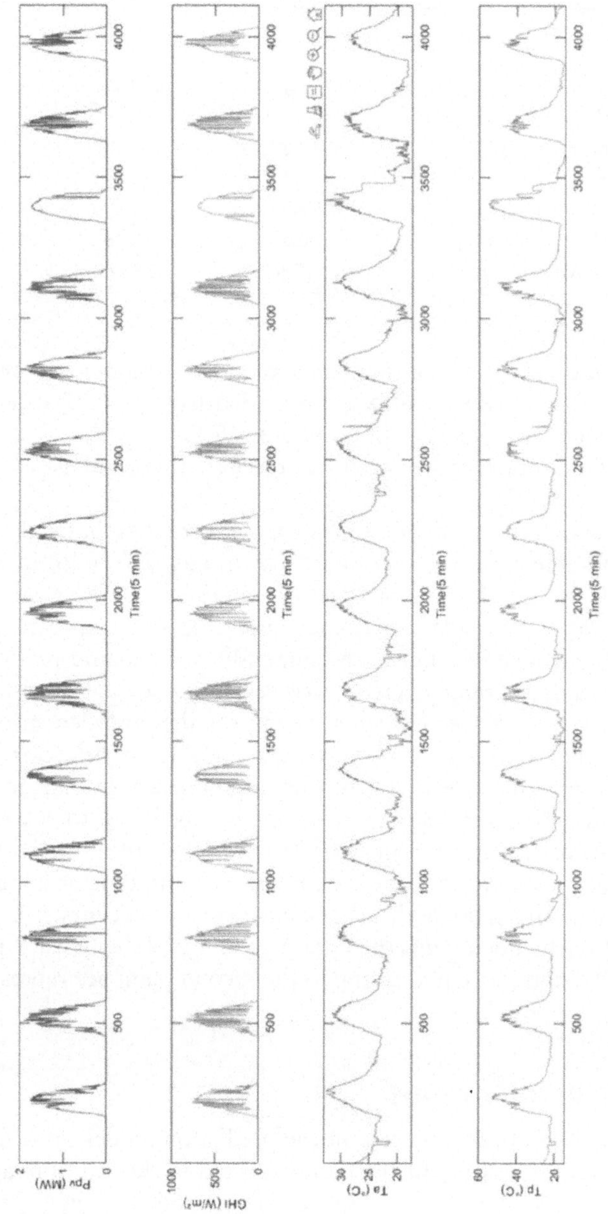

Figure 3.6 Behavior of database 1 variables from December 18 to December 31, 2016.

Table 3.3 Description of the variables present in database 2

Variables present in the database	*Unit of measure*	*Minimum value*	*Maximum value*	*Mean*
PV power generated (*Ppv*)	[MW]	0	8.36	3.55
Global horizontal irradiance (*GHI*)	[W/m²]	0	1281	511.80
Ambient temperature (*Ta*)	[°C]	4.85	43.70	26.75
Relative humidity (*Rh*)	[%]	5.82	70.35	23.61
Wind speed (*Ws*)	[m/s]	0.07	4.98	2.14
Wind direction (*Wd*)	[°]	19.52	350.60	160.50

horizontal irradiance, ambient temperature, relative humidity, wind speed, and wind direction. These data were recorded from July 29 to December 10, 2020, with a temporal resolution of 5 minutes.

A general description of the variables present in this database is presented in Table 3.3.

Figure 3.7 shows the behavior of the different variables that make up database 2 for the time period from August 7 to August 20, 2020. In this case, similar conclusions to the previous case are obtained regarding the behavior of photovoltaic power, solar irradiance, and ambient temperature. In the case of the wind speed variable, an eminently variable and random character is observed, although there is a certain behavior pattern that shows that this variable increases its value during the day and decreases during the night. In the case of the wind direction variable, its behavior is totally random and a certain pattern of behavior cannot be clearly defined. Finally, the relative humidity variable shows the opposite behavior to power, irradiance, ambient temperature, and wind speed. When these variables reach their maximum value during the day, relative humidity is at its minimum value, and then during the night, the opposite process occurs.

In the rest of the document, the data pertaining to the first PV plant are defined as ($D1$) and those pertaining to the second plant are represented by ($D2$).

3.3.2 Data preprocessing

To guarantee a good performance of the prediction models, it is necessary to perform a preprocessing of the data collected in the initial phase of the study. This can improve the convergence of the models, reduce the computational cost, and increase the quality and accuracy of the forecasts to be carried out. Preprocessing may include several important phases such as outlier correction, missing data handling, filtering of night values, correlation analysis to define possible inputs to the prediction model, change of the

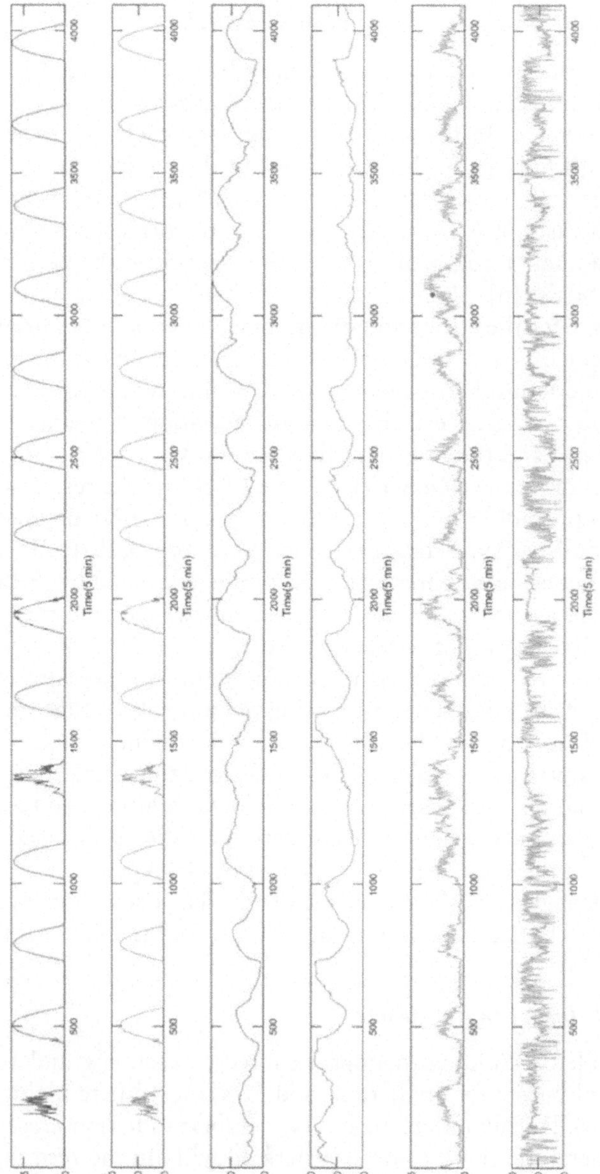

Figure 3.7 Behavior of database 2 variables from December 7 to December 20, 2020.

time resolution, and data normalization. A description of each of the above phases is presented in the following sections.

3.3.2.1 Correction of outliers and missing data

Outliers are values that differ significantly from the remaining values of the time series and can be produced mainly by errors in the sensors or measurement equipment. In the case of missing data, their presence is mainly associated with failures or disconnections in the equipment during a certain period of time, which causes the loss of information in that period of time. In both cases, it is extremely important to correct these problems to improve data quality and achieve a positive impact on the performance of prediction models [16].

In this work, for the correction of outliers, a method was initially implemented that consists of correcting the values that are outside the logical limits of each time series. Subsequently, an algorithm was used that considers outliers as values that exceed the standard deviation of the mean by more than three times and then replaces those values with the non-outlier closest to its position in the time series. Figure 3.8 shows the result of applying the described procedure to the *GHI* time series of the *D*1 database for the period of time between February 7 and February 10, 2016. In this case, the initial data has been represented, the outliers detected by the algorithm (8 in this case), as well as the clean final data derived from the process of detection and correction of outliers.

To solve the problems related to missing data, a procedure is applied, based on filling the missing data by using the moving average method and a sliding window of a given length. In this work, a window size equal to 12 measurements was used, which corresponds to a time period of one hour. However, for larger data absences, the window resolution can be extended. Figure 3.9 shows the missing value correction for the *Ta* time series of data set *D*1 during a three-day period of the month of June 2016. It can be seen that the implemented method works properly. The corrections which were made appear to be accurate and fit the behavior of the time series.

3.3.2.2 Filtering of night values

A recommended technique to improve forecast accuracy and reduce the computational burden of prediction models is to eliminate nighttime measurements [16]. During nighttime hours, the solar radiation presents a zero value, and therefore, the PV power generated will also be zero during that time period. However, the night hours vary according to the seasons of the year and this aspect should be considered when eliminating these measurements [56]. It is best to perform a preliminary inspection of the data and precisely define the values that can be eliminated, so as not to affect the quality of the data.

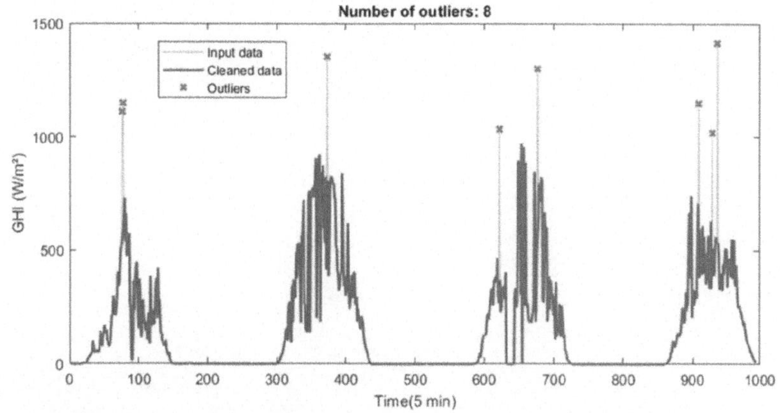

Figure 3.8 Outlier correction in the *GHI* time series of the *D1* data set.

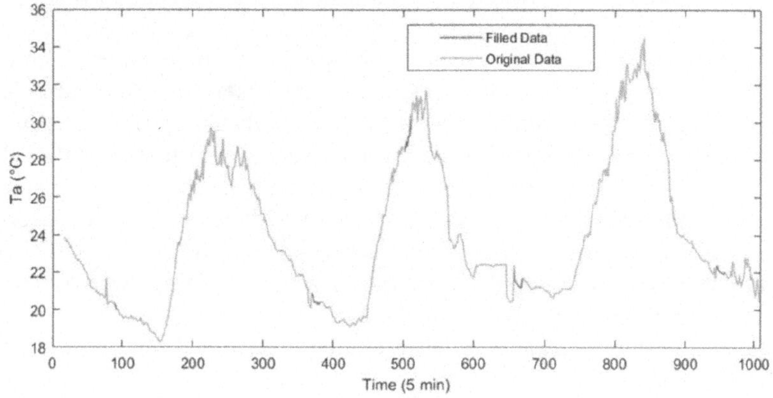

Figure 3.9 Correction of missing data in the *Ta* time series of data set *D1*.

Figure 3.10 shows the behavior of the *Ppv* time series for the time period August 10 to August 13, 2020. The first case represented corresponds to the original *Ppv* series and the second case corresponds to that series after filtering the night values. It can be observed that the applied method maintains the quality and adequate behavior of the filtered data and eliminates a large number of zero value measurements, which are unnecessary for the prediction.

3.3.2.3 Correlation analysis

Correlation analysis is a very effective technique for determining the degree of linear relationship between two specific variables. The linear correlation coefficient can take values in the range [1]. A value between 0 and 1 indicates a positive correlation between the variables, i.e., there is a direct

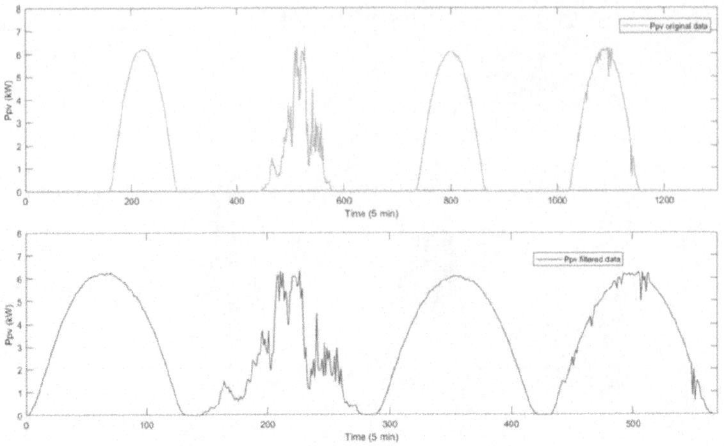

Figure 3.10 Filtering of nighttime values in the *Ppv* time series of data set *D*2: (a) Represents the original series and (b) Represents the filtered series.

relationship between them. On the other hand, a value between 0 and –1 is equivalent to a negative correlation; i.e., there is an inverse relationship between the variables analyzed. Correlation coefficient values equal to or close to zero indicate that there is no (or a very weak) linear relationship between the variables. Values close to 1 or –1 indicate that there is a strong relationship between the variables, in either the direct or inverse direction [57]. In this study, the Pearson correlation coefficient (PCC) method [33] is used to determine the degree of correlation of each variable, with respect to the target variable to be predicted (*Ppv*).

The Pearson correlation coefficient can be determined for two vectors x and y by using the Equation 3.20:

$$PCC = \frac{\frac{1}{n}\sum_{i=1}^{n}(x_i - \bar{x})(y_i - \bar{y})}{\sqrt{\frac{1}{n}\sum_{i=1}^{n}(x_i - \bar{x})^2}\sqrt{\frac{1}{n}\sum_{i=1}^{n}(y_i - \bar{y})^2}} \tag{3.20}$$

where $\bar{x}$ and $\bar{y}$ are the mean of each vector, x_i and y_i represent the values of the variables for observation i, and n corresponds to the number of elements in the sample.

Figures 3.11 and 3.12 show the Pearson correlation values between the different variables that make up the *D*1 and *D*2 databases, respectively. Figure 3.11 shows that the *GHI* variable has the highest level of correlation (0.9616) with respect to the *Ppv* variable. It is followed in that order by the variables *Tp*, with a correlation of 0.8725, and *Ta*, with a value of 0.5464. Since the three variables mentioned correlate adequately with the output variable, it is decided to use these three variables (*GHI*, *Tp*, and *Ta*) as possible inputs to the prediction model for data set *D*1.

In the case of Figure 3.12, it can be seen that there are a larger number of variables involved in the correlation analysis. As in the previous case, the variable most strongly correlated with *Ppv* is *GHI,* with a value of 0.9592. It is followed by the variable *Ta,* with a correlation of 0.4426, and *Ws,* with a value of 0.3437. On the other hand, the variable *Rh* presents a weak and negative correlation with respect to *Ppv,* with a value of −0.2576, and the variable *Wd* also presents a negative correlation and is very close to zero (−0.0902). Therefore, for data set *D2*, the variables *GHI*, *Ta*, and *Ws* are defined as possible inputs to the prediction model.

3.3.2.4 Change of the temporal resolution of the data

In this work, the data obtained were measured with a high temporal resolution (5 min). However, the objective of the research is the prediction of PV energy for a one-hour horizon and for the following day. Therefore, in some cases, it is necessary to change the temporal resolution of the data to 30 min or 1 hour. The method used is based on averaging the values of each interval and taking these average values to reconstruct the time series. Figure 3.13 shows the results obtained by changing the temporal resolution of the *GHI* variable in the *D*1 data set. As can be seen, as the temporal resolution changes, data smoothing is achieved, and fluctuations are reduced. However, sometimes some useful information may be lost, regarding phenomena that occur in relatively short time intervals.

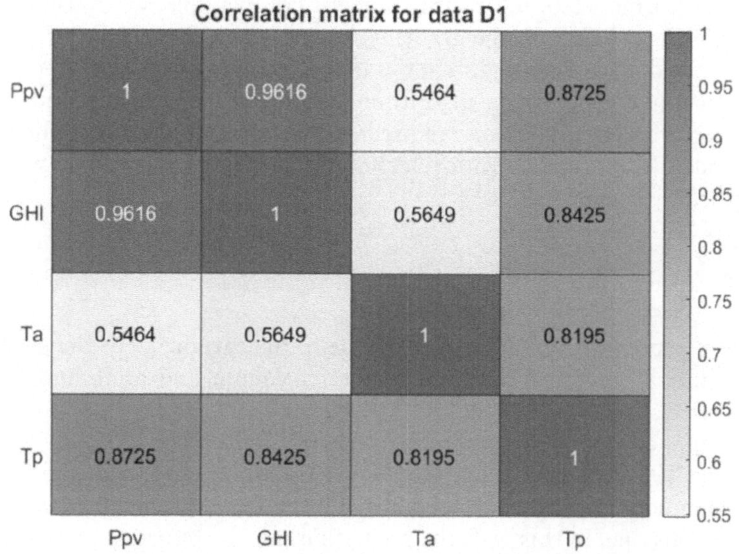

Figure 3.11 Correlation matrix for dataset D1.

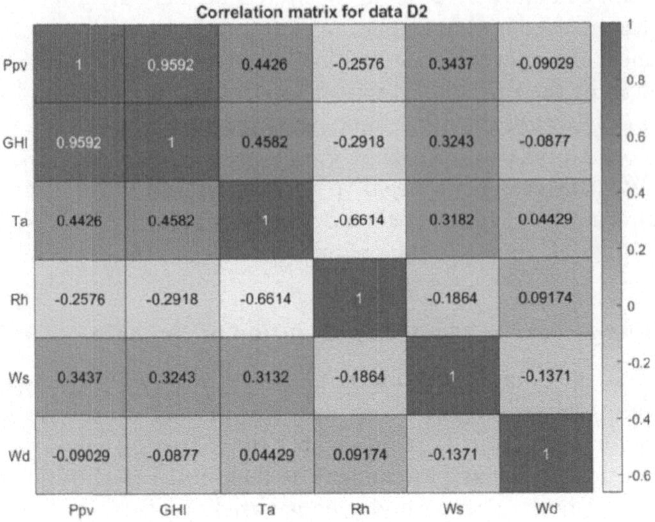

Figure 3.12 Correlation matrix for dataset *D2*.

3.3.2.5 Data normalization

The normalization of data is considered an aspect of great importance when working with time series that present different scales of values. As can be seen in Tables 3.2 and 3.3, the ranges of minimum and maximum values of each variable differ. These differences in scale can have a negative impact on the performance of forecasting models. For example, when using training algorithms based on the gradient descent technique, differences in the scale of the data can cause difficulties in the learning process and reduce the speed of convergence of the algorithm. To solve these problems and ensure good performance and quality in predictions, normalization is applied [16].

A widely used normalization technique is Min-Max scaling, which is described by Equation 3.21:

$$x' = \frac{x - x_{min}}{x_{max} - x_{min}} \tag{3.21}$$

where x' represents the normalized value of the variable, x is the observed value, and x_{max}, x_{min} correspond to the maximum and minimum values, respectively.

The application of this technique allows rescaling of all the variables in the range [0, 1] and reducing the dispersion among the data. Finally, the model predictions must be denormalized by means of a reverse procedure to the previous one. In this way, the predictions can be returned to their original scale and appropriate comparisons can be made with the real values.

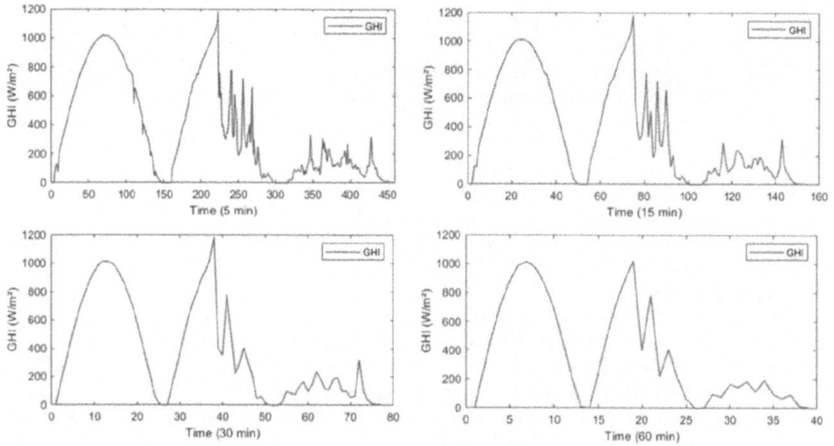

Figure 3.13 Change in temporal resolution of the *GHI* variable of data set *D1*.

3.3.3 Proposed methodology for data analysis and preprocessing

Figure 3.14 presents the logical and sequential explanation of the proposed methodology for data preprocessing in the form of pseudocode. The algorithm presented covers all the steps described above, from data collection and visualization to data quality improvement using different techniques, as well as the definition of the most relevant inputs for the forecasting models and their normalization. In this way, the quality of the available data is improved and a satisfactory performance of the prediction models is obtained.

3.4. DEVELOPMENT OF THE PREDICTIVE MODEL

This section discusses the partitioning of the available data into training and test sets. In addition, several consistent metrics are defined to evaluate the performance of the models and the quality of their predictions. Next, the general structure of the proposed prediction model is defined, and the fit of its most relevant hyperparameters obtained using the BOA algorithm is shown. Then, a partial description of the programming performed for the creation of the model is given. Finally, the general methodology proposed for PV power prediction is presented.

3.4.1 Division of the data

The division of the available historical data is an important aspect in the creation of a given forecast model. In the phase prior to model training, the

Algorithm 1. *Data analysis and preprocessing*

```
1: Obtain PV plant data
2: Graphing data to detect anomalies and trends
3: Determine possible minimum and maximum values for each time series
4: Follow steps (5-11) to eliminate out of range measurements
5: for i=1: length (time_serie) do
6:     if time_serie(i)< minimum possible value
7:             time_serie(i) = minimum possible value
8:     elseif time_serie(i) > maximum possible value
9:             time_serie(i) = maximum possible value
10:    end if
11: end for
12: Detect and replace outliers in each time series using step 13
13:        time_serie = filloutliers (time_serie, fillmethod, findmethod)
14: Detect and replace missing values in each time series using step 15
15:        time_serie = fillmissing (time_serie, movmethod, window)
16: Convert the database to table format and follow steps 17and 18 to filter the nightly values
17: Define start and end time of the solar day
18: Use the date column and take for each time series the values between the start and end
    time of each day
19: Calculate the Pearson correlation coefficient of each variable with respect to PV power
    using equation (20)
20: Select the variables with the highest correlation coefficient as inputs to the prediction
    model
21: Define the time resolution required for the data
22: for j=1: length (time_serie) do
23:    if temporary resolution required = temporary resolution original
24:            keeping each time series the same
25:    else determine the average value for each time interval
26:           reconstruct time series for new resolution
27:    end if
28: end for
29: Determine the minimum and maximum value of each time series
30: Normalize the data using equation (21)
```

Figure 3.14 Pseudocode for data analysis and preprocessing. (**the programming of the algorithm was developed in MATLAB*)

data are divided into several independent sets. In this work, the criterion shown in Figure 3.15 for the division of the historical data is applied.

The training data set is used to train the prediction model so that it learns the relationships between the data and is able to recognize patterns and trends in the time series. Within the training set itself, most of the data is used to train the model, and a small part is used to validate the learning using the cross-validation method. The test data set is used to evaluate the predictive and generalization capability of the developed model against a set of new data that were not part of the training process.

3.4.2 Performance analysis

Several metrics recommended in the literature are used to evaluate the performance of the prediction model developed. Among these are the mean absolute error (MAE), the mean absolute percentage error (MAPE), the root mean square error (RMSE) and the coefficient of determination (R^2) [58]. The metrics are defined by Equations 3.22–3.25:

$$MAE = \frac{1}{n}\sum_{i=1}^{n}\left|y_p - y_m\right| \tag{3.22}$$

$$RMSE = \sqrt{\frac{1}{n}\sum_{i=1}^{n}\left(y_p - y_m\right)^2} \tag{3.23}$$

$$MAPE = \frac{1}{n}\sum_{i=1}^{n}\left|\frac{y_p - y_m}{y_m}\right| \times 100\% \tag{3.24}$$

$$R^2 = 1 - \frac{\sum_{i=1}^{n}\left(y_p - y_m\right)^2}{\sum_{i=1}^{n}\left(y_m - y_{avg}\right)^2} \tag{3.25}$$

where y_p represents the value predicted by the prediction model, y_m represents the measured value of the PV power, y_{avg} is the average of the observed values and n corresponds to the number of observations.

The meaning of these metrics is described below. The MAE measures the average absolute difference between the actual values and those predicted by the model. The RMSE estimates the error by the square root of the average of the squared differences between the actual values and the model predictions. It is a suitable metric for analyzing large errors. In the case of the MAPE, it is considered a widely useful standard metric and is very fair for determining the accuracy of predictions. The coefficient of determination R^2 measures the degree of linear correlation between actual values and predictions and its value can range from zero to one. In practice, small values of MAE, MAPE and RMSE, and values close to one of the R^2, are desired to ensure good quality in the predictions made [58].

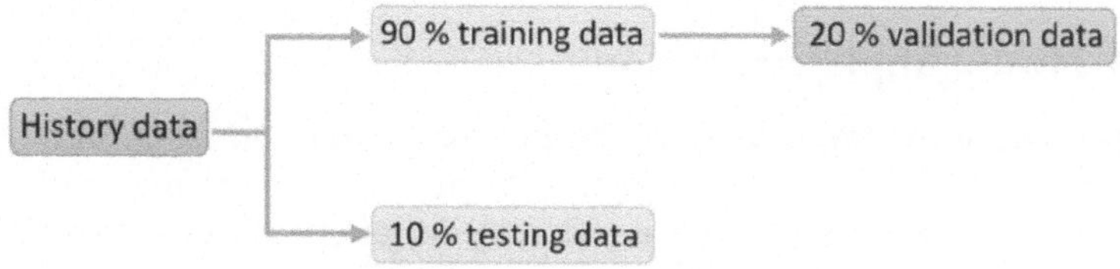

Figure 3.15 Division of available historical data.

3.4.3 Structure of the predictive model

For the prediction of the power generated in plants *D*1 and *D*2, a structure similar to the one shown in Figure 3.16 is used. The internal structure of both models is the same, differing only in the inputs used. The structure of each model contains a sequence input layer, a BiLSTM layer with a certain number of hidden units, a dropout layer, a fully connected layer, and a regression layer at the output.

To avoid the risk of overfitting the prediction model, the dropout layer is added. By means of this layer, memory units are randomly removed from the neural network with a probability that can vary between 0.1 and 1.0. The addition of the dropout layer reduces the error rate in the forecasts and increases the robustness of the developed model [53]. The objective of the work is the prediction of the *Ppv* for the hour-ahead and the day-ahead. In the case of the hour-ahead prediction, the measurements were taken with a time resolution of 30 minutes for both data sets. In this case, six delays from each input time series are used as inputs to the prediction model. That is, the value of each time series at the current time and at five previous time instants is used. The number of lags is determined by autocorrelation analysis.

For the prediction of the day-ahead, it should be taken into account that, by eliminating the nighttime measurements and working only with the data between 7:00 am and 19:00 pm, the resulting days have a duration of 12 hours. Therefore, in the prediction of the next day (12 hours-ahead), an hourly time resolution is used for the historical data. For this type of prediction, 24 delays of each time series were used as inputs to the prediction model. An interesting aspect to note is that, despite using a large number of inputs to the prediction model (24 inputs for the hour-ahead prediction and 96 inputs for the day-ahead prediction), the computation time does not increase considerably. It could be seen that the computational cost of the model is mostly related to the internal structure of the network and the adjustment of its hyperparameters.

The adjustment of the hyperparameters of the prediction model is performed using the BOA algorithm. There is another group of hyperparameters that are manually adjusted according to values recommended in the literature. Table 3.4 shows the values of the different hyperparameters of the BiLSTM model for the two time horizons analyzed.

3.4.4 Model implementation

The procedure for implementing the *Ppv* prediction model is represented in the form of pseudocode, as shown in Figure 3.17; the elements of major importance for the construction and training of the proposed model are represented.

Some important issues related to the training of the prediction model must be addressed. In neural network models, an optimization algorithm is

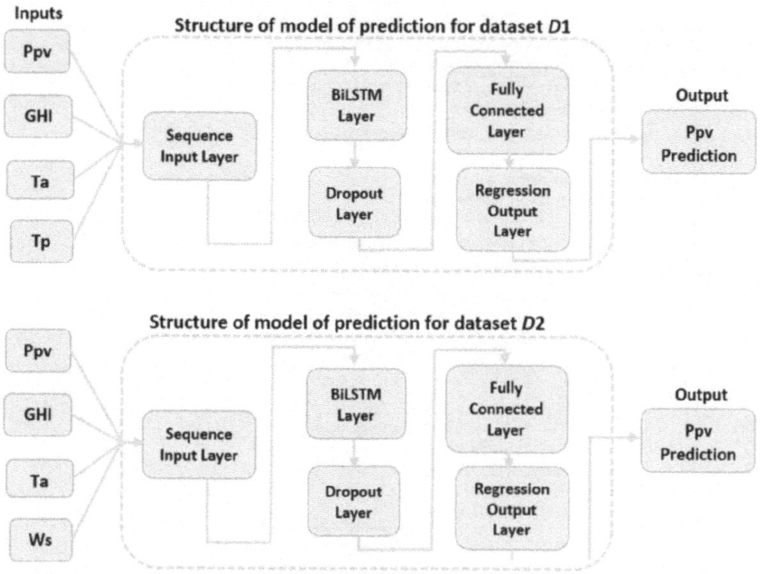

Figure 3.16 General structure of the prediction models for data sets *D1* and *D2*.

Table 3.4 Adjustment of the hyperparameters of the BiLSTM model for the prediction of *Ppv*

Type of adjustment	*Hyperparameter to adjust*	*Fit obtained (One-hour-ahead prediction)*	*Fit obtained (Day-ahead prediction)*
Through BOA	Number of hidden layers	1	1
	BiLSTM units in hidden layers	90	120
	Initial Learn Rate	0.005	0.01
	Mini-batch size	64	128
	Regularization factor L_2	0.0001	0.0001
Manually	Learn rate schedule	Per parts (periods)	Per parts (periods)
	Optimization algorithm used for training	Adaptive moment estimation (Adam)	Adaptive moment estimation (Adam)
	Dropout value	0.2	0.2
	Number of training epochs	200	200
	Learn rate drop period	125	125
	Learn rate drop factor	0.5	0.5
	Gradient Threshold	1	1

Algorithm 2. *Model building and training*

```
1: Define the layers of the BiLSTM model using steps 1-6
2:        layers = [ sequenceInputLayer (number of features)
3:        bilstmLayers (number of hidden units)
4:        dropoutLayer (dropout value)
5:        fullyConnectedLayer (number of responses)
6:        regressionLayer];
7: Define the training options by steps 8-18
8:        options = trainingOptions ('adam', . . .
9:                       'MaxEpochs', number of epochs, . . .
10:                      'GradientThreshold', value, . . .
11:                      'InitialLearnRate', value, . . .
12:                      'LearnRateSchedule', 'piecewise', . . .
13:                      'LearnRateDropPeriod', value, . . .
14:                      'LearnRateDropFactor', value, . . .
15:                      'Verbose', value, . . .
16:                      'MiniBatchSize', value, . . .
17:                      'L2Regularization', value, . . .
18:                      'Plots', 'training-progress', . . .
19: Perform model training using step 20
20:       model = trainNetwork (XTrain, Ytrain, layers, options);
21: Perform PV power forecasting using step 22
22:       YPred = predict (model, XTest);
```

Figure 3.17 Pseudocode for building and training the prediction model. (**the programming of the algorithm was developed in MATLAB*)

used to minimize the error rate. The effectiveness of a specific optimization algorithm is given by its convergence speed and its generalization capability. Several optimization algorithms can be used in training deep learning models, including adaptive moment estimation (Adam), stochastic gradient descent with momentum (SGDm), and root mean square propagation (RMSprop). In this work, the Adam algorithm, recommended for its high computational efficiency and easy tuning, was used. Another important aspect in model training is the number of epochs or iterations performed by the prediction model. In this case, 200 training epochs were used; this value was determined by the trial and error method. In this case it was observed that, around this value, the validation error practically stops decreasing and behaves in a stable way, so it is not necessary to continue increasing the number of iterations of the model [59]. Other aspects related to the mini-batch size, initial learning rate and L_2 regularization were addressed in the hyperparameter optimization section.

3.4.5 General methodology for PV power prediction

PV power prediction encompasses a set of steps and procedures that must be integrated to obtain a prediction model that performs favorably. Figure 3.18 shows the general methodology proposed for the prediction of PV power generated by a BiLSTM model with Bayesian hyperparameter optimization.

In the initial phase of the methodology, the necessary historical data are obtained, the analysis and preprocessing of this data are performed and then the data are divided into the training and test sets. In the next phase,

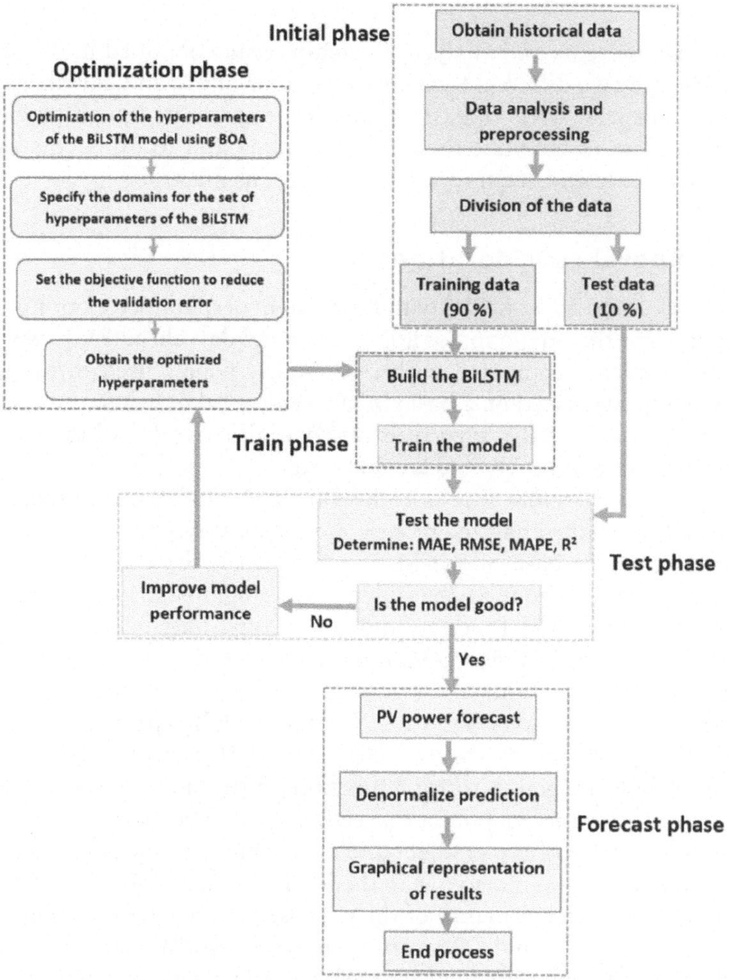

Figure 3.18 Proposed methodology for PV power prediction.

the architecture of the prediction model is defined, the optimization of its most relevant hyperparameters is performed through BOA and its training is carried out. Subsequently, the performance of the proposed model is tested using the test set data and its quality is determined. If the results obtained are not adequate, the hyperparameters of the model are modified and it is trained again. Finally, the trained model is used to make the next hour's or next day's *Ppv* forecast. The predictions are then denormalized, to restore them to their original scale; the results obtained are plotted and the process is completed.

3.5 RESULTS AND DISCUSSION

This section presents and analyzes the main results obtained in the prediction of *Ppv* for two different time horizons (hour-ahead and day-ahead) and different data sets (*D*1 and *D*2). The results are presented in graphical and numerical form to facilitate their understanding and to draw conclusions about the quality and accuracy of the predictions made.

3.5.1 General considerations

In this work, a BiLSTM model with Bayesian hyperparameter optimization is proposed for the prediction of hour-ahead and day-ahead PV power. The proposed model is compared with two models widely used in the literature: a first model based on a basic MLP-type neural network and a second model based on a NARX type recurrent neural network. All models were implemented in MATLAB R2021b software.

The proposed architecture for the BiLSTM model was discussed in Section 3.4.3. All the analyzed models use a similar structure for the inputs and output. In the case of the MLP neural network, a structure with a hidden layer, 40 neurons in the hidden layer, sigmoidal logarithmic type hidden layer activation function and linear type output layer activation function was used. Levenberg-Marquardt was used as the learning algorithm. The minimum value of the gradient was set to 0.0001 and the maximum number of iterations was 1000. For the NARX type recurrent neural network, an architecture with a hidden layer, 36 neurons in the hidden layer, hyperbolic tangent activation function for the hidden layer, and linear activation function for the output layer was used. The learning algorithm, the minimum value of the gradient and the maximum number of iterations were adjusted in the same way as in the MLP neural network.

To evaluate the performance of the proposed model and the comparison models, data sets (*D*1 and *D*2) corresponding to two PV plants of different installed capacity and different geographical location were used. In all cases the predictions were made for two different time horizons (hour-ahead and day-ahead).

3.5.2 Results obtained in the prediction of the PV power with one hour-ahead

The prediction of the hour-ahead PV power is very useful to ensure proper operation of power grids. Figure 3.19 shows a comparison of the actual *Ppv* values recorded during one week and the predictions made by the BiLSTM, MLP, and NARX models for a prediction horizon of one hour-ahead. It should be clarified that all studies performed use data from the test set.

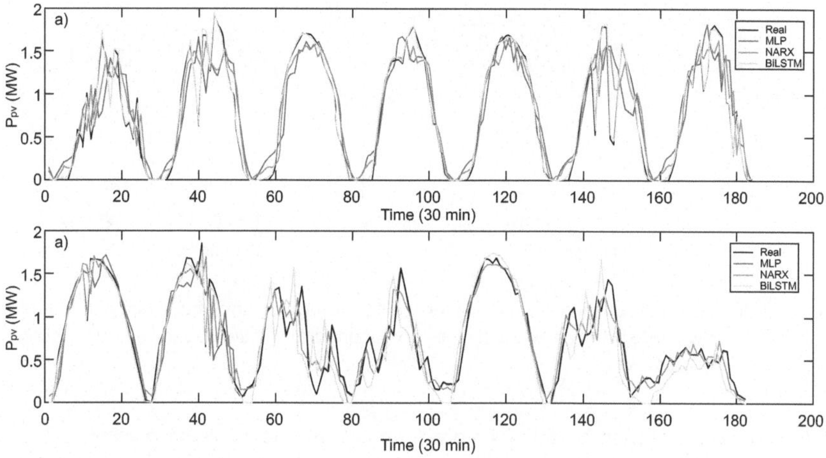

Figure 3.19 Prediction of one-hour-ahead *Ppv*: (a) Prediction for data set *D1* and (b) Prediction for data set *D2*.

To demonstrate the level of adaptability and robustness of the proposed prediction model, the study is performed on days of different behavior and on two different data sets. As can be seen, the predictions of the proposed model adequately fit the real PV power patterns during the whole time period analyzed. In the case of the MLP and NARX models, their predictions do not fit with good accuracy to the actual behavior patterns of the PV power time series. These differences are more significant on days with high power variability due to the effect of cloudiness.

Figure 3.20 shows a comparison between the actual average PV power during the week analyzed and the average of the predictions made by each model. In this case, similar conclusions are obtained as in the previous case, since in both data sets the superiority of the BiLSTM model over the MLP and NARX models is evident. It can be observed that the proposed model behaves favorably in the prediction of photovoltaic power one hour-ahead and shows good stability and accuracy in its predictions for the analyzed time period.

Table 3.5 shows the average prediction errors made by each model during the time period analyzed. In both data sets, the BiLSTM model presents the lowest values of MAE, RMSE, and MAPE and the highest value of the coefficient of determination R^2, which indicates a high quality of its predictions. The second place in accuracy is occupied by the recurrent neural network of the NARX type. In third place is the MLP neural network, which shows the most unfavorable indicators and the greatest uncertainties in prediction.

Figure 3.21 shows the scatter plots between the actual *Ppv* values and the predictions of the BiLSTM, MLP and NARXmodels. The straight

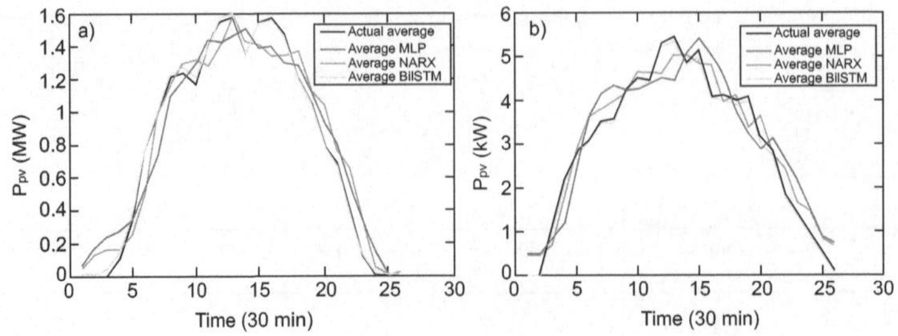

Figure 3.20 Comparison between the average actual power and the average prediction of each model for one week: (a) Results obtained for dataset *D1* and (b) results obtained for dataset *D2*.

Table 3.5 Errors of each model in the prediction of *Ppv* one hour-ahead using databases *D1* and *D2*

		Prediction models		
Dataset used	*Metrics*	*MLP*	*NARX*	*BiLSTM*
D1	MAE (kW)	116.01	102.82	18.27
	RMSE (kW)	187.43	168.68	26.61
	MAPE (%)	9.82	9.04	1.02
	R^2	0.9262	0.9318	0.9982
D2	MAE (kW)	0.51	0.46	0.09
	RMSE (kW)	0.59	0.54	0.11
	MAPE (%)	10.24	9.73	1.08
	R^2	0.9236	0.9298	0.9973

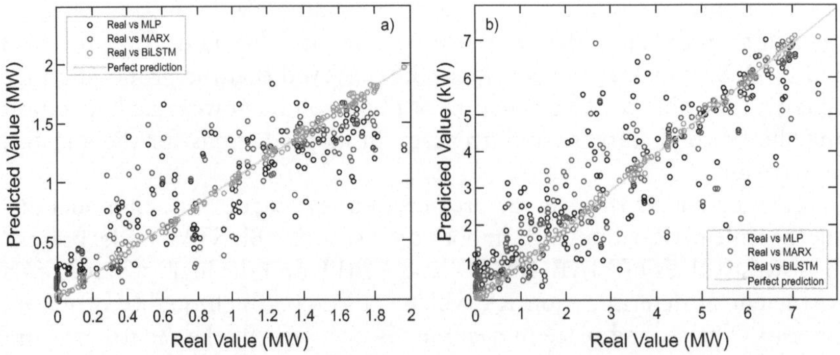

Figure 3.21 Scatter plots between actual values and predictions from MLP, NARX, and BiLSTM models using data sets *D1* (case a) and *D2* (case b).

line represents a perfect prediction. As can be seen in the two cases analyzed, most of the predictions of the BiLSTM model are concentrated on the perfect prediction line or very close to it. This indicates that there is a high level of correlation between the actual values and the predictions of this model. For the MLP and NARX models, a greater dispersion of the observations is evident, which is equivalent to a lower level of correlation between the predictions and the actual values and therefore a lower quality of the predictions.

Table 3.6 presents the values of MAE, RMSE, MAPE, and R^2 for three different types of days (sunny, partly sunny, and cloudy) and for both data sets. All models analyzed show superior performance (lower MAE, RMSE, MAPE and higher R^2 values) on a sunny day. The opposite is true for cloudy days and the performance of the prediction models is much lower. The partly sunny day can be considered as an intermediate case in the prediction. It can also be seen that the BiLSTM model shows higher accuracy and stability in predictions on different types of days compared to the NARX and MLP models.

The significant differences in the accuracy of the predictions for the different types of days are mainly due to the variability of the weather conditions. On sunny days there is no cloud cover and therefore no abrupt variations in solar irradiance incident on the PV plant occur. Under these conditions the prediction models can perform more favorably. However, as the incident cloudiness over the PV plant increases, fluctuations in the power output occur, and prediction becomes more difficult.

The results obtained in the prediction of the one-hour-ahead PV power show that the BiLSTM prediction model with Bayesian hyperparameter optimization is a powerful tool in time series prediction. Its particular structure allows it to extract deep features and learn complex nonlinear patterns. Therefore, it performs satisfactorily in predicting the PV power of the hour-ahead. In the case of the NARX and MLP models, the quality of their predictions is much lower compared to the proposed model. These two models are structurally simpler and require less computational cost. However, when faced with large and complex time series, they do not have sufficient learning capacity to achieve stable and high-quality predictions.

3.5.3 Results obtained in the prediction of the day-ahead PV power

Day-ahead *Ppv* predictions are also of great importance, especially in matters related to power system planning and operation. Figure 3.22 shows the results obtained in the day-ahead PV power prediction from the proposed model and the comparison models. As in the hour-ahead prediction, two different data sets are used and a time period of one week is analyzed. As can be seen in comparison to the one-hour-ahead prediction, in this case the

Table 3.6 Behavior of the prediction errors of each model for a prediction horizon of one hour-ahead in different types of days

		Results of the prediction models for dataset D1			*Results of the prediction models for dataset D2*		
Type of day	*Metrics used*	*MLP*	*NARX*	*BiLSTM*	*MLP*	*NARX*	*BiLSTM*
Sunny	MAE (kW)	72.72	63.02	14.19	0.38	0.36	0.06
	RMSE (kW)	100.03	91.36	19.46	0.48	0.45	0.08
	MAPE (%)	5.89	5.67	0.34	5.17	4.96	0.32
	R^2	0.9656	0.9679	0.9991	0.9693	0.9711	0.9993
Partly sunny	MAE (kW)	105.29	92.68	17.67	0.49	0.44	0.07
	RMSE (kW)	192.14	174.54	23.49	0.70	0.62	0.09
	MAPE (%)	9.94	8.02	1.01	8.82	7.32	1.25
	R^2	0.9187	0.9294	0.9986	0.8970	0.9173	0.9981
Cloudy	MAE (kW)	196.02	164.27	31.98	0.67	0.63	0.11
	RMSE (kW)	272.40	240.45	41.24	0.87	0.78	0.13
	MAPE (%)	17.21	15.46	1.98	18.27	17.03	2.27
	R^2	0.8428	0.8537	0.9951	0.8338	0.8493	0.9930

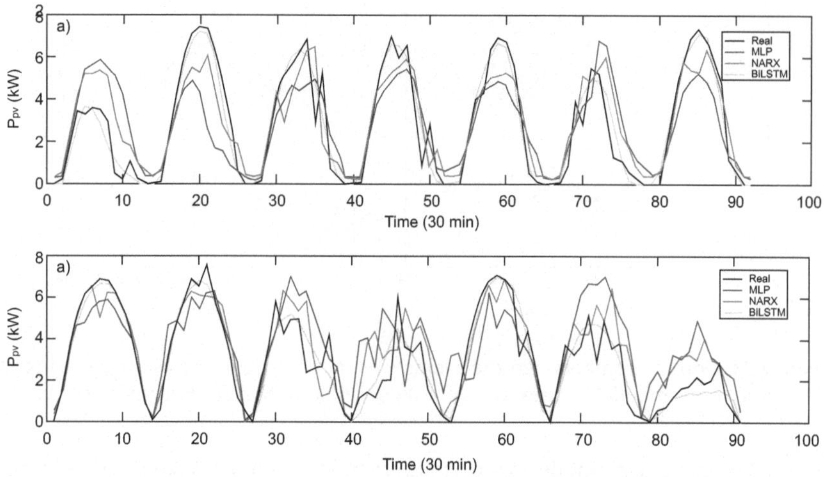

Figure 3.22 *Ppv* prediction for the day-ahead: (a) Prediction for data set *D1* and (b) prediction for data set *D2*.

accuracy of all models has decreased for both data sets as expected. Again, the BiLSTM model shows the best performance. The proposed model is able to fit the general patterns of behavior of the power time series, even though there are some days whose prediction is very difficult due to the high variability of weather conditions. In the case of the MLP and NARX models, their predictions have worsened significantly and their quality is poor. The predictions of these two models show significant differences with respect to the real power values recorded in both PV plants.

Figure 3.23 shows a comparison between the actual average PV power during the week analyzed and the average of the predictions made by each model for a day-ahead prediction horizon. The results obtained show that the average prediction of the BiLSTM model adequately adjusts to the average real power during the analyzed week. This reaffirms the effectiveness of this model for day-ahead *Ppv* prediction. In the case of the MLP and NARX models, they do not show a good performance and large differences with respect to the real power average are observed.

Table 3.7 shows the values of several metrics to determine the quality of the predictions made by each model for the time horizon of the day-ahead. Evidently the accuracy of the models has decreased with increasing prediction horizon. In both data sets the BiLSTM model shows the best indicators, followed by the NARX model and the MLP model. All models show slightly better performance for the first data set. This may be associated with two fundamental issues. First, data set *D*1 has a duration of one year and set *D*2 only has measurements of 132 days. The length of the time series can be an influential factor in the training quality of a prediction

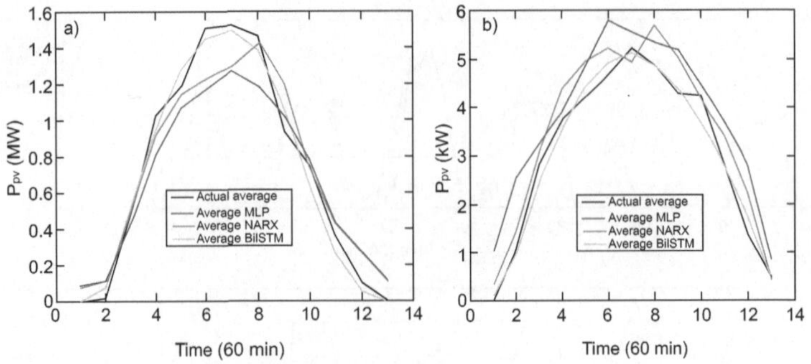

Figure 3.23 Comparison between the average actual power and the average prediction of each model for one week: a) Results for dataset *D1* and (b) Results for dataset *D2*.

Table 3.7 Errors of each model in predicting day-ahead *Ppv* using databases *D1* and *D2*

		Prediction models		
Dataset used	*Metrics*	*MLP*	*NARX*	*BiLSTM*
D1	MAE (kW)	331.34	260.61	86.31
	RMSE (kW)	416.96	335.70	137.68
	MAPE (%)	36.43	27.84	10.19
	R^2	0.6614	0.7224	0.9352
D2	MAE (kW)	1.47	0.96	0.44
	RMSE (kW)	1.82	1.31	0.63
	MAPE (%)	37.12	28.78	10.96
	R^2	0.6486	0.6934	0.9295

model. On the other hand, the time period selected for dataset *D2* has a higher variability in the behavior of PV power, which makes its prediction even more difficult.

Figure 3.24 shows the scatter plots between the actual *Ppv* values and the predictions of each model for the following day. In this case, a larger scatter is observed between the predictions of each model and the actual *Ppv* power values, indicating that the quality of the predictions has degraded with increasing prediction horizon. The BiLSTM model again shows the best fit with respect to the perfect prediction line.

Table 3.8 shows the prediction errors made by each model for different types of days. As in the next hour prediction the best results are obtained for the sunny day and the most pessimistic results are obtained for the cloudy day. Again, the superiority of the BiLSTM model compared to the NARX and MLP models is evident for all test conditions analyzed.

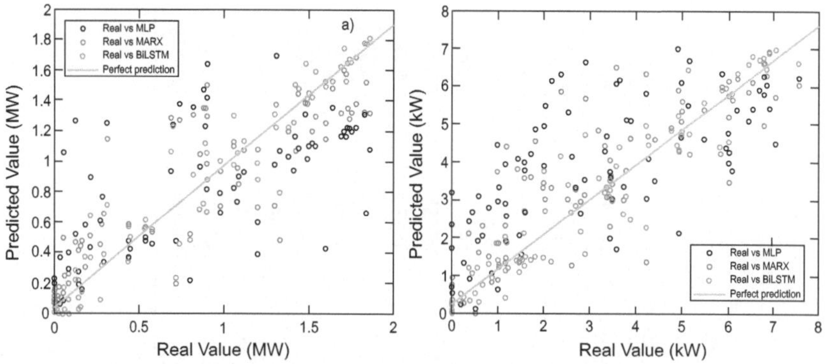

Figure 3.24 Scatter plot between actual values and predictions from MLP, NARX, and BiLSTM models using data sets *D1* (case a) and *D2* (case b).

The general analysis of the predictions made for the day-ahead shows that as the prediction horizon increases, the quality and accuracy of the results decreases for all the models analyzed. The proposed prediction model has demonstrated its superiority in all the tests performed. Moreover, it has been able to adapt with good accuracy to different data sets and different operating conditions. Even in adverse weather conditions where the comparison models showed totally erroneous results, the BiLSTM model was able to generally capture the behavior pattern of the original PV power series and produce an accurate forecast.

3.6 CONCLUSIONS

Solar photovoltaic energy is a renewable energy source that has been widely used in recent years. However, the power generated in a PV installation depends on several meteorological factors that cause its behavior to be dynamic, variable, and intermittent. In order to solve these problems and achieve a successful integration of this energy source into power grids, an accurate forecast of the PV power generated is required. In this study, a BiLSTM deep learning model was proposed to forecast the PV power of the hour-ahead and day-ahead. To obtain better performance of the forecasting model, a data preprocessing algorithm was implemented. This procedure improved the quality of the available historical data and increased the accuracy of the predictions. The adjustment of the most relevant hyperparameters of the proposed model was performed by means of a Bayesian optimization algorithm. In this way, the performance of the model and its generalization capacity could be improved.

The predictions of the BiLSTM model were compared with an MLP-type neural network model and a NARX type recurrent neural network model.

Table 3.8 Behavior of the prediction errors of each model for a day-ahead prediction horizon on different day types

		Results of the prediction models for dataset **D1**			*Results of the prediction models for dataset* **D2**		
Type of day	*Metrics used*	*MLP*	*NARX*	*BiLSTM*	*MLP*	*NARX*	*BiLSTM*
Sunny	MAE (kW)	218.43	143.65	49.67	0.70	0.43	0.19
	RMSE (kW)	263.04	175.59	80.98	0.84	0.67	0.23
	MAPE (%)	18.93	14.85	5.26	19.56	16.27	5.42
	R^2	0.8472	0.9041	0.9878	0.8374	0.8955	0.9814
Partly sunny	MAE (kW)	289.82	242.55	82.97	0.88	0.55	0.26
	RMSE (kW)	354.76	316.83	121.52	1.07	0.74	0.32
	MAPE (%)	26.28	22.86	10.26	27.11	23.07	10.81
	R^2	0.8085	0.8442	0.9789	0.7934	0.8324	0.9702
Cloudy	MAE (kW)	507.51	410.45	128.09	1.69	1.09	0.49
	RMSE (kW)	641.68	493.91	191.43	1.83	1.23	0.63
	MAPE (%)	69.28	56.49	19.43	73.15	60.02	20.16
	R^2	0.5846	0.6214	0.9389	0.5714	0.6006	0.9276

Different consistent metrics recommended in the specialized literature were used to determine the accuracy of the predictions. The models were tested using two historical data sets belonging to different PV installations. The performance of the models was evaluated for two prediction horizons and for different types of days (sunny, cloudy, and partly sunny). In all cases analyzed, the BiLSTM model obtained the best results and showed higher robustness, accuracy, and adaptability in the predictions. Even in adverse weather conditions where the MLP and NARX models showed unfavorable performance, the BiLSTM model was able to adequately adjust to the actual PV power patterns. The analysis of the results obtained for each available dataset and for the different forecast horizons studied demonstrated that the BiLSTM model is a powerful tool for PV power prediction.

Future work will be focused on further improving the performance of the proposed prediction model; for this, some hybrid model architectures such as BiLSTM-CNN can be applied so that temporal and spatial features can be extracted from the time series. The use of meteorological forecasts can also be useful to improve the accuracy of the predictions, especially for the day-ahead forecast horizon. In addition, larger data sets and data from other geographic locations can be used in the study, which will provide more conclusive evidence on the effectiveness of the proposed method.

It is also planned to use the proposed forecasting method for on-line applications, with a probabilistic approach that takes into account possible uncertainties in the prediction. In this sense, one-hour-ahead on-line forecasts are extremely important for real-time monitoring. Day-ahead forecasts are suitable for generation dispatch and the definition of reserve margins in the power grid. Although the results obtained with the BiLSTM model are promising, there is room for further improvement in prediction accuracy and real-time efficiency. One possible avenue for future research would be to explore new technologies such as artificial intelligence (AI), the Internet of Things (IoT), and robotics to optimize the prediction of PV generation. AI could be used to analyze large amounts of historical data and detect patterns and trends that can help to more accurately predict the power generated at a PV installation. IoT could be used to collect data from sensors and devices connected to solar power plants, allowing a better understanding of the environment and conditions in real time. Robotics, meanwhile, could be used in tasks related to inspecting and maintaining solar panels and obtaining accurate measurements of solar radiation. Future research can explore how these technologies can be integrated into the proposed predictive models to improve their performance and adaptability to different operating conditions, allowing for more efficient management of solar photovoltaic energy.

REFERENCES

1. Lotta, M. C.; Pye, S.; Dodds, P. E. Quantifying the co-impacts of energy sector decarbonization on outdoor air pollution in the United Kingdom. *Energy Policy*, 2017, vol. 101, pp. 42–51, doi: 10.1016/j.enpol.2016.11.028.
2. Abdel-Nasser, M.; Mahmoud, K. Accurate photovoltaic power forecasting models using deep LSTM-RNN. *Neural Computing & Applications*, 2019, vol. 31, pp. 2727–2740, doi: 10.1007/s00521-017-3225-z.
3. Zang, Haixiang, et al. Day-ahead photovoltaic power forecasting approach based on deep convolutional neural networks and meta learning. *International Journal of Electrical Power & Energy Systems*, 2020, vol. 118, p. 105790, doi: 10.1016/j.ijepes.2019.105790.
4. Gandoman, F. H.; Abdel Aleem, S. H. E.; Omar, N.; Ahmadi, A.; Alenezi, F. Q. Short-term solar power forecasting considering cloud coverage and ambient temperature variation effects. *Renew Energy*, 2018, vol. 123, p. 793e805, doi: 10.1016/j.renene.2018.02.102.
5. Gaëetan, M.; Izumi, K. *Trends. In: Photovoltaic Applications*. Paris, France: IEAPVPS; 2019.
6. REN21.2022. *Renewables 2022 Global Status Report* (Paris: REN21 Secretariat), ISBN 978-3-948393-04-5.
7. Jones, Dave. Ember *Global Electricity Review* 2021, march 2021, www.ember-climate.org/global-electricity-review-2021.
8. Ismail, Abdul Muhaimin, et al. Progress of solar photovoltaic in ASEAN countries: A review. *Renewable and Sustainable Energy Reviews*, 2015, vol. 48, pp. 399–412, doi: 10.1016/j.rser.2015.04.010.
9. Cheng, Lilin, et al. Short-term solar power prediction learning directly from satellite images with regions of interest. *IEEE Transactions on Sustainable Energy*, 2021, vol. 13, no. 1, pp. 629–639, doi: 10.1109/TSTE.2021.3123476.
10. Liu, Luyao, et al. Prediction of short-term PV power output and uncertainty analysis. *Applied Energy*, 2018, vol. 228, pp. 700–711, doi: 10.1016/j.apenergy.2018.06.112.
11. Kumar, Dhivya Sampath, et al. Review of power system impacts at high PV penetration Part II: Potential solutions and the way forward. *Solar Energy*, 2020, vol. 210, pp. 202–221, doi: 10.1016/j.solener.2020.08.047.
12. Rodríguez, Fermín, et al. Very short-term temperature forecaster using MLP and N-nearest stations for calculating key control parameters in solar photovoltaic generation. *Sustainable Energy Technologies and Assessments*, 2021, vol. 45, p. 101085, doi: 10.1016/j.seta.2021.101085.
13. Das, Utpal Kumar, et al. Forecasting of photovoltaic power generation and model optimization: A review. *Renewable and Sustainable Energy Reviews*, 2018, vol. 81, pp. 912–928, doi: 10.1016/j.rser.2017.08.017.
14. Akhter, Muhammad Naveed, et al. Review on forecasting of photovoltaic power generation based on machine learning and metaheuristic techniques. *IET Renewable Power Generation*, 2019, vol. 13, no 7, pp. 1009–1023, doi: 10.1049/iet-rpg.2018.5649.
15. Wang, Kejun; Qi, Xiaoxia; Liu, Hongda. A comparison of day-ahead photovoltaic power forecasting models based on deep learning neural network. *Applied Energy*, 2019, vol. 251, p. 113315, doi: 10.1016/j.apenergy.2019.113315.

16. Alkhayat, Ghadah; MEHMOOD, Rashid. A review and taxonomy of wind and solar energy forecasting methods based on deep learning. *Energy and AI*, 2021, vol. 4, p. 100060, doi: 10.1016/j.egyai.2021.100060.
17. Ammar, Rim Ben; Ammar, Mohsen Ben; Oualha, Abdelmajid. Deep learning and optimization algorithms-based PV power forecast for an effective hybrid system energy management. *International Journal of Renewable Energy Research (IJRER)*, 2022, vol. 12, no 1, pp. 97–108, doi: 10.20508/ijrer.v12i1.12608.g8382.
18. Alonso-Montesinos, J.; Batlles, F. J.; Portillo, C. Solar irradiance forecasting at one-minute intervals for different sky conditions using sky camera images. *Energy Conversion and Management*, 2015, vol. 105, pp. 1166–1177, doi: 10.1016/j.enconman.2015.09.001.
19. Dolara, Alberto; Leva, Sonia; Manzolini, Giampaolo. Comparison of different physical models for PV power output prediction. *Solar Energy*, 2015, vol. 119, pp. 83–99, doi: 10.1016/j.solener.2015.06.017.
20. Hossain, Monowar, et al. Application of extreme learning machine for short term output power forecasting of three grid-connected PV systems. *Journal of Cleaner Production*, 2017, vol. 167, pp. 395–405, doi: 10.1016/j.jclepro.2017.08.081.
21. Lago, Jesus, et al. Short-term forecasting of solar irradiance without local telemetry: A generalized model using satellite data. *Solar Energy*, 2018, vol. 173, pp. 566–577, doi: 10.1016/j.solener.2018. 07.050.
22. Carriere, Thomas, et al. A novel approach for seamless probabilistic photovoltaic power forecasting covering multiple time frames. *IEEE Transactions on Smart Grid*, 2019, vol. 11, no 3, pp. 2281–2292, doi:10.1109/TSG.2019.2951288.
23. Harty, Travis M., et al. Intra-hour cloud index forecasting with data assimilation. *Solar Energy*, 2019, vol. 185, pp. 270–282, doi: 10.1016/j.solener.2019.03.065.
24. Das, Subhra. Short term forecasting of solar radiation and power output of 89.6 kWp solar PV power plant. *Materials Today: Proceedings*, 2021, vol. 39, pp. 1959–1969, doi: 10.1016/j.matpr.2020.08.449.
25. Agoua, Xwégnon Ghislain; Girard, Robin; Kariniotakis, George. Short-term spatio-temporal forecasting of photovoltaic power production. *IEEE Transactions on Sustainable Energy*, 2017, vol. 9, no 2, pp. 538–546, doi: 10.1109/TSTE.2017.2747765.
26. Alfadda, Abdullah, et al. Hour-ahead solar PV power forecasting using SVR based approach. 2017 IEEE Power & Energy Society Innovative Smart Grid Technologies Conference (ISGT). IEEE, 2017, pp. 1–5, doi: 10.1109/ISGT.2017.8086020.
27. Essam, Yusuf, et al. Investigating photovoltaic solar power output forecasting using machine learning algorithms. *Engineering Applications of Computational Fluid Mechanics*, 2022, vol. 16, no. 1, pp. 2002–2034, doi: 10.1080/19942060.2022.2126528.
28. Guermoui, Mawloud, et al. Forecasting intra-hour variance of photovoltaic power using a new integrated model. *Energy Conversion and Management*, 2021, vol. 245, p. 114569, doi: 10.1016/j.enconman.2021.114569.

29. Rodríguez, Fermín, et al. Forecasting intra-hour solar photovoltaic energy by assembling Wavelet based time-frequency analysis with deep learning neural networks. *International Journal of Electrical Power & Energy Systems*, 2022, vol. 137, p. 107777, doi: 10.1016/j.ijepes.2021.107777.
30. Pan, Mingzhang, et al. Photovoltaic power forecasting based on a support vector machine with improved ant colony optimization. *Journal of Cleaner Production*, 2020, vol. 277, p. 123948, doi: /10.1016/j.jclepro.2020.123948.
31. Rafati et al. High dimensional very short-term solar power forecasting based on a data-driven heuristic method. *Energy*, vol. 219, p. 119647, 2021, doi: 10.1016/j.energy.2020.119647.
32. M. A. Hassan, N. Bailek, K. Bouchouicha and S. C. Nwokolo. Ultra-short-term exogenous forecasting of photovoltaic power production using genetically optimized non-linear autoregressive recurrent neural networks. *Renewable Energy*, vol. 171, pp. 191–209, 2021, doi: 10.1016/j.renene.2021.02.103.
33. M. S. Hossain and H. Mahmood. Short-term photovoltaic power forecasting using an LSTM neural network and synthetic weather forecast *IEEE Access*, vol. 8, pp. 172524–172533, 2020, doi: 10.1109/ACCESS.2020.3024901.
34. Akhter, Muhammad Naveed, et al. An hour-ahead PV power forecasting method based on an RNN-LSTM model for three different PV plants. *Energies*, 2022, vol. 15, no 6, p. 2243, doi: 10.3390/en15062243.
35. Elsaraiti, Meftah; MERABET, Adel. Solar power forecasting using deep learning techniques. *IEEE Access*, 2022, vol. 10, pp. 31692–31698, doi: 0.1109/ACCESS.2022.3160484.
36. Massaoudi, Mohamed, et al. An effective hybrid NARX-LSTM model for point and interval PV power forecasting. *IEEE Access*, 2021, vol. 9, pp. 36571–36588, doi: 10.1109/ACCESS.2021.3062776.
37. Ren, Xiaoying, et al. Quad-kernel deep convolutional neural network for intra-hour photovoltaic power forecasting. *Applied Energy*, 2022, vol. 323, p. 119682, doi: 10.1016/j.apenergy.2022.119682.
38. Wojtkiewicz, Jessica, et al. Hour-ahead solar irradiance forecasting using multivariate gated recurrent units. *Energies*, 2019, vol. 12, no. 21, p. 4055, doi: 10.3390/en12214055.
39. Zhen, Hao, et al. Photovoltaic power forecasting based on GA improved Bi-LSTM in microgrid without meteorological information. *Energy*, 2021, vol. 231, p. 120908, doi: 10.1016/j.energy.2021.120908.
40. Bou-Rabee, Mohammed A., et al. BiLSTM network-based approach for solar irradiance forecasting in continental climate zones. *Energies*, 2022, vol. 15, no. 6, p. 2226, doi: 10.3390/en15062226.
41. Sabri, Mohammed; El Hassouni, Mohammed. Predicting photovoltaic power generation using double-layer bidirectional long short-term memory-convolutional network. *International Journal of Energy and Environmental Engineering*, 2022, pp. 1–14, doi: 10.1007/s40095-022-00530-4.
42. Sharadga, Hussein; Hajimirza, Shima; Balog, Robert S. Time series forecasting of solar power generation for large-scale photovoltaic plants. *Renewable Energy*, 2020, vol. 150, pp. 797–807, doi: 10.1016/j.renene.2019.12.131.
43. Jia, Pengyun, et al. Short-term photovoltaic power forecasting based on VMD and ISSA-GRU. *IEEE Access*, 2021, vol. 9, pp. 105939–105950, doi: 10.1109/ACCESS.2021.3099169.

44. Akhter, Muhammad Naveed, et al. A hybrid deep learning method for an hour ahead power output forecasting of three different photovoltaic systems. *Applied Energy*, 2022, vol. 307, p. 118185, doi: 10.1016/j.apenergy.2021.118185.
45. Li Gangqiang, et al. Recurrent neural networks based photovoltaic power forecasting approach. *Energies*, 2019, vol. 12, no 13, p. 2538, doi: 10.3390/en12132538.
46. Pang, Zhihong; Niu, Fuxin; O'Neill, Zheng. Solar radiation prediction using recurrent neural network and artificial neural network: A case study with comparisons. *Renewable Energy*, 2020, vol. 156, pp. 279–289., doi: 10.1016/j.renene.2020.04.042.
47. Sherstinsky, Alex. Fundamentals of recurrent neural network (RNN) and long short-term memory (LSTM) network. *Physica D: Nonlinear Phenomena*, 2020, vol. 404, p. 132306, doi: 10.1016/j.physd.2019.132306.
48. Hochreiter, Sepp; Schmidhuber, Jürgen. Long short-term memory. *Neural Computation*, 1997, vol. 9, no 8, p. 1735–1780, doi: 10.1162/neco.1997.9.8.1735.
49. Jung, Yoonhwa, et al. Long short-term memory recurrent neural network for modeling temporal patterns in long-term power forecasting for solar PV facilities: Case study of South Korea. *Journal of Cleaner Production*, 2020, vol. 250, p. 119476, doi: 10.1016/j.jclepro.2019.119476.
50. Lin, Wenshuai, et al. Multi-step prediction of photovoltaic power based on two-stage decomposition and BILSTM. *Neurocomputing*, 2022, vol. 504, pp. 56–67, doi: 10.1016/j.neucom.2022.06.117.
51. Salehinejad, Hojjat, et al. Recent advances in recurrent neural networks. arXiv preprint arXiv:1801.01078, 2017, doi: 10.48550/arXiv.1801.01078.
52. Sultana, Nahid, et al. Bayesian optimization algorithm-based statistical and machine learning approaches for forecasting short-term electricity demand. *Energies*, 2022, vol. 15, no. 9, p. 3425, doi: 10.3390/en15093425.
53. Michael, Neethu Elizabeth, et al. Short-term solar irradiance forecasting based on a novel Bayesian optimized deep Long Short-Term Memory neural network. *Applied Energy*, 2022, vol. 324, p. 119727, doi: 10.1016/j.apenergy.2022.119727.
54. Martinez-cantin, Ruben. Bayesian optimization with adaptive kernels for robot control. In *2017 IEEE international conference on robotics and automation (ICRA)*. IEEE, 2017. p. 3350–3356, doi: 10.1109/ICRA.2017.7989380.
55. https://dkasolarcentre.com.au/download?location=alice-springs
56. Alzahrani, Ahmad, et al. Solar irradiance forecasting using deep neural networks. *Procedia Computer Science*, 2017, vol. 114, pp. 304–313, doi: 10.1016/j.procs.2017.09.045.
57. Zhou, Yi, et al. Prediction of photovoltaic power output based on similar day analysis, genetic algorithm and extreme learning machine. *Energy*, 2020, vol. 204, p. 117894, doi: 10.1016/j.energy.2020.117894.
58. Ahmed, Razin, et al. A review and evaluation of the state-of-the-art in PV solar power forecasting: Techniques and optimization. *Renewable and Sustainable Energy Reviews*, 2020, vol. 124, p. 109792, doi: 10.1016/j.rser.2020.109792.
59. Harrou, Fouzi; Kadri, Farid; Sun, Ying. *Forecasting of photovoltaic solar power production using LSTM approach. Advanced statistical modeling, forecasting, and fault detection in renewable energy systems*, London, UK: IntechOpen, vol. 3, 2020.

Chapter 4

Bifacial PV technology

Performance analysis and applications

Birinchi Bora, Arup Dhar, Deepak Yadav, and Mugala Naveen Kumar

4.1 INTRODUCTION

4.1.1 What is bifacial technology?

A bifacial solar photovoltaic cell (BSPV) is capable of absorbing photons from both front and rear sides; therefore it is capable of generating more power compared to its mono-facial counterpart. BSPV cells absorb global radiation from the front side and reflected diffused (mostly) radiation from the rear side. Figure 4.1 shows the structure and configuration of bifacial and mono-facial photovoltaic cells.

The working principle of the BSPV cells is similar to the conventional mono-facial solar cells except for the fact that the rear side also consists of a passivated coating having anti-reflection properties and the back metal contact is in the form of fingers and is similar to the front metal contact. The anti-reflection coating on the rear side enhances the photon absorption from that aside and generates electron-hole pairs (EHPs). All the charge carriers generated near the front region move toward the junction, and the EHPs are separated such that the electrons are pushed toward the negative contact and the holes are pushed toward the positive contact. From the metal contacts, the charge carriers move through the external circuit, and the work is done. Figure 4.2 shows the schematic diagram of bifacial photovoltaic technology.

In contrast to the standard PV module, the bifacial PV module consists of glass on both the front and rear sides instead of one glass on the front side and a back sheet on the rear side. The total irradiance on the backside is mostly from the diffused light reflected from the surroundings and the light passing through the module, possibly due to the cell gaps. This allows for the module to produce more power than the mono-facial counterpart, without engaging additional area and utilizing the same amount of semiconductor material. Figure 4.3 shows the configuration of mono-facial and bifacial PV modules.

DOI: 10.1201/9781003623168-6

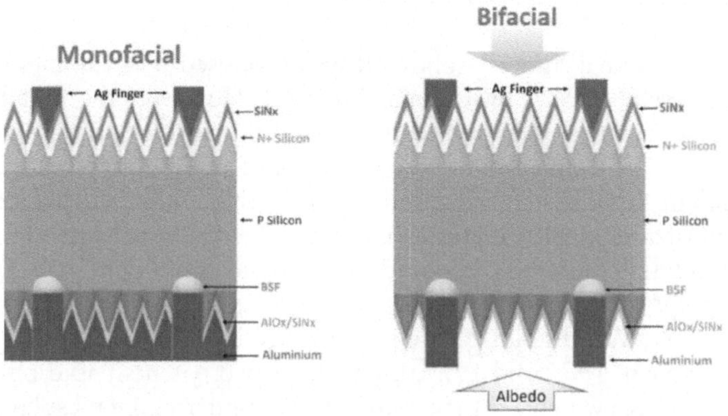

Figure 4.1 Structure and configuration of bifacial and mono-facial photovoltaic cells.

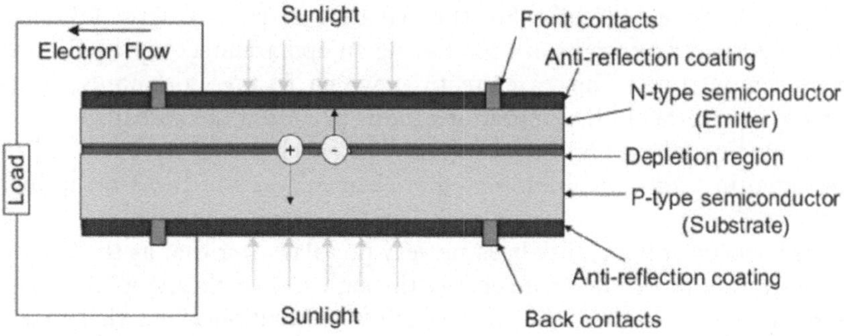

Figure 4.2 Schematic of BPV technology.

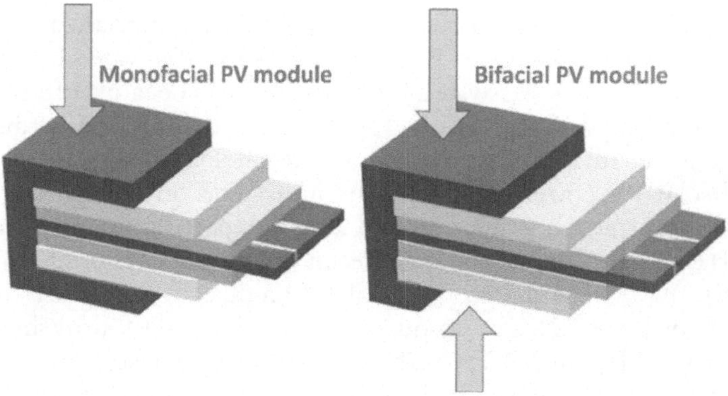

Figure 4.3 (A) A bifacial PV module and (B) a standard mono-facial PV module.

4.2 HISTORY OF BIFACIAL SOLAR CELL

The first theoretical bifacial solar cell was proposed in a Japanese patent dated October 4, 1960, by Hiroshi Mori [1]. However, the first bifacial solar cells and panels were fabricated during the Soviet Space Program in the Salyut 3 (1974) and Salyut 5 (1976) at the LEO military space station [2]. In Salyut 3, small experimental panels with a total cell surface of 24 cm^2 demonstrated a 34% increase in energy generation per satellite revolution due to Earth's albedo compared to the mono-facial panels at that time [3, 4]. Except for these implementations in Russian satellite systems, the bifacial solar PV technology remained mostly ignored over the next two decades. Later in 1976 and 1977, Luque filed two patents based on microelectronic devices which are the precursors for modern bifacial solar cells [5–7]. These solar cells were based on both the npp+ and the pnp structures. The development in bifacial solar cell research increased resulting in three important theses authored by Andrés Cuevas (1980), Javier Eguren (1981), and Jesús Sangrador (1982). For the first time in 1976, Cuevas fabricated Luque's patented bifacial solar cell having an npn architecture [8]. Eguren's thesis demonstrated npp+ architecture, which is now commonly known as back surface field [9]. Sangrador's thesis proposed the so-called vertical multi-junction edge-illuminated solar cell in which p+nn+ were stacked and connected in series and illuminated at their edges, and these being high-voltage cells required no surface metal grid to extract the current [10].

The concept of bifaciality became very popular after that as the demand for generating more and more energy through renewable energy increased. More research was focused on cost-effective solutions, and their use in several niche applications, such as noise barriers, parking shades, and fences increased during the 1990s and early 2000s. Bifacial PV technology caught the global market momentum in 2010 when solar PV manufacturers like Sanyo and Yingli Solar started commercializing bifacial modules using state-of-the-art PV technologies like Passivated Emitter and Rear Contact (PERC), Passivated Emitter with Rear Locally Diffused Cells (PERL), Passivated Emitter Rear Totally diffused (PERT), Interdigitated Back Contact (IBC), and heterojunction technology (HJT). Many large bifacial solar PV power plants have been installed throughout the globe, which demonstrated a gain of around 20% compared to mono-facial SPV plants [11]. As per the International Technology Roadmap for Photovoltaic (ITRPV), the market share of bifacial PV modules will increase to more than 35% by 2028 [12]. The introduction of the bifacial PV workshop "bifi-PV-workshop" [13] in 2012 brought together R&D, industry, and investors under one roof and enabled them to work together in making the bifacial PV technology bankable. Most remarkable results were demonstrated with n-PERT bifacial PV technologies, but the introduction of PERC+ revolutionized the PV module market and enabled the widespread availability of bifacial PV modules with prices at par with its mono-facial counterpart. The reason behind the acceptance of PERC+ modules by the manufacturers is that the production process flow is almost the same except for the rear metallization pattern compared to the PERC modules. The n-PERT and TOPCon technologies having a higher bifaciality factor will dominate the PV market in the near future as PERC technology is approaching its efficiency limits and more and more PERC Tier1 producers are also activating their n-type road maps, including Jinko, Canadian, and JA Solar [14]. With the introduction of high-quality n-PERT and TOPCon technology, the Levelized Cost of Electricity (LCOE) can decrease even faster due to increased bifaciality (up to 95%) and reduced degradation (no light-induced degradation (LID) and lower light and elevated temperature-induced degradation (LeTID)). Further, they inherit some unique features like lower temperature coefficients and better low-light performance.

Bifacial PV modules are built with either frame or frameless configurations as most of the modules consist of a double glass architecture. The junction boxes are kept shallow in size to avoid any possibility of shadows at the rear side of the module. As bifacial PV modules absorb light from both front and rear, the current produced is higher compared to the mono-facial modules, in which the number of busbars is increased and cut cell technology is incorporated to reduce resistive power losses in the connecting ribbons. Another important development in the bifacial module technology is the filling up of inter-cell space with white reflectors replicating the mono-facial PV module conditions where the reflected radiation is being used by the cells again. In the case of bifacial PV modules, it is circumstantial that the presence of the white reflectors in between the cell gap will enhance the power compared to the condition when the lights are being allowed to pass through and get reflected from the rear surface in the form of albedo.

Figure 4.4 shows various possible configurations in which the bifacial PV modules can be installed. The maximum bifacial gain in terms of energy can be obtained from the conventional fixed tilt configuration shown in Figure 4.4 with optimized height and spacing between the panels. The horizontal installations as shown in Figure 4.4 can be used in car parking and other applications where PV modules are used for shading purposes. The vertical configuration shown in Figure 4.4 is used for fencing, sound barrier, etc. The least LCOE can be achieved with the single-axis tracking

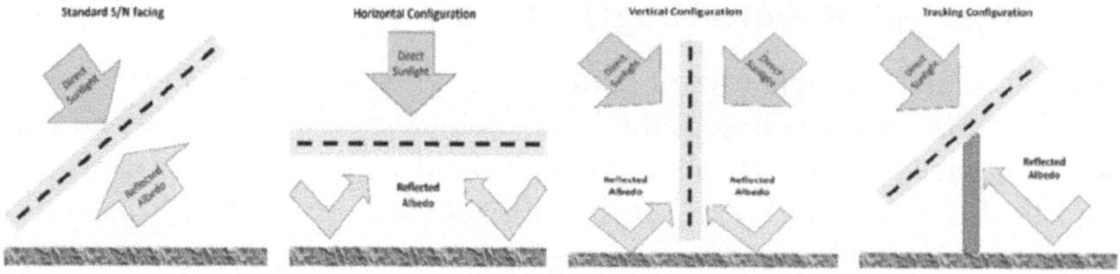

Figure 4.4 Possible installation configuration for bifacial modules.

system as shown in Figure 4.4 as they not only can generate the highest amount of energy but also can demand periodic maintenance.

The energy generation from the bifacial PV modules depends on the rear side irradiation which ultimately depends on the following factors:

- Albedo of the ground surface,
- Mounting height of the modules above the ground,
- Spacing between the rows,
- The ratio between diffuse horizontal irradiance (DHI) and global horizontal irradiance (GHI).

The higher precision of the energy generation estimation can be done if all the dependent parameters are known with higher accuracy. Proper energy estimation will play a key role in making bifacial PV technology bankable.

4.3 APPLICATIONS

The bifacial PV modules system can be used for various terrestrial applications:

4.3.1 Solar power plants

As lower LCOE is preferred by the PV market, the utilization of sunlight from both sides is enabled by the use of bifacial PV modules. The energy generation from the bifacial solar photovoltaic (SPV) power plant is quite different from the mono-facial one and depends on the installation configuration as shown in Figure 4.4. For utility-grade applications, maximum possible energy generation and minimum LCOE are the key requirements. To achieve these, the site survey is done properly and precise information regarding the parameters required for the energy generation is collected. It is observed that the fixed tilt configuration with optimized parameters enhancing the rear irradiation generates the highest energy in the non-tracking configuration. If financial support is available then minimum LCOE is achieved by a single-axis tracking system which then again demands periodic maintenance.

4.3.2 Agro-photovoltaic/agrivoltaic

Agrivoltaic and multiple uses of land is a hot topic of discussion nowadays. The definition of agrivoltaic is the simultaneous use of land for the production of electricity through photovoltaic modules and agricultural production. In this direction, Next2sun in collaboration with TOTAL in France has developed a very innovative technology using vertical mounting systems [15] on dual-use land. Apart from conventional silicon-based solar cell technology, semitransparent photovoltaics were also used for agrivoltaic, either with spectral semitransparency with selective use of wavelengths [16] or with regional semitransparency by splitting the portion of the received solar rays between the PV panel and the crop below [17]. Figure 4.5 shows the vertical installations of bifacial PV modules at the National Institute of Solar Energy (NISE), Gurgaon, India.

4.3.3 Buildings integrated with bifacial PV

Bifacial PV technology is also employed in Building-Integrated Photovoltaics (BIPV) [18], such as vertical facade integration [19] and shades [20]. There are many advantages to this application. Firstly, they not only generate electricity but also function as conventional building materials. In addition, they are less exposed to soiling losses, and therefore, huge cleaning costs can be reduced. Moreover, the orientation effect is not as important as in conventional mono-facial PV modules, which implies that they can face any orientation, including east and west. It can also serve as thermal insulation and a noise barrier to the building.

Figure 4.5 Vertically installed bifacial PV power plant at NISE.

4.3.4 Noise barriers

With the increasing population, the land cost is also increasing in the cities, which puts an economic barrier to solar PV installations in urban areas. In such conditions, the bifacial PV modules can be installed as noise barriers along highway and railway tracks and have a huge potential for cost-effective renewable energy generation. In 1997, the first bifacial PV noise barriers were installed and put into use in Switzerland [21]. Nowadays, noise barriers with bigger capacities are being built, for example, 730 kWp was installed in Italy in 2009, and a series of bifacial PV noise barriers ranging from 1 MWp to 2.065 MWp were installed in Germany [22].

However, the use of bifacial photovoltaic modules as a noise barrier requires an increase in the thickness of the module to be able to absorb noise and withstand any stones from the vehicle's movement. These can also cast a shadow on the rear side of the bifacial module which may reduce the energy yield [23]. Two solutions can be considered to reduce the losses due to shading, firstly placing the bifacial module cells away from the module frame and secondly increasing the number of bypass diodes [22].

4.3.5 Floating bifacial PV power plants

Floating bifacial photovoltaic solar power is a conception in which a solar photovoltaic system is installed directly on a body of water (Figure 4.6), rather than on the ground or the roofs of buildings [24].

This type of installation is generally composed of a floating platform to keep the photovoltaic system above the water and a mooring system to keep the panels in the same position and prevent them from rotating or drifting away [25]. Additionally, for this application, more robust bifacial PV modules, such as those with polymers, are required to tolerate salty environments. The cables and connectors to extract the electricity from the solar photovoltaic installation and transport it to the shore are also to be adapted to the floating conditions. This means that the cables have to be properly coated with waterproof material because they generally have to pass through water to reach the land [26]. PV modules installed on water bodies can be very advantageous compared to PV systems installed on land in terms of savings in land costs, reduction of water evaporation, enhancement of water quality, minimizing the effect of dust, and lowering the PV module temperatures due to inherited water-cooling effect [27]. The bifacial floating photovoltaic system receives reflected irradiation from water. However, the albedo of water bodies is very low compared to the normal albedo of the soil [28]. A comparison between mono-facial and bifacial floating modeling and experimental data was performed, as mentioned in reference [29]. The installations had capacities of 3.84 kW for mono-facial and 4.14 kW for bifacial modules. The results show that the bifacial PV had a bifacial gain of 4.5% in Frankfurt (Germany) and 7.3% in Catania

Figure 4.6 A picture of floating PV power plant.

(Italy). The data regarding floating bifacial PV are fewer and, to the best of the authors' knowledge, there are not many large commercial installations made in the world. However, this is expected to change, as the cost of bifacial PV is falling rapidly. Currently, a 2.83 MW floating solar PV farm, with about 6900 bifacial modules has been installed in Ratchathani, Thailand [30] from which an energy production of 4440 MWh is expected.

4.4 ADVANTAGES AND DISADVANTAGES OF BIFACIAL SOLAR PHOTOVOLTAICS OVER MONO-FACIAL PHOTOVOLTAICS

4.4.1 Advantages

(i) Enhanced power generation: As bifacial solar panels can absorb light from both the front and the rear sides, it produces more power compared to their mono-facial counterparts and ultimately increases the power generation density. More power can be produced in the same amount of area which makes solar photovoltaic technology become cost economic. Some manufacturers claim that bifacial solar panels can generate up to 30% more energy than conventional mono-facial solar panels. This higher efficiency translates into less space per watt.

(ii) Durability: As bifacial solar panels are covered with tempered glass from both sides, they're often more durable. The tempered glass is weather resistant, UV resistant, and can withstand high temperatures and strong winds. As a result of this, the durability of bifacial solar panels increases.

(iii) Aesthetically pleasing: Bifacial modules come in many designs, including frameless. The wide variety of designs makes them usable for BIPV applications, which demand good aesthetics and power generation at the same time.

(iv) Capability to work with diffused light: The bifacial modules work well with diffused light also, and therefore, one doesn't need to worry much about the orientation of the modules.

(v) Better reliability: As most of the bifacial modules come with a frameless design, the solar cells are less likely to suffer from potential induced degradation (PID) as the most prominent path for the leakage current to flow is through the metallic frames. Furthermore, bifacial panels without a metal frame don't require grounding.

(vi) Longer warranty: As bifacial PV modules are made up of high-quality solar cells like PERC, TOPCon, and HJT, the lifespan of these modules is generally more than 30 years as claimed by the manufacturers.

4.4.2 Disadvantages

(i) High initial cost: The use of high-quality solar cells and the existence of glass on both sides increase their cost compared to its mono-facial counterpart. This increases the overall cost of the project.

(ii) Higher installation cost: As bifacial solar panels are heavier due to the presence of glass on both sides of the panel, the transportation and handling costs increase, which finally increase the installation costs.

(iii) Higher cell temperature: The presence of glass on the back side is believed to trap heat which increases the cell temperature. Many studies are in progress to quantify the temperature of bifacial solar cells, but the presence of glass makes it difficult. Non-contact temperature measurements are preferred for the bifacial PV modules.

4.5 CURRENT BIFACIAL SOLAR CELL AND MODULE TECHNOLOGIES

Bifacial solar photovoltaics is an enhanced technology promising improved energy generation by virtue of its capability to absorb light from both sides in the PV module. In this chapter, the operating principles of the fundamental bifacial solar photovoltaic cell will be explained and various technological designs will be discussed. Various PV module configurations for bifacial applications will also be explained.

4.5.1 Bifacial solar cell technologies

In conventional mono-facial solar cells, light in the form of photons is absorbed only from the front side, whereas in bifacial solar cells, the photons are absorbed from both the front and rear sides. This additional absorption of photons from the rear side increases the EHP generation, and simultaneously, the power generation of the solar cell increases. Bifacial solar photovoltaic configurations have been under investigation since the 1960s [1].

The enhanced power generation increases the power density of the PV modules compared to that of its mono-facial counterpart, and therefore, reduces the land requirement for generating a similar amount of energy. Another significant advantage of the bifacial solar cells is the reduction in absorption of the infrared part of the solar spectrum due to the absence of the full-area aluminum metallization at the rear side of the solar cells. However, thermal insulation on the rear side of the PV module compensates for this advantage where glass is present on the rear side. Another advantage of bifacial configuration is the reduced recombination at the silicon-Al interface near the back surface field of the conventional mono-facial solar cells. Some of the currently used solar cell technologies in bifacial modules are given below:

i) **PERC bifacial solar cell**

Al-BSF solar cells were produced on a large scale as the solar photovoltaic market expanded significantly in the last decade. These solar cells have some drawbacks like rear side surface recombination near the silicon and full-area aluminum metallization, partial absorption of the infrared part of the solar spectrum, and the low carrier lifetime of the p-type silicon. Passivated Emitter Rear Contact (PERC) solar cell technology addressed the rear surface recombination and infrared absorption significantly by introducing rear side passivation and localized metal contacts at the rear side of the solar cell. As per the International Technology Roadmap for PV (ITRPV), PERC solar cells shared 50% of the worldwide PV industry

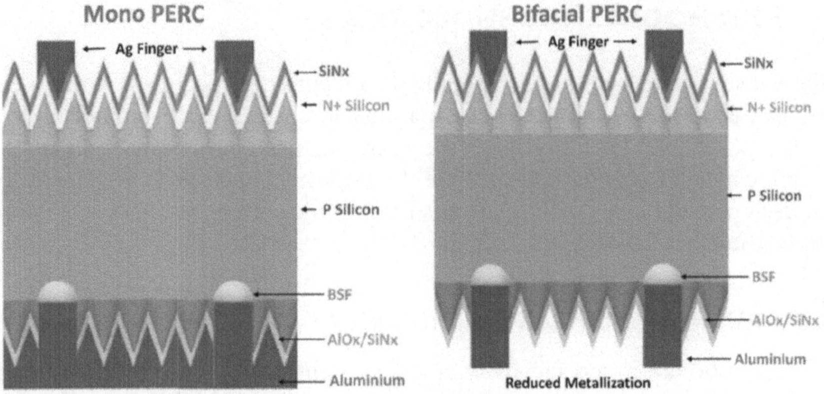

Figure 4.7 Bifacial PERC solar cell.

in 2019 and are expected to reach approximately 80% within the coming years [31].

PERC technology is already reaching its theoretical efficiency limit of 22.5 %, and in this scenario, to enhance the power output, bifacial PERC solar cells have been introduced and are also popularly known as PERC+ [32]. The manufacturing process also doesn't need much change as can be seen in Figure 4.7. The structure of the bifacial PERC (PERC+) is shown in Figure 4.7. It is very much similar to the PERC solar cell structure which has p-type silicon as the base wafer, N+ doped emitter at the front and above that, a layer of SiNx is present as a passivation cum anti-reflection coating (ARC) layer, and finally, the metal fingers are present on the front side for the collection of generated electrons. On the rear side, a stacked layer of SiNx and AlOx is present for passivation through which metal contacts are laser grooved. The difference between mono-facial and bifacial PERC is that the full-area rear Al screen-print in mono-facial PERC is replaced with an Al finger grid screen design in bifacial PERC. The Al metal fingers with a pitch identical to the local laser grooved opening pitch must be aligned to ensure the overlap of Al fingers and laser contact opening introduced through the rear passivation stack of AlOx/SiNy [33]. Finally, the local Al-BSF is formed during the firing process, with Al in direct contact with the silicon wafer.

With bifacial PERC+ cells, Al paste consumption decreases up to 0.2 g per wafer [34]. Optimization of the rear stacking layer consisting of AlOx/SiNy is very critical from both the passivation and anti-reflection points of view. Also, the design of the rear metal grid has to be such that minimum series resistance loss is incurred.

At present, bifacial p-type Czochralski-grown PERC+ are produced with an average efficiency above 21.5%, whereas Trina Solar has announced a certified efficiency of 23.39% for a 252 cm^2 PERC+ cell with nine-busbar technology using standard manufacturing equipment [35]. At the module level, SolarWorld initiated the mass production of PERC+ glass/glass bifacial modules in 2015. Since then, various companies, such as Neo Solar Power Energy Corporation, Trina Solar, and LONGi Solar, have followed Solar World's technology route to offer commercial PERC+ PV modules.

In the near future, the integration of the Tunnel Oxide Passivated Contact (TOPCon) technology with the PERC+ architecture is expected in which the passivating contact can be created with a thin interfacial oxide and a highly doped polysilicon layer on the top. TOPCon solar cells are expected to dominate the PV industry after the era of PERC solar cells. There are various limitations in today's p-type wafers, which are creating barriers to efficiency enhancement. One of the most significant issues with p-type wafers is their low bulk carrier lifetime, which plays a dominant role in the performance of the solar cell. Another prominent issue is the light-induced degradation caused by boron-oxygen complexes. P-type wafers are also highly sensitive toward metal impurities.

ii) **n-PERT bifacial solar cell**

The limitations of p-type wafer-based solar cells are driving the PV industry's attention toward high-efficiency n-type solar cells, including n-PERT solar cells due to their high bulk carrier lifetime and potential of bifaciality factor of up to 95% [36].

Figure 4.8 shows the architecture of n-PERT bifacial solar cell structure which typically consist of n-type base wafer and P+ boron doped emitter at the front side which is also passivated mostly with SiO_2/SiN_x stacking

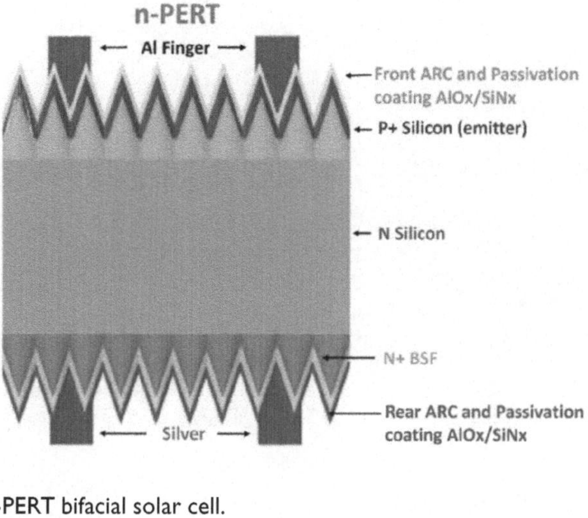

Figure 4.8 N-PERT bifacial solar cell.

layer. An N+ phosphorous-doped back surface field cell (BSF) covers the rear side completely and is passivated by a SiNx dielectric layer. The basic difference between PERC+ and PERT solar cells is that the rear BSF is present locally in the case of PERC+ while the BSF covers the complete rear side which eliminates the requirement of laser grooving and therefore decreases the series resistance loss significantly. The concept of n-PERT solar cells was first developed in 2002 at the University of New South Wales [37]. Interuniversity Microelectronics Centre (IMEC), in collaboration with Jolywood, recently reported a 23.2% efficient fully screen-printed bifacial n-PERT cell, which has a bifaciality coefficient above 80%.

iii) **Heterojunction bifacial solar cell**

Figure 4.9 presents a typical bifacial heterojunction (HJT) solar cell having n-type monocrystalline silicon as the base absorber material. On both sides of the absorber, a thin (~5 nm) intrinsic amorphous silicon [a-Si:H(i)] is deposited with plasma-enhanced chemical vapor deposition (PECVD). This layer provides high-quality passivation and obstructs surface recombination at the edges of the base absorber. This significantly enhances the effective lifetime of the solar cell. Now, the n- and p-type doped a-Si:H layers are applied to opposite sides of the wafer, respectively, to form electrical

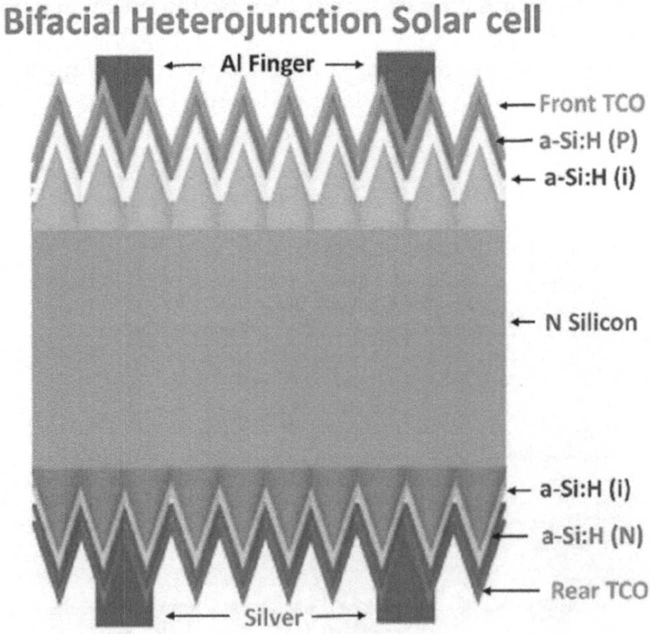

Figure 4.9 Bifacial HJT solar cell.

contacts to the electrons and holes in the wafer. This in-situ doped a-Si:H layer facilitates efficient carrier selectivity. To enhance lateral transport of the collected carriers toward the metallic grids, transparent conductive oxides are sputtered on top of the doped a-Si:H layers on both sides of the wafer. In HJT solar cell technology, the efficiency is significantly enhanced by virtue of both *surface passivation* and *carrier selectivity.*

The HJT solar cells possess high Voc (typically around 750 mV) owing to high-quality chemical passivation by the a-Si:H(i) layer. Efficiencies greater than 25% have already been demonstrated for large-area HJT in integration Interdigitated Back Contact (IBC) technology [38]. Bifacial HJT solar cells offer significant benefits of low-temperature coefficient, compatibility with thin wafers, and most importantly, high bifaciality factor (typically > 90%). In terms of limitations, bifacial HJT suffers from parasitic absorption losses in TCO and a-Si:H layer, which creates a trade-off between the Jsc and Voc [39].

iv) **Bifacial perovskite-heterojunction silicon tandem solar cells**

The highest efficiency achieved in a single-junction silicon solar cell is 26.7% with the integration of IBC and HJT concepts by Kaneka in Japan [40]. Still, the efficiency remains lower than the theoretical efficiency limit of 29.43% due to thermalization and transmission losses happening in single-junction solar cells [41]. To overcome these major drawbacks, tandem solar cells came into existence. These have multi-junction configurations with various optical band gaps to harvest the entire solar spectrum range. Recently, four-terminal perovskite-heterojunction silicon tandem solar cells with spectral albedo have been reported with an efficiency of around 30% [42]. Figure 4.10 shows the structure of bifacial perovskite-heterojunction silicon tandem solar cells.

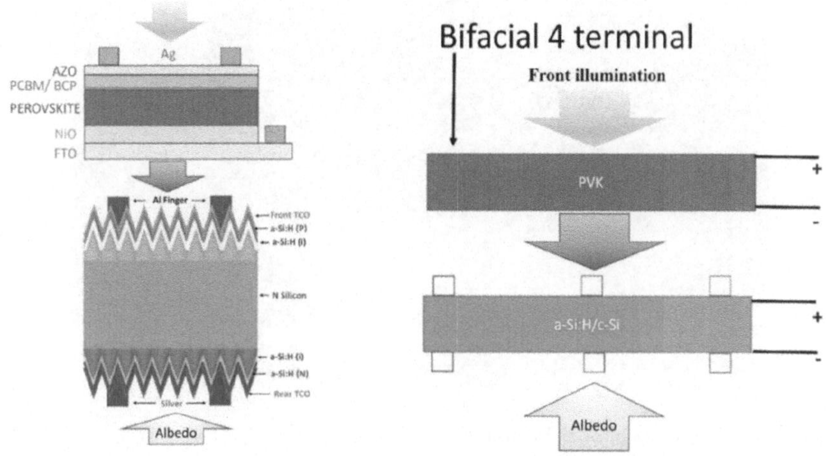

Figure 4.10 Bifacial perovskite-heterojunction silicon tandem solar cells.

4.5.2 Bifacial solar module technologies

Bifacial PV modules have either glass or transparent polymer back sheet at the rear side to enable the absorption of light from the rear side of the PV modules. The most common practice is to use glass at the rear side of bifacial modules, which is supported by the aluminum frames, but in some cases, these bifacial PV modules can be made frameless by increasing the rigidity of the adhesion between the glass-glass architecture.

The frameless PV modules enjoy the advantage of high insulation resistance, which enhances their reliability against PID when used for high-voltage power generation. This also reduces the weight of the PV modules, which eases the transport and installation of PV modules. Figure 4.11

Figure 4.11 Frameless bifacial PV modules.

shows a frameless bifacial PV module. Another advancement in the bifacial PV module technology which is penetrating the market significantly is the flexible module technology. This Light-Flexible High-Efficiency Module uses advanced composite encapsulation materials that are a perfect match for curved modules weighing 70% less than the products of its kind. Since they are light and flexible, they can be used in a variety of locations, such as a rooftop with low load-bearing capacity or the exterior of a building. Additionally, it is convenient to install these modules.

4.6 PERFORMANCE MEASUREMENT OF THE BIFACIAL MODULE

4.6.1 Performance measurement of the bifacial module in indoor conditions

Specific measurement procedures are demonstrated in IEC TS 60904-1-2 to characterize the power of bifacial PV modules that are capable of generating power from both the front and rear surfaces. The characterization of the bifacial PV modules includes three important steps:

a) Measurement of bifaciality factor at standard test conditions (STC).
b) Determination of rear-irradiance-driven power gain yield, BiFi.
c) Output power determination at 10% and 20% rear irradiance with respect to the front.

The performance measurement of the bifacial module in indoor conditions can be measured in two different ways as per IEC 60904-1-2 [43].

i) Solar simulator with adjustable irradiance levels for single-side illumination.

Solar simulators used for measuring bifacial PV modules are capable of providing irradiance levels of typically more than or equal to 1200 W/m^2. The simulator's non-uniformity of irradiance could be below 2% and could remain below this value at irradiance levels used for the characterization of bifacial devices.

a) Bifaciality factors measurement at STC.

Measurement of the front side is taken at STC following the conditions defined in IEC 60904-7 [44] while blocking the rear side of the module with black cloth/ paper so that the light incident on the rear side is less than 3 W/m^2. Similarly, the measurement at STC is taken from the rear side while covering the front side. This way, all the electrical parameters are obtained from the front and rear sides individually at STC. Now the ratio of three crucial parameters is considered for bifaciality factor calculations:

$\varphi Pmax$ = Ratio of rear to front side maximum power (*Pmax*),
φVOC = Ratio of rear to front side open-circuit voltage (*Voc*),
φI_{sc} = Ratio of rear to front side short circuit current (φI_{sc}).

The least value among the above three parameters is considered to be the bifaciality factor, φ . Values of φ typically range from 75% to 95% for n-PERT bifacial modules, from 60% to 70% for p-type PERC bifacial modules, and >90% for HJT bifacial modules.

b) Determination of rear irradiance-driven power gain yield, BiFi.

The power output of the bifacial module is determined by exposing the front side of the bifacial PV module at G_f=1000 W/m^2 and various rear side irradiance G_r, especially at different lower irradiance levels.

G_r = From 0 to 100 W/m^2, from 100 to 200 W/m^2, more than 200 W/m^2

For solar simulators having only one light source, power output can be calculated using the equivalent irradiance G_E method described by the equation:

$$G_E = 1000\ W/m^2 + \varphi * G_r$$

Now, the Pmax vs. G_r is plotted for different rear irradiance and the slope calculated is defined as X_i

c) Output power determination at 10% and 20% rear irradiance with respect to the front.

Now the calculated X_i is used to determine the performance of the PV module for two rear irradiance levels i.e., 10% and 20%:

$$\text{Pmax } X_i\ 100 = Pmax\text{, STC} + X_i * 100\ W/m^2$$

$$\text{Pmax } X_i\ 200 = Pmax\text{, STC} + X_i * 200\ W/m^2$$

ii) Solar simulator with adjustable irradiance levels for double-side illumination.

A solar simulator, as defined in IEC 60904-9 [45], with the additional capability to simultaneously illuminate the bifacial device on both sides could be used. Such simulators were able to provide irradiance at different levels on both sides. The non-uniformity of irradiance was below 2 % on both sides, at the irradiance levels used for the characterization of bifacial devices. For simulators having two light sources with adjustable irradiance levels, the front side can be illuminated with G_f = 1000 W/m² and the rear side with at least two irradiance levels (100 W/m² and 200 W/m²). The nameplate power output may be defined at STC on both the front and rear sides individually and at 1000 W/m² on the front side and 100 W/m² and 200 W/m² on the rear side.

4.6.2 Performance measurement of the bifacial module in outdoor conditions

The performance measurement of the bifacial module can also be done in outdoor conditions using an I-V tracer. In addition to the measuring equipment as per IEC 60904-1, at least two additional PV reference devices, as described in IEC 60904-2, are required to measure the irradiance level on the rear side and the rear side irradiance non-uniformity. Their spectral responsivity should be as close as possible to the PV module under test. It should be ensured that the non-uniformity of irradiance on the rear side must be below 10% during outdoor measurement under natural sunlight. To determine the non-uniformity, at least five reference devices shall be placed across the rear side of the PV module as shown in Figure 4.12 and their mean value should be considered. At present, there is no standard procedure to translate the I-V data of the bifacial module to STC. However, the translation can be done using the procedure defined in IEC 60891. The module temperature in the outdoor condition should be measured using either a non-contact temperature sensor or the Voc data. Research is still required to establish the procedure for estimating the bifacial module temperature and the translation of I-V.

4.6.3 Bifacial PV module nameplate rating

As of now, there is no clarity on the nameplate rating of the bifacial PV modules. Some of the manufacturers use conventional front STC results, whereas some manufacturers add some power based on the assumed contribution from the rear. Technischer Überwachungsverein (TÜV) Rheinland has proposed specific bifacial standard test conditions of 1000 W/m² for front side irradiance and 135 W/m² for rear side irradiance [44]. The definition of rear side irradiance is as follows:

- Albedo factor: 0.21 (light soil)

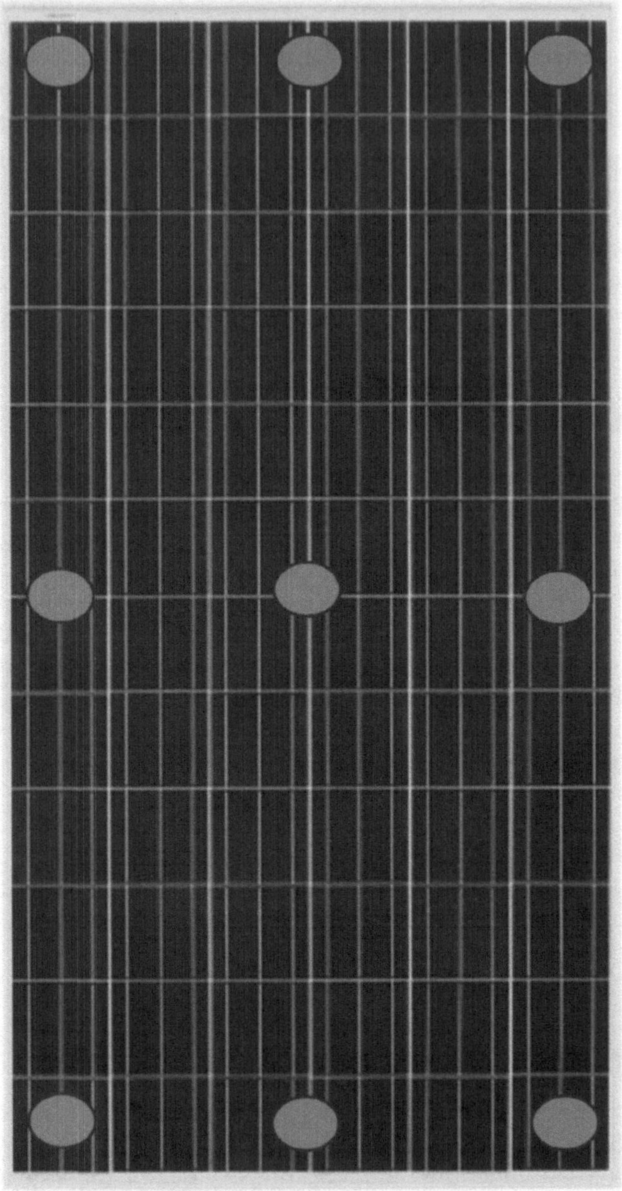

Figure 4.12 Position of irradiance measurement device to check the non-uniformity.

- Clearance height: 1 m
- Inclination angle: 37°
- Front side irradiance:1000 W/m^2

The nominal output power at BSTC of the bifacial module is then measured with an equivalent irradiance of GE = 1000 W/m² + φ * 135 W/m².

As per IEC 60904-1-2, the nameplate rating of bifacial module can be given as follows:

Bifaciality coefficient	*Front side irradiance (W/m²)*	*Rear side irradiance (W/m²)*	*Equivalent irradiance (W/m²)*	*Pmax (W)*	*BiFi , ref (W/W/ m²)*	$P_{max\ Bifi\text{-}DUT}$
X	1000	0	1000	P1		Z1
		100		P2		Z2
		200		P3		Z3

Based on the reflecting surface used in the field, the bifacial PV power plant size should be calculated for estimating the energy output.

4.7 PERFORMANCE AND RELIABILITY TESTING OF THE BIFACIAL MODULE

The performance and reliability of PV modules can be measured in indoor and outdoor conditions. For the indoor reliability testing of PV modules, people use IEC 61215 standards for design and qualification checking and IEC 61730 for safety testing.

4.7.1 Testing procedure as per IEC 61215: 2021 [46]

i) **Visual inspection (Module Quality Test (MQT) 1):** The visual defects of PV modules can be inspected under the illumination of 1000 lux. For qualifying the visual inspection test as per IEC 61215, the following visual defects should not be observed during the inspection:
   a) Broken, cracked, or torn external surfaces, including superstrates, substrates, frames, and junction boxes;
   b) Bent or misaligned external surfaces, including superstrates, substrates, frames, and junction boxes;
   c) A crack in a cell, the propagation of which could remove more than 10% of that cell's area from the electrical circuit of the module;
   d) Bubbles or delaminations forming a continuous path between any part of the electrical circuit and the edge of the module;
   e) Loss of mechanical integrity, to the extent that the installation and/or operation of the module would be impaired.

ii) **Maximum power determination (MQT 2):** The purpose of this test is to measure the performance of the module at the maximum power point. The following apparatuses are required for this test:

a) Either natural light source or radiant light source as per IEC 60904-9.
b) I-V tracing unit as per IEC 60904-1 for measuring current-voltage characteristics.
c) A PV reference device for measuring the in-plane irradiance.
d) Module temperature sensor.

For measuring the performance, the spectral response of the PV reference device, test sample as per IEC 60904-8, and spectrum of the light are required to estimate the mismatch factor as per IEC 60904-7. The MMF should be included while measuring the in-plane irradiance. The maximum power determination should be done at a temperature between 20 °C and 50 °C and an irradiance between 700 W/m^2 and 1100 W/m^2. The standard does not mention about the measurement of the bifacial module with both sides exposed simultaneously. However, one can measure it by covering the other side.

iii) **Insulation test** (MQT 3): This test is used to check the resistance between the solar cell and the module frame. The test procedure for the bifacial module is the same as a mono-facial module. The system voltage should be applied for 2 minutes to measure the resistance of the PV module. The passing criteria of insulation resistance will be 40 MΩ m^2.
iv) **Measurement of temperature coefficients** (MQT 4): The purpose of this is to measure the effect of temperature on the short circuit current, open-circuit voltage, and maximum power at 1000 W/m^2. For measuring the temperature coefficient, a procedure as per IEC 60891 can be used. The short circuit current, open-circuit voltage, and power should be plotted against the temperature, and the linear coefficient should be estimated. From the slope of the linear plot, the temperature coefficient can be estimated using the electrical parameter at STC.
v) **Performance of bifacial module at STC (MQT 6.1) and low irradiance condition** (MQT 7): The performance of the bifacial module at STC should be measured using the procedure mentioned in IEC TS 60904-1-2. For the measurement at STC, in the bifacial module, 1000 W/m^2 on the front side and 100 W/m^2 and 200 W/m^2 on the rear side were used for estimating the performance. The performance of the bifacial module was measured at a low irradiance of 200 W/m^2 and 25 °C using the procedure as mentioned in IEC TS 60904-1-2. The procedure for measuring the performance at low irradiance conditions is the same as STC; however, the irradiance is 200 W/m^2 instead of 1000 W/m^2.
vi) **Outdoor exposure test** (MQT 8): The main purpose of this study is to assess the ability of the module to withstand exposure to outdoor conditions. The bifacial module should be installed at a latitude of ±5°. A

suitable fixed resistive load or MPPT load can be used to operate near the maximum point. A uniform reflective surface may be used for this test. The in-plane irradiance should be measured on the front side, and the test should be done for a total exposure of 60 kWh/m^2.

vii) **Hot spot endurance test (MQT 9):** The main purpose of this test is to check the ability of the module to withstand reverse bias hot spot heating effects. The procedure for mono-facial and bifacial modules is the same. For bifacial module hot spot testing, light source as per IEC 60904-1-2 should be used and the backside should be covered if single-side illumination is used.

viii) **UV preconditioning test (MQT 10):** The purpose of this test is to check the susceptibility of the PV module to UV light degradation. The procedure for a mono-facial and bifacial module is the same. The total UV dose should be an irradiation of at least 15 kWh/m^2 at 60 ± 5°C. For bifacial modules, the rear side also should be exposed to UV light.

ix) **Thermal cycling test (MQT 11):** The main purpose of this test is to check the withstanding capability of the PV module with repeated changes in temperature. The module needs to be installed in a climatic chamber with a 5 N weight and the temperature stress should range from −40 °C to 85 °C. For one test sequence, the number of cycles is 50 and for other test sequences it is 200. The applied current needs to be estimated for the bifacial C-Si module as per IEC 61215-1-1: 2021.

x) **Humidity freeze test (MQT 12):** The main purpose of this test is to check the withstanding capability of the PV module at high temperature and humidity followed by subzero temperature. The module needs to be installed in a climatic chamber and temperature and humidity stress from 85 °C, 85% RH to −40 °C should be applied with no humidity control. Totally, ten cycles of stress need to be given over the module to check the capability. The applied current should be 50% of the current estimated for the bifacial C-Si module as per IEC 61215-1-1: 2021 for the thermal cycling test.

xi) **Damp heat test (MQT 13):** The main purpose of this test is to check the capability of the PV module to withstand long-term exposure to humidity. The module needs to be installed in a climatic chamber and the temperature and humidity stress should be set at 85 °C, 85% RH for 1000 cycles. The module should be short-circuited during the stress.

xii) **Robustness of terminations (MQT 14):** The main purpose of this test is to check the withstanding capability of terminations during normal assembly or handling operations. There are two types of robustness of terminations test: retention of the junction box on the mounting surface and cord anchorage test. The test procedure for the bifacial module is the same as the mono-facial module.

xiii) **Wet leakage current test (MQT 15):** The purpose of this test is to evaluate the insulation resistance of the solar cell and the frame of the PV module during wet conditions. Usually, system voltage is applied for 2 minutes in the module with the shorted positive and negative terminal and the module frame. The test procedure for bifacial will be the same as that for mono-facial module. Water needs to be sprayed properly on the rear surface of the module also. The passing criteria for large-size modules is that the insulation resistance should be more than 40 $M\Omega.m^2$.

xiv) **Static mechanical load test (MQT 16):** The purpose of this test is to evaluate the ability to withstand a static mechanical load. The test load should be equal to the design load multiplied by the safety factor. The test load should be maintained consecutively on each side of the module for 1 hour and the cycles should be repeated three times.

xv) **Hail test (MQT 17):** The purpose of this is to evaluate the hail impact withstanding capability of PV modules. Ice balls of the size mentioned in the IEC 61215 are used. They are thrown over the different regions of the PV module at a specified speed. After the impact, the power degradation and cracks need to be checked in the module.

xvi) **Bypass diode test (MQT 18):** There are two types of bypass diode tests: bypass diode thermal test and bypass diode functionality test. The purpose of the thermal test is to check the adequacy and reliability of the diode to reduce the damage caused due to hot spots. A current of 1.25 times the short circuit current should be applied with a pulse of less than 1 ms. For bifacial modules, the applied current should be 1.25 times the short circuit current including the rear side irradiance as defined in IEC 61215-1. The purpose of the bypass diode functionality test is to check the functionality of the diode after giving stress over the module. For this purpose, the I-V tracer is used to measure the current-voltage characteristics of the PV module at different conditions as mentioned in IEC 61215-2.

xvii) **Stabilization test (MQT 19):** The purpose of this test is to stabilize the PV module electrically. The performance of the module needs to be measured repeatedly after exposing the module with different procedures as mentioned in IEC 61215-2. The stabilization of the PV module is technology specific and defined as follows:

$$\frac{Maximum\,power - Minimum\,power}{Average\,power} < z$$

where z is the technology-specific value.

The light-induced degradation can be performed in both indoor and outdoor conditions. The module should be in the MPPT conditions during the stabilization procedure. The initial stabilization should be done to verify the nameplate of the PV module after repetitive exposure to light. There are two passing criteria defined in IEC 61215-1 (gate 1 and gate 2). Gate 1 is for initial stabilization and gate 2 is for estimating the module degradation. There is another stabilization procedure to put the defects causing the boron-oxygen LID into the regenerate state.

xviii) **Cyclic (dynamic) mechanical load test (MQT 20):** The purpose of this is to check the withstanding capability of PV module components in the dynamic mechanical load. It is required to check the integrity of the module components after giving stress. The test procedure is the same for both the mono-facial and bifacial modules.

xix) **Potential induced degradation test (MQT 21):** The purpose of this test is to check the withstanding design capability of the PV module at system voltage under high humidity and high temperature. The module needs to be put in a climatic chamber and the system voltage defined by the module manufacturer should be applied at 85 °C temperature and 85% relative humidity. Stress is applied for 96 hours. The test procedure is the same for both the mono-facial and bifacial modules.

xx) **Bending test (MQT 22):** The purpose of this is to check the bending capability of the flexible module as specified by the manufacturer. The module needs to roll from a flat position to the specified radius of curvature of a cylinder. There should not be any crack in the module during visual inspection, and the module should qualify for the wet leakage test.

4.7.2 Testing procedure as per IEC 61730 standards [47]

At present, for the bifacial module, there is no IEC standard available for safety testing. So, the IEC 61730:2016 is modified in light of the IEC 61215: 2021 for bifacial module testing and the details are given below:

i) **Visual inspection (Module Safety Test (MST) 01):** As per IEC 61730, visual inspection needs to be done as per IEC 61215-2, and other safety aspects should be carried out as per MST 01 of IEC 61730-2. The inspection criteria are the same for both mono-facial and bifacial modules.

ii) **Performance at STC (MST 02):** The purpose of this test is to verify *I*sc and Voc of PV modules after stabilization as per IEC 61215 with respect to the nameplate value declared by the manufacturer. The passing criterion was *I*sc and Voc within tolerances declared by the manufacturer after stabilization.

iii) **Maximum power determination (MST 03):** The purpose of this test is to verify the electrical characteristics as per MQT 02 of IEC 61215-2. There should not be any abnormal behavior in the I-V curve with respect to the initial I-V curve as per MST 02.
iv) **Insulation thickness test (MST 04):** The purpose of this test is to verify the insulation thickness of the PV module as per different classes of PV modules specified in IEC 61730-1. The thickness needs to be checked for both sides of bifacial PV modules.
v) **Durability of markings (MST 05):** The purpose of this test is to check the durability of markings in the PV modules. The inspection is done by rubbing the hand with a piece of cloth soaked with water and petroleum spirits.
vi) **Sharp edge test (MST 06):** The purpose of this test is to check the sharp edges and burs in the PV module, which may damage the conductors and injure the user. This test should be done during inspection only.
vii) **Bypass diode functionality test (MST 07):** The bypass diode functionality test for bifacial PV modules needs to be done as per MQT 18.2 of IEC 61215 as mentioned earlier.
viii) **Accessibility test (MST 11):** The purpose of this test is to check the protection against access to the hazardous live parts in the PV module. The test fixture should be as per IEC 61032. The resistance between the live part and the test fixture should not be less than 1 MΩ.
ix) **Cut susceptibility test (MST 12):** The purpose of this test is to check the withstanding capability of polymeric materials used in the PV module during handling PV module installation and maintenance to avoid electric shock. However, for a glass-to-glass bifacial module, this test is not applicable.
x) **Continuity test of equipotential bonding (MST 13):** The main purpose of this test is to check the resistivity of accessible conductive parts like the metallic frame in PV modules. The resistance between conductive components should be less than 0.1 Ω.
xi) **Impulse voltage test (MST 14):** The main purpose of this is to check the high-voltage insulation withstanding capability of the PV module. The module needs to be covered with conductive metallic foil with the best possible contact. The negative terminal of the impulse voltage generator should be connected to the foil and the positive terminal should be connected to shorted output terminals of the PV module. The surge impulse voltage should be applied as per IEC 60060-1. There should not be any visual defects, dielectric breakdown, and failure in the insulation test after applying surge impulse voltage to qualify the test.
xii) **Insulation test (MST 16):** The purpose of this test is to check the insulation between the conductive component of the PV module and the

frame. The applied voltage should be as per IEC 61730 for different classes of PV modules. The voltage should be applied between the shorted output terminal and the frame. The insulation resistance should be more than 40 MΩ.m^2.

xiii) **Wet leakage current test (MST 17):** The purpose of this test is to check the insulation resistance between the conductive component and the frame during wet conditions. The voltage should be applied between the shorted output terminal and frame as per IEC 61730-1 for different classes of PV modules. The insulation resistance during the wet leakage test should be more than 40 MΩ.m^2.

xiv) **Temperature test (MST 21):** The purpose of this test is to estimate the maximum reference temperature for different components of the PV module during its operation. There are two types of methods to do this test: the outdoor method and the indoor method. In the outdoor method, the following parameters should be measured during the test: temperature of PV components, ambient temperature, in-plane irradiance, and wind speed. The environmental conditions during the test should be ambient temperature ranging from 20 °C to 45 °C and irradiance ranging from 700 W/m^2 to 1000 W/m^2. In indoor conditions, the intensity during the test should be 1000 W/m^2.

xv) **Hot spot endurance test (MST 22):** The purpose of this test is to check the ability of the module to withstand reverse bias hot spot heating effects. The test is the same as mentioned earlier for the hot-spot endurance test as per IEC 61215.

xvi) **Fire test (MST 23):** The purpose of this test is to check the fire resistance properties of PV modules to fire originating from an outside source. The fire testing of the PV module can be done using ENV 1187 and ANSI/UL 1703.

xvii) **Ignitability test (MST 24):** The purpose of this test is to check the ignitability of PV modules using a direct flame. For this test, the standard ISO 11925-2 is used. The testing should be done at an environmental temperature of 23 ± 5 °C and relative humidity of 50% ± 20%. The ignitability test should be done on the surface and edge of the PV module. The flame application time for this test is 15 S.

xviii) **Bypass diode thermal test (MST 25):** The purpose of this is to check the adequacy and reliability of the diode in reducing the damaging effect of hot spots. For the bifacial module, the applied current should be 1.25 times the short circuit current including the rear side irradiance as defined in IEC 61215-1. The test procedure is the same as the bypass diode thermal test of IEC 61215.

xix) **Reverse current overload test (MST 26):** The purpose of this test is to check the withstanding capability of the PV module at overload reverse current. Due to the applied reverse current in the PV module, the components get heated up and dissipate heat. The current should be 135% of the maximum over the current rating i.e., the fuse rating of the PV module. The module needs to be covered with white tissue paper during the test, and there should not be any flaming or charring in the tissue after the test.

xx) **Module breakage test (MST 32):** The purpose of this test is to check the risk of physical injuries if the module is broken during the installation. An impactor of weight 45.5 ± 0.5 kg is used to impact the module from a height of 300 mm from the surface of the PV module. After the stress, the module should not be separated from the frame. There should not be any breakage in the module, and if there is any breakage, then the size of the opening should not be more than 76 mm diameter, and the particle ejected from the module surface should be less than 65 cm^2.

xxi) **Screw connections test (MST 33):** There are two types of screw connections test: test for general screw connections (MST 33a) and test for locking screws (MST 33b). The purpose of this test is to check the reliability of screw connections used in the PV module. In case of general screw connections test, the screws and nuts that are likely to be tightened by users should be tightened and loosened five times at a particular torque specified in the standard. There should not be any damage to the screw during the test, and the screw should still be usable. The purpose of locking screw testing is to check the reliability after applying a specified torque to the screw. The torque should be applied to the screw for 1 minute each both in clockwise and anticlockwise directions at 25 °C. The screw should not be loose after the test.

xxii) **Static mechanical load test (MST 34):** The purpose of this test is to evaluate the ability to withstand a static mechanical load. The test procedure is the same as the test procedure mentioned in MQT 16 in IEC 61215-2.

xxiii) **Peel test (MST 35):** The purpose of this test is to check the durability of the adhesion of different polymeric materials in the PV module. The test is not applicable to glass-to-glass bifacial modules. The test needs to be done as per ISO 5893. The polymeric material needs to be peeled from the module surface at 90° ± 10° at a rate of 50 mm/min ± 5 mm/min. The test needs to be performed at different positions of the PV module, and the mean value of peel off force should be estimated.

xxiv) **Lap shear strength test (MST 36):** The purpose of this test is to check the durability of the adhesion of glass-to-glass assemblies. The test is mainly applicable to glass-to-glass bifacial modules. The test needs to be done as per ISO 4587. Glass-to-glass samples need to be prepared as per the design criteria of the module. Ten samples need to be tested before giving stress, and ten samples need to be tested after a weathering test like UV exposure. The loss in the mean breaking force of ten samples before applying stress to the mean breaking force of ten samples after weathering stress should be less than 50%.

xxv) **Materials creep test (MST 37):** The purpose of this test is to check the creep or adhesion capability of PV module materials after operating at the highest temperature that a module will face in the field. The module needs to be installed in an environmental chamber and the temperature should be set at 105 °C ± 5 °C for 200 h. Depending on the class of the PV module, the creepage and clearance distances should meet the criteria as per IEC 61730-1.

xxvi) **Robustness of terminations test (MST 42):** The main purpose of this test is to check the withstanding capability of terminations during normal assembly or handling operations. The test procedure is the same as the MQT 14.1 and MQT 14.2 in IEC 61215-2.

xxvii) **Thermal cycling test (MST 51):** The main purpose of this test is to check the withstanding capability of the PV module at repeated changes in temperature. The test procedure is the same as the MQT 11 in IEC 61215-2.

xxviii) **Humidity freeze test (MST 52):** The main purpose of this test is to check the withstanding capability of the PV module at high temperature and humidity followed by subzero temperature. The test procedure is the same as the MQT 12 in IEC 61215-2.

xxix) **Damp heat test (MST 53):** The main purpose of this test is to check the withstanding capability of the PV modules during long-term penetration of humidity. There are two different standard durations used in this test: 1000 h and 200 h. The test procedure is the same as the MQT 13 in IEC 61215-2.

xxx) **UV test (MST 54):** The purpose of this test is to check the susceptibility of the PV module to UV light degradation. In this test, two different standard doses are used: 15 kWh/m^2 and 60 kWh/m^2. The test procedure is the same as the MQT 10 in IEC 61215-2.

xxxi) **Cold conditioning (MST 55):** The purpose of this test is to check the withstanding capability of the PV module at sub-zero temperature. A climatic chamber should be used, and the module should be installed in the chamber. The temperature of the chamber should be set at −40 °C for 48 h. There should not be any visual defects, and the insulation resistance test should be passed by withstanding the stress.

xxxii) **Dry heat conditioning (MST 56):** The purpose of this test is to check the withstanding capability of the PV module at high temperatures. A climatic chamber should be used, and the module should be installed in the chamber. The temperature of the chamber should be set at 105 °C ± 5 °C, which is less than 50% RH for 200 h. There should not be any visual defects, and the insulation resistance test should be passed by withstanding the stress.

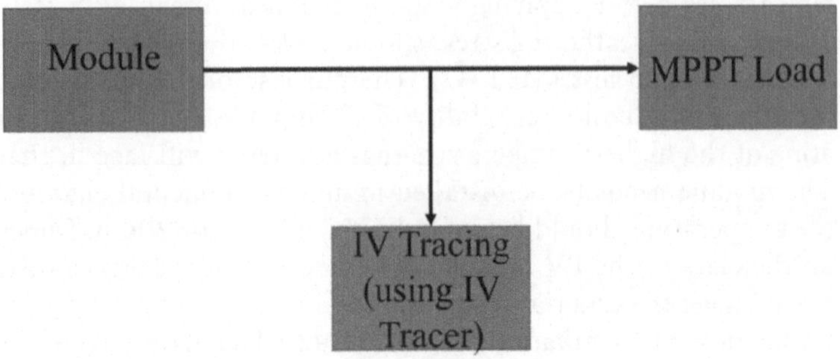

Figure 4.13 Block diagram of single module exposure for long-term testing.

4.7.3 Performance and reliability test of the bifacial module in outdoor conditions

The performance and reliability test of the bifacial PV module in outdoor conditions can be performed in the following configurations: single module exposure with MPPT load, PV modules as a PV power plant

i) **Single module exposure with Maximum Power Point Tracker (MPPT) load**

For performance and reliability analysis on a single module basis, the following configurations can be used to expose the PV module in the field. Figure 4.13 shows the configuration of single module testing in outdoor conditions.

The following parts are required for this configuration:

- Module mounting structure,
- Reference solar cell for irradiance measurement (in-plan and albedo),
- Temperature sensors for the module and ambient temperature measurement,
- Humidity sensor,
- Datalogger,
- I-V tracer,
- MPPT load/micro-inverter/DC-DC optimizer with a fixed load,
- Wind sensor.

Module needs to be connected with MPPT load. The load may be an electronic load or one can use a DC-DC optimizer with the fixed load. The module mounting structure can be either with a fixed tilt equal to the latitude of the test site or with a tracking facility. The temperature sensor should be placed near the mounting structure in shadow conditions for measuring the ambient temperature. For module temperature, it is not possible to

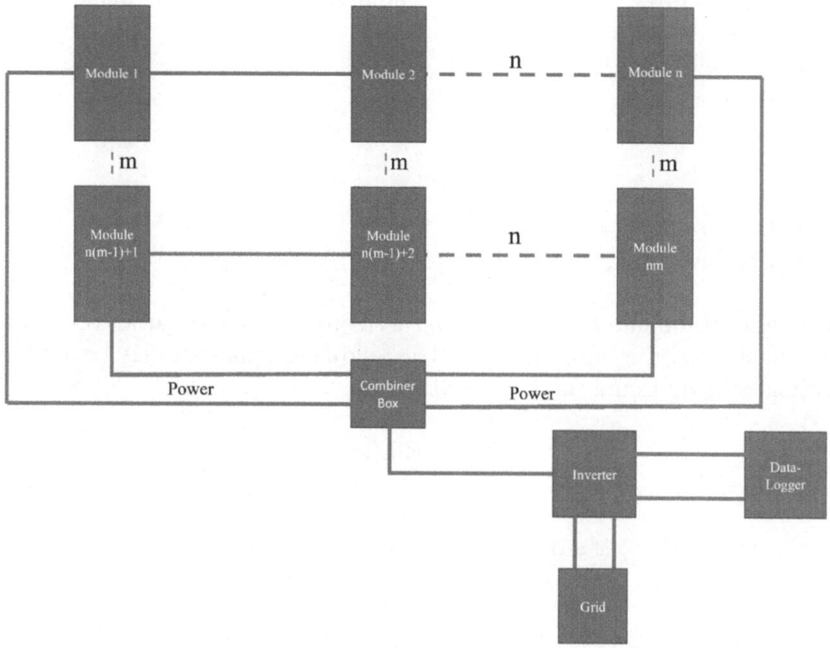

Figure 4.14 Block diagram of PV module power plant for long-term testing.

stick the temperature sensor in the backside of the bifacial module as it will block the radiation on the rear side. For this, one can measure and put the temperature sensor in the glass. There is one more procedure to estimate the module temperature, which is by measuring the Voc, ambient temperature, and in-plane irradiance and wind sensors.

The module connected to the MPPT load may be connected through a relay system, and this can be used to connect and disconnect the module from the load and I-V tracer. From the I-V data, the degradation of the PV module can be estimated. The details are given in the successive section.

ii) **PV modules as a PV power plant**

For performance and reliability analysis on power plant basis, the following configurations can be used to expose the PV module in the field. Figure 4.14 shows the configuration that can be used for performing long-term testing of bifacial PV modules as power plant mode.

The following parts are required for this configuration:

- Module mounting structure,
- Reference solar cell for irradiance measurement (in-plan and albedo),

- Temperature sensors for the module and ambient temperature measurement,
- Humidity sensor,
- Datalogger,
- I-V tracer,
- Inverter,
- Wind sensor,
- Grid connection,
- Combiner box.

The bifacial module PV power plant needs to be installed at an optimum tilt angle as per the site conditions. The following equations may be used to estimate the optimum tilt angle [48].

$$\beta_o = x \times L + y$$

$$x = 0.86 - 0.57 \times A \times \exp(-E/H)$$

$$y = 4.5 + 62 \times A \times \exp(-E/H)$$

$$\text{If } \beta_o \geq 90^\circ, \beta_{opt} = 90^\circ \text{ and if } \beta_o < 90^\circ, \beta_{opt} = \beta_o$$

Where β_{opt} is the optimal tilt angle, L is the latitude, E is the elevation, H is the module height, and A is the albedo.

The number of PV modules in series of the bifacial PV power plant can be decided based on the system voltage of the PV power plant. At least two strings in parallel should be used to install the PV power plant. One can use reflective surfaces for enhancing the albedo effect of bifacial PV module performance. Before installation, it is always better to measure the reflectivity of the surface to be used. The irradiance-measuring sensor should be used to measure the in-plane irradiance and irradiance in the backside. For measuring the temperature, the procedure mentioned may be used.

iii) **Procedure for performance degradation and reliability analysis**

The following equipment are required for performance and reliability analysis in outdoor conditions:

- I-V tracer,
- Electroluminescence (EL) camera with DC power supply,
- Infra-red (IR) camera,
- Insulation tester.

Before installation of the bifacial PV module for long-term analysis, it is always recommended to baseline the measurement of the module in indoor conditions. The I-V data of the bifacial module need to be measured at STC and at 10% and 20% albedo effect. However, this measurement should be as per the albedo effect of the reflective surface to be used in the field. The electroluminescence (EL) imaging of the module needs to be done for both sides at 0.1 *I*sc and *I*sc. The infrared thermal imaging should be done at MPPT condition using DC power supply. The insulation resistance testing and wet leakage current testing of the PV module need to be done before installation. If an I-V tracer is used to measure the performance of a PV module in the outdoor condition, then a proper translation procedure should be used to translate the I-V data to STC as per IEC 60891.

It will be good if the performance of the PV module measured by I-V testing, electroluminescence (EL) imaging, infrared (IR) imaging, insulation testing and wet leakage testing can be done once every week. This will give proper data for measuring the degradation of the module. The continuous logging of other environmental parameters is required to check in a particular interval of time. In case of bifacial PV power plants, there is no standard procedure for estimating the performance ratio. For estimating the performance ratio of the bifacial module, the following equation can be used:

$$Performance\,ratio = \frac{\left[\dfrac{Energy\,output\,in\,kWh}{PV\,power\,plant\,size}\right]}{\left[\dfrac{In\,plan\,irradiance\,in\,kWh + Rear\,side\,irradiance\,in\,kWh}{1000\dfrac{W}{m2}}\right]}$$

The PV power plant size should be based on the rating of the PV module given with respect to the albedo effect of the reflective surface and both the front side irradiance and back side irradiance should be used.

The degradation can be estimated for both sides of the bifacial module individually. For this, the performance of the front side should be measured by covering the rear side and vice versa. The following equation can be used to estimate the degradation rate of bifacial module:

$$Power\,degradation\,rate = \frac{\begin{array}{l}[Initial\,Pmax(front\,or\,rear\,side)\,at\,STC\\ -Final\,Pmax(front\,or\,rear\,side)\,at\,STC\end{array}}{\begin{array}{l}Initial\,Pmax(front\,or\,rear\,side)\,at\,STC\\ \times Year\,of\,exposure\end{array}} \times 100\%$$

The bifacial PV module is a new technology that exists in the market. The uses of bifacial module technology are increasing day by day. However, there are many issues that need to be addressed through research for this particular technology to become even more efficient.

REFERENCE

1. Radiation energy transducing device, Hiroshi Mori, Hayakawa Denki Kogyo KK, Patent no. US 3278811, filing date 3 October 1961.
2. Bordina, N.M.; Golovner, T.M.; Zadde, V.V.; Zaitseva, K.N.; Landsman, A.P.; Streltsova, V.I. Operation of a thin silicon photoconverter under illumination on both sides, *App. Sol. Energy*, vol. 11, no. 6, pp. 12–19, 1975.
3. Grigorieva, G, M.; Kagan, M.B.; Zviagina, K.N.; Kulicauscas, V.; Kreinin, L.; Bordina, N.; Eisenberg, N. Future of bifacial si solar cells for space application, 23rd European Photovoltaic Solar Energy Conference, Valencia, Spain, 1–5 September 2008.
4. Geliotekhnika, Letin, V.A.; Kagan, M.B.; Nadorov, V.P.; Zayavlin, V.R. Bifacial solar arrays of russian space crafts, 28th IEEE Photovoltaic Specialists Conference - 2000 (Cat. No.00CH37036), Anchorage, AK, USA, pp. 1067–1070, 2000, doi: 10.1109/PVSC.2000.916071.
5. Luque, A. Procedure for the photovoltaic conversion of solar energy and device for the put into practice of this procedure, Patent no. ES 453575A1, filing date 24 November 1976
6. Luque, A. Procedure for obtaining bifacial solar cells, Patent no. ES 458514A1, filing date 5 May 1977
7. Luque, A. Double-sided solar cell with self-refrigerating concentrator, Patent no. US 4169738, filing date 21 November 1977
8. Luque, A.; Cuevas, A.; Ruiz, J.M. Double-sided n+-p-n+ solar cell for bifacial concentration, *Solar Cells*, vol. 2, no. 2, pp. 151–166, 1980.
9. Luque, A.; Cuevas, A.; Eguren, J. Solar cell behaviour under variable surface recombination velocity and proposal of a novel structure. *Solid State Electron.*, vol. 21, no. 5, pp. 793–794, 1978, doi:10.1016/0038-1101(78)90014-X
10. Sangrador, J.; Sala, G. Two- dimensional analysis of the collection efficiency of vertical multijunction solar cells. International Electron Devices Meeting, pp. 318–321, 1979, doi:10.1109/IEDM.1979.189613
11. Ishikawa, N.; Nishiyama, S. World first large scale 1.25 MW bifacial PV power plant on snowy area in Japan, Presented at the 3rd Bifi PV Workshop, Miyazaki, Japan, 2016
12. International Technology Roadmap for Photovoltaic (ITRPV) – 2017 Results, http://www.itrpv.net
13. WWW.bifiPV-workshop.com
14. N-Type Expansions to Benefit from 2020 Downturn-Driven PV, https://www.pv-tech.org/news/n-typeexpansions-to-benefit-from-2020-downturn-driven-pv

15. Ghgh Thompson, E.P.; Bombelli, E.L.; Shubham, S.; Watson, H.; Everard, A.; D'Ardes, V.; Schievano, A.; Bocchi, S.; Zand, N.; Howe, C.J.; et al. Tinted semi-transparent solar panels allow concurrent production of crops and electricity on the same cropland. *Adv. Energy Mater.*, vol. 10, p. 2001189, 2020.
16. Zhi, L.; Yano, A.; Marco, C.; Yoshioka, H.; Kita, I.; Ibaraki, Y. Shading and electric performance of a prototype greenhouse blind system based on semi-transparent photovoltaic technology. *J. Agric. Meteorol.*, vol. 74, pp. 114–122, 2018.
17. Joge, T.; Araki, I.; Takaku, K.; Nakahara, H.; Eguchi, Y.; Tomita, T. Advanced applications of bifacial solar modules, 3rd World Conference on Photovoltaic Energy Conversion, pp. 2451–2454, 2003.
18. Soria, B.; Gerritsen, E.; Lefillastre, P.; Broquin, J-E. A study of the annual performance of bifacial photovoltaic modules in the case of vertical facade integration. *Energy Sci Eng.*, vol. 4, pp. 52–68, 2016.
19. Hezel, R. Applications of bifacial solar cells. *Novel Prog. Photovoltaics Res Appl.*, vol. 11, pp. 549–556, 2003.
20. Attoye, D.E.; Hassan, A. A review on building integrated photovoltaic façade customization potentials. *Sustainability*, vol. 9, p. 2287, 2017.
21. World's first bifacial sound insulation system near Wallisellen/ Zurich, Switzerland, 10 kWp pilot system. https://www.tnc.ch, 1997
22. Faturrochman, G.J.; de Jong, M.M.; Santbergen, R.; Folkerts, W.; Zeman, M.; Smets, A.H.M. Maximizing annual yield of bifacial photovoltaic noise barriers. *Sol Energy*, vol. 162, pp. 300–305, 2018.
23. de Jong, M.M.; van den Donker, M.N.; Verkuilen, S.; Folkerts, W. Self-shading in bifacial photovoltaic noise barriers, Proceedings of the 32nd European Photovoltaic Solar Energy Conference and Exhibition Munich, Germany, vol. 53, pp. 2732–2734, 20–24 June 2016.
24. Faturrochman, G.J.J.; de Jong, M.M.M.; Santbergen, R.; Folkerts, W.; Zeman, M.; Smets, A.H.M.H.M. Maximizing annual yield of bifacial photovoltaic noise barriers. *Sol. Energy*, vol. 162, pp. 300–305, 2018.
25. Sharma, P.; Muni, B.; Sen, D. Design parameters of 10 KW floating solar power plant conference. *Int. Adv. Res. J. Sci. Eng. Technol.*, vol. 2, pp. 85–89, 2016.
26. Rodrigues, I.S.; Ramalho, G.L.B.; Medeiros, P.H.A. Potential of Floating Photovoltaic Plant in a Tropical Reservoir in Brazil. *J. Environ. Plan. Manag.*, vol. 63, pp. 2334–2356, 2020.
27. Sahu, A.; Yadav, N.; Sudhakar, K. Floating photovoltaic power plant: A review. Renew, sustain. *Energy Rev.*, vol. 66, pp. 815–824, 2016.
28. Ponce, V.M.; Lohani, A.K.; Huston, P.T, Surface Albedo and water resources: Hydro climatological impact of human activities, *J. Hydrol. Eng.*, vol. 2, pp. 197–203, 1997.
29. Widayat, A.A.; Ma'arif, S.; Syahindra, K.D.; Fauzi, A.F.; Setiawan, E.A.; Adhi Setiawan, E. Comparison and optimization of floating bifacial and monofacial solar PV system in a tropical region, 9th International Conference on Power Science and Engineering (ICPSE), London, UK, pp. 66–70, 23–25 October 2020.
30. First Floating Solar Farm in Thailand with UBE. https://www.baywa-re.es.

31. "International Technology Roadmap for Photovoltaic (ITRPV) - Results 2019," April 2020. http://itrpv.vdma.org.
32. Deline, C.; Peláez, S. A.; Marion, B.; Sekulic, B.; Woodhouse, M.; Stein, J.; Bifacial, P.V. System performance: Separating fact from fiction, presented at the IEEE-PVCS-46, Chicago, 2019.
33. Romijn, I.; et al. Bifacial cells, presented in Bifacial Photovoltaics: Technology, Applications and Economics. IET, 2019.
34. Dullweber, T.; Schmidt, J. Industrial silicon solar cells applying the passivated emitter and rear cell (PERC) concept—a review. *IEEE J Photovolt.*, vol. 6, no. 5, pp. 1366–1381, 2016.
35. Dullweber, T.; et al. Present status and future perspectives of bifacial PERC+ solar cells and modules, *Japanese J. Appl. Phys.*, vol. 57, no. 8S3, 2018, 'Trina Solar press release,' March 06, 2020. https://www.trinasolar.com.
36. Tous, L.; et al. *Progress on Bifacial Ni/Ag Plated NPERT Cells for Module Fabrication with SWCT*, Brussels, Belgium, pp. 439–443, 2018.
37. Zhao, J. H.; Wang, A. H.; Altermatt, P. P.; Green, M. A.; Rakotoniaina, J. P.; Breitenstein, O. High efficiency PERT cells on n-type silicon substrates. Conference Record of the Twenty-Ninth IEEE Photovoltaic Specialists Conference, 2002., New Orleans, LA, USA, 2002, pp. 218–221, doi: 10.1109/PVSC.2002.1190495.
38. Yamamoto, K.; et al. Record-breaking efficiency back-contact heterojunction crystalline Si solar cell and module. In Proc. 33rd Eur. Photovolt. Sol. Energy Conf. Exhib, 201–204.
39. Chavali, R. V. K.; et al., Multiprobe characterization of inversion charge for self-consistent parameterization of HIT cells, *IEEE J. Photovoltaics*, vol. 5, no. 3, pp. 725–735, May 2015, doi: 10.1109/JPHOTOV.2014.2388072.
40. Kunta, Y. et al., Silicon heterojunction solar cell with interdigitated back contacts for a photoconversion efficiency over 26%, *Nat. Energy.*, vol. 2, pp. 1–8, 2017.
41. Shockley, W.; Queisser, H. J. Detailed balance limit of efficiency of pn junction solar cells. *J. Appl. Phys.*, vol. 32, pp. 510–519, 1961.
42. Kim, S., Trinh, T.T., Park, J. et al., Over 30% efficiency bifacial 4-terminal perovskite-heterojunction silicon tandem solar cells with spectral albedo. *Sci. Rep.*, vol. 11, p. 15524, 2022. https://doi.org/10.1038/s41598-021-94848-4.
43. IEC TS 60904-1-2:2019, Photovoltaic devices - Part 1-2: Measurement of current-voltage characteristics of bifacial photovoltaic (PV) devices.
44. Herrmann, W.; Schweiger, M.; Bonilla Castro, J. Performance characteristics of bifacial PV modules and power labeling, presented at the 4th Bifi PV Workshop, Konstanz, Germany, 2017
45. IEC 60904-9:2020, Photovoltaic devices - Part 9: Classification of solar simulator characteristics
46. IEC 61215-2:2021, Terrestrial photovoltaic (PV) modules - Design qualification and type approval - Part 2: Test procedures
47. IEC 61730-2:2016, Photovoltaic (PV) module safety qualification - Part 2: Requirements for testing
48. Sun, X.; Khan, M. R.; Deline, C.; Alam, M. A. Optimization and performance of bifacial solar modules: A global perspective. *Appl. Energy*, vol. 212, pp. 1601–1610, 2018, doi: 10.1016/j.apenergy.2017.12.041.

Chapter 5

Performance degradation of photovoltaic module

Amandeep Singh Makhija and Shabbir S. Bohra

5.1 INTRODUCTION

The development of sustainable energy has been given strong emphasis in UN SDG-7, and it is something that the world is getting closer to. By 2030, enhanced international cooperation will be achieved in order to facilitate public access to clean energy research and technology, including renewable energy, energy efficiency, and new, cleaner fossil fuels, as well as to encourage investment in energy infrastructure and clean energy technologies [1]. The energy demand continues to rise worldwide, and according to International Energy Agency's (IEA) 2019 international energy outlook reference case, the world's energy consumption would rise by about 50% between 2018 and 2050, with Asia seeing the biggest rise. Due to the considerable downsides of conventional fossil fuels and the escalation of irreversible environmental issues, the development trend of the world's energy has entered a critical transition phase. A crucial component of the development of energy is electricity. Hydropower, wind power, solar power, and other sources are all part of the typical renewable energy generation strategy. Among them, solar power is thought to be the most promising power generation method due to its availability, universality, reproducibility, and lack of pollution. The hydropower and wind power are other popular renewable resources at some places.

The most direct and effective way to use solar energy is through SPV power generation [2]. The SPV has become increasingly popular in recent years because its energy production costs are lower; it has no maintenance cost, it is pollution free, and it has government support. Around 3.4 million exajoules (EJ) of energy are delivered to Earth. Earth's solar energy potential is about 3,500 GW, even at 10% efficiency. As of 2018, 425 GW of solar energy has been produced utilizing various PV systems. By 2050, it is anticipated that global solar power generation would have increased about 20 times to 1.8 TW due to improvements in SPV technologies. According to estimates, SPV systems will be supplying around 25% of the world's total energy requirements by 2030. The officials have implemented new laws and offered subsidies to encourage the use of SPV installations. By offering

DOI: 10.1201/9781003623168-7

subsidies, businesses and homeowners will be encouraged to switch to independent solar energy and lessen their dependency on grid electricity [3].

Making PV-generated electricity cost-effective is the primary problem faced by the industry at the moment. Determining the factors that can lower the cost of energy generated by PV is therefore crucial. To make PV technology a financially viable source of energy, the long-term durability of PV modules is therefore crucial. The failure mode and degradation process of solar panels can be checked to gauge their reliability when used in field. It has been observed that operating a PV system outside in tropical conditions under higher voltage bias can have a significant impact on the system's performance [4].

5.2 DEGRADATION MODES IN PV MODULES

Solar panel life and power generation products are affected by several factors over time, including climate, type of module technology, and tracking systems. A progressive decrease in solar panel production over time is called the degradation rate. The inaccuracy in calculating the degradation rate of the photovoltaic power generation system increases the financial risk [5]. The projected service life for c-Si SPV modules is 25–30 years, with annual power degradation of 0.5%–1.0%. However, it has been claimed that installed modules incur a 0.5%–10% annual power degradation rate. The field-installed c-Si SPV deteriorates for a variety of reasons. Exposure to a variety of cyclic temperatures combined with operations at high temperatures [6] and exposure to changing weather, depending on the local climate, might lead to accelerated material deterioration [7]. The cyclic temperature loading causes thermo-mechanical fatigue in the solder joints in the c-Si SPV module interconnects. The accumulation of power degradation in the SPV modules is largely a result of operations under a variety of currents and voltages. The degradation rate is also affected by the ultraviolet (UV) light that strikes SPV cells. The combined effects of these elements on the deterioration of SPV operating in tropical conditions are critical. The performance of the module is lowered and its power output is decreased by additional factors related to the packaging material, connectivity, solder joint, adhesion, delamination, moisture buildup, and semiconductor device temperature difficulties, which raises concern about SPV dependability[6]. Figure 5.1 depicts several important factors affecting SPV performance. Among all, weather conditions (specifically tropical climate) and the PV module packaging quality have a significant effect on the degradation.

The next section gives a summary of the degradations or faults related to SPV cells/modules, including corrosion of front grid fingers, EVA discoloration, snail trails, hotspots, bubble formation, de-lamination, LID, LeTID, PID, damage in solar cell, panel back sheet, and junction box failure.

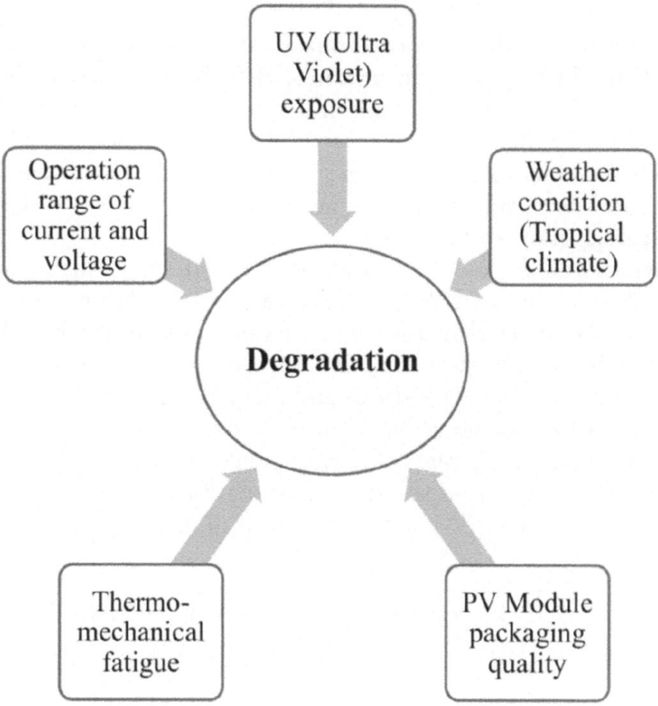

Figure 5.1 Factors responsible for SPV degradation.

5.2.1 Corrosion of front grid fingers

After prolonged exposure to high temperatures and humidity, the delamination effect causes moisture to enter through EVA and corrode the metal components. Ingress of moisture can penetrate PV modules more easily because of the EVA's comparatively high diffusivity [6]. The corrosion often occurs near the edges of cells and at cell breakage.

The output terminals, fingers, bus bars, cell, and string interconnects get corroded, which in turn reduces the module's output power due to increase in series resistance (R_s) and decrease in fill factor (FF). In a detailed field investigation, interconnected fracture and corrosion caused 90% of the modules to fail [4]. In a study on a 5 MW PV system, corrosion of the metallization and interconnects reduces annual power by 1.5–2.5% [8]. Figure 5.2b depicts the corrosion of the fingers grid of a 1 MW canal top plant in Chandrasan, India, which has been operational for the past 10 years.

5.2.2 EVA (ethylene-vinyl acetate) discoloration

This defect is mainly due to poor encapsulant quality, and long-term exposure of PV module to high temperatures, UV exposure, and humidity [9]. Acetic acid may develop as a result of incompatible additions in the EVA. Photo-oxidative bleaching is the cause of these occurrences, in which oxygen bleaches the discolored areas in the presence of UV radiation [10]. In that case, the encapsulant absorbs more light in the visible high-energy range, turning yellow or brown [7].

This defect would cause the transmittance to decline, which would then cause the short circuit current (I_{sc}) to decrease. The I_{sc} decomposition range can be 6%–8% lower than the nominal value, and it can be 10%–13% lower for entire discoloration [11]. Due to non-uniform discoloration, performance decreases from 1%–10%, and a considerable reduction in FF is seen [4, 10]. EVA discoloration on 100% of modules was noted due to outdoor exposure of 22 years, in an instance [12]. The vertical path of Browning Effect is seen on the surface of the PV module in Figure 5.2a, taken at a 1 MW canal top plant in Chandrasan, India.

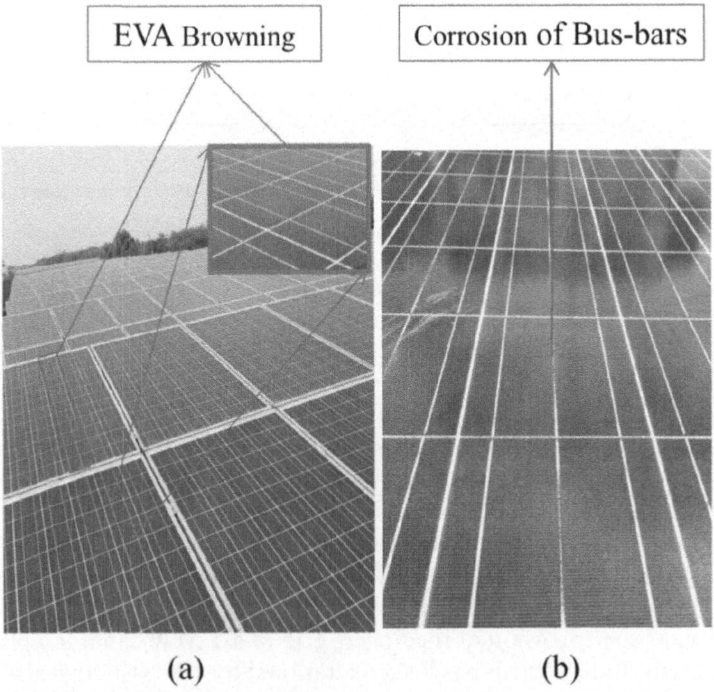

Figure 5.2 Defects observed at canal top SPV project, Chandrasan, India. (a) EVA browning and (b) fingers corrosion.

5.2.3 Snail trails

The snail trails result from the oxidation of the silver grid, which is caused by the entry of moisture, oxygen, carbon dioxide, and other chemicals through cell gaps and cracks [7]. The front contacts' silver paste can develop a grayish hue at cell edges or cracks, creating the appearance of a snail trail above the cell. They can appear on the margins of solar cells as well as near micro cracks running through them [9]. The snail trails can be seen with the naked eye (Figure 5.3a). When a crack is present, depending on the climatic conditions, snail trails start to form after 3–12 months of operation.

The presence of snail trail lowers the irradiance transmittance to cells, which lowers the I_{sc} and impacts the panel's performance.

5.2.4 Hotspot in cell

The solar cell in PV module is connected in series in order to increase module voltage. If a short circuit occurs and a particular PV cell malfunctions, the voltage will reverse, becoming equal to and the opposite of that of the remaining cells in the series. That particular portion of the cell gets heated

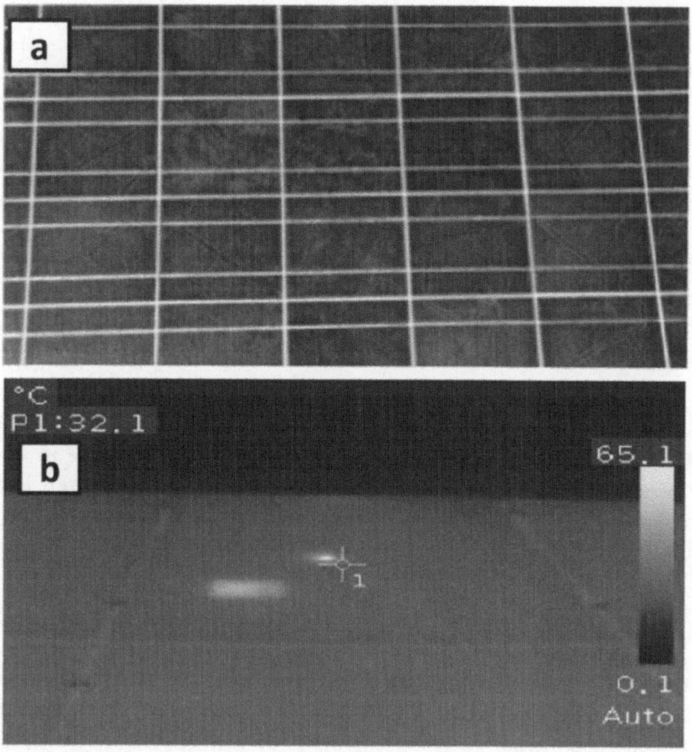

Figure 5.3 (a) Snail trails [9], (b) hot spots[7].

and turns into darker spot, which is known as a hotspot (Figure 5.3b.). Hot spots result from this defective solar cell's burden on the other cells and comparatively high heat-dissipation sites [11]. The PV materials get affected due to elevated temperature spots, and other modes of degradation are accelerated by it. Hot patches can turn into burning spots and causes the browning of the encapsulant [7]. The following situations result in hotspots:

a) Any damaged or shaded solar cells within the module.
b) High PV module temperatures and irradiation.

To avoid the effect of the hot spot, two/three bypass diodes are used for 72-cell PV module. The bypass diode bypasses the current from the hot spot area to another non-hot spot area, reducing the open circuit voltage (V_{oc}) [9].

It was discovered that after 130 Sun hours, the hot spot declined the performance of the copper indium selenium (CIS) module by more than 20% and the performance of the c-Si module by 60% [4].

5.2.5 Bubble formation in front panel and back sheet

The detachment of the two layers, EVA-glass and EVA-back sheet, is a possible reason for both bubble formation and delamination. The impacted area of the bubble is less than that of the delamination, which is how they differ from one another. During field operation, the temperature of module, and hence, the cells rises; it causes certain gases to be released from the back of the solar cell as a result of a chemical reaction and creates a void space or bubble. The back sheet's deterioration, such as delamination and bubble formation, can locally reduce thermal conductivity, leading to hot patches that can eventually worsen solar cells' performance [4]. The bubbles on the back and the front side of PV modules are shown in Figure 5.4

In a survey, only a small number of modules (13% of the total examined modules) exhibited bubbles and delamination in the back sheet, and they were all older than ten years. All of the damaged modules are in the composite zone's hot and humid zone, both of which have operated in a high relative humidity (RH). It may be argued that this means that excessive humidity encourages the growth of bubbles [12].

5.2.6 Delamination

Delamination happens when the EVA loses its adhesive strength in conditions of high humidity and high temperature, particularly around the ribbon [11]. In contrast to the solar cells' blue anti-reflective coating, the encapsulant delamination on their top surface is visible as white patches

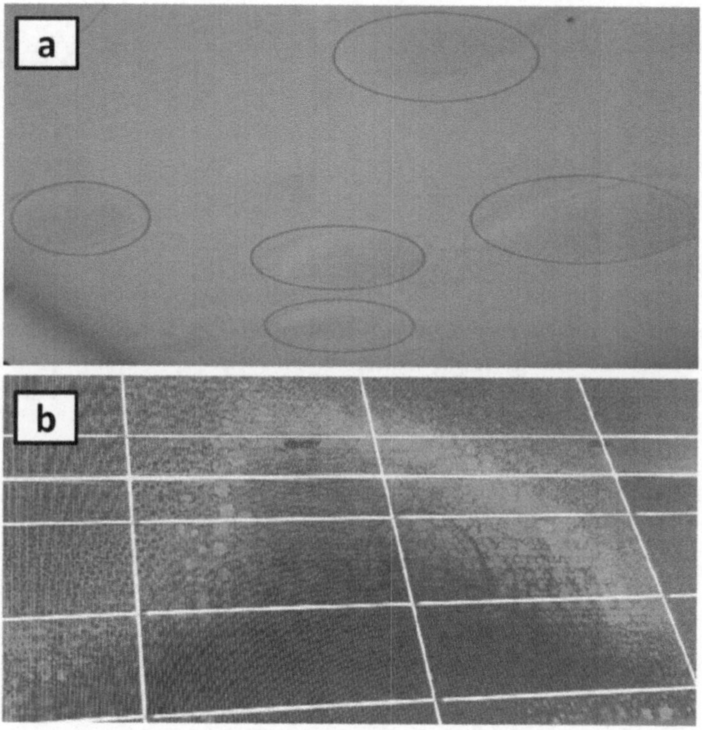

Figure 5.4 Bubbles formation in PV module. (a) Back panel [9] and (b) front side [11].

or spots on the solar cells (Figure 5.5). The transparent conductive oxides (TCOs) in contact with glass could create sodium ions as a result, which affects the TCO and causes it to degrade quickly [4]. In the afflicted zone, delamination may encourage moisture infiltration and buildup, which may result in physical corrosion. Delaminated areas have less heat conductivity, which could cause hot spots to emerge. It is the most frequent reason for PV modules to degrade when exposed for an extended period to high temperatures and humidity [9].

As a result of the delaminated section reflecting a large portion of the incident light, the affected cell's light input is reduced, which lowers its I_{sc} and performance losses up to 4% [7].

5.2.7 Light-induced degradation (LID)

The LID occurs when crystalline modules are exposed to the sun light for the first few hours. There were minute amounts of oxygen in the molten silicon during the Czochralski process. Due to the influence of light exposure, positively charged oxygen dimers may spread throughout the silicon lattice

Figure 5.5 Patches of delamination on the PV front panel [11].

and combine with acceptors of Boron dopant. The Boron-Oxygen (B-O) complexes create energy levels in the silicon lattice and can trap electrons and holes that would otherwise be lost due to the PV effect. The LID is therefore associated with both B-O concentrations.

The LID loss, which can range from 1% to 3%, is commensurate to wafer manufacturing quality (or even more). Initial power loss in amorphous silicon (a-Si) modules is up to 15%, whereas, in c-Si cells, it ranges from 3% to 6% [7].

5.2.8 Light and elevated temperature-induced degradation (LeTID)

The probability of LeTID occurring in PV modules with p-type passivated emitter rear contact cells is the highest. It has subsequently been found that it also occurs, albeit to a lower extent, in other varieties of crystalline Si cells. Unlike LID, it occurs at temperatures over 50°C. Although one of the most frequent causes is the diffusion of hydrogen into the cell's bulk, which has already happened as a result of the deposition of the rear-side passivation layer. The precise mechanism of this phenomenon is still unknown. The degradation in V_{oc} was observed after prolonged exposure to high temperatures, which was most likely caused by an increase in rear-surface recombination, which can happen even in the absence of light [13]. During normal operation, the LeTID degradation may saturate and occasionally even rebound. Complete degradation is anticipated to take ten years, whereas recovery takes much longer and may extend longer than the module's lifetime [7].

The performance loss due to the LeTID varies from 1% to 4% and records up to 16% are even reported [7].

5.2.9 Potential-induced degradation (PID)

In c-Si cells, polarization and PID have been identified as significant failure mechanisms. A high negative voltage exists between the PV module's cells and the aluminum frame, which is earthed for safety, when the modules are operating with a negative potential to Earth. The effect becomes more pronounced toward the negative polarity of the PV array, where the potential can exceed half of the array's voltage. The outcome is that the electrons from the PV module's parts can separate, move with the electric field, and eventually leave through the aluminum frame. If preventative steps are not followed, the result is an increase in the module's charge (polarization), which changes its characteristic curve and, as a result, its power. Additionally, the encapsulation (often referred to as EVA) and the anti-reflective coating (ARC) allow the sodium ions to go from the glass plate to the cell. These ions reduce the output by effectively opening a shunt channel throughout the cell. The effect is stronger and gradually becomes cumulative when the cell is operated at a higher negative potential. Contrary to corrosion and normal aging-related degeneration, it has been found that this polarization can typically be reversed; therefore, it is not a permanent effect [14].

The plant may have PID if any of the following symptoms are present:

a. The maximum voltage-to-open-circuit voltage (V_m/V_{oc}) ratio and the V_{oc} have steadily decreased compared to the datasheet values.
b. The influence of the PID was readily apparent due to increased local shunting brought on by surface recombination and a decline in the FF.

The PID is caused by a variety of elements, including the architecture of a panel, its high potential with respect to the ground, and environmental elements like humidity and temperature [11]. The PID exists in three levels in PV modules i.e., system, module, and solar cells. PID is highlighted by darker cells in an electroluminescence (EL) image that have accumulated close to the frame. With infrared thermography (IR-T), the heat produced by the shunted cells can also be observed [7].

PID decreases shunt resistant (R_{sh}), lowers the maximum power (P_{max}) and V_{oc} of the module, and subsequently decreases the overall PV plant yield [15]. The module degradation rate varies from 1% to 4% per annum, up to 20% per annum, and reaches a complete power loss of up to almost 100% [7]. The PID in a solar power plant has a significant impact on the PR on a large scale. A power loss of up to 30% has been observed in the majority of the panels when measured in the string's negative area [16]. After 9 years of outdoor exposure in Malaysia, a degradation rate of 61%, 66%, and 73% was observed at the negative end of the c-Si PV module as a result of the combined effects of outdoor exposure and voltage of 300, 400, and 600 V, respectively [4]. Three key factors that affect the performance of PV modules are an increase in R_s, a fall in the cell's R_{sh}, and ARC degradation [17, 18]. As shown in Figure 5.6, the deterioration of modules in the string follows a pattern, with the negative side being the most affected than the positive side [19]. Figure 5.7 shows the several dark cells/spots which are typically detected in the PID-affected module caused due to the shunting of solar cell.

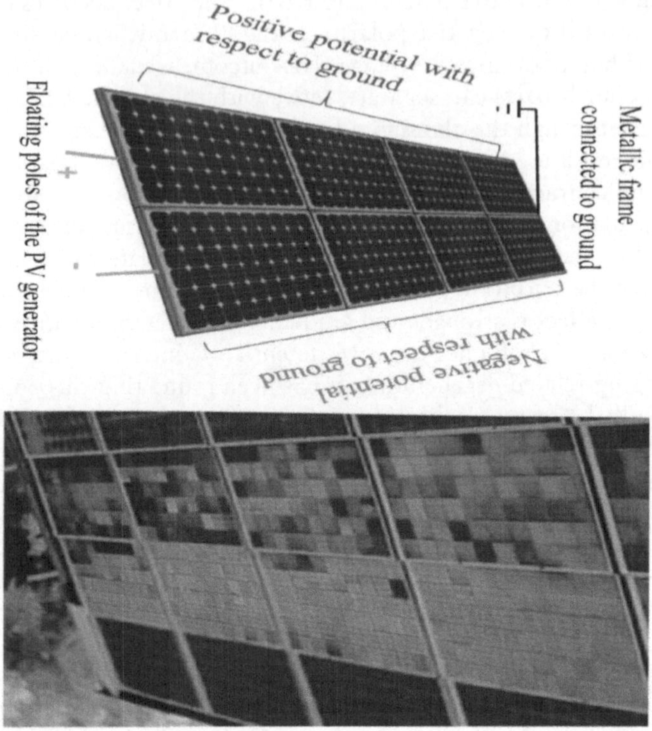

Figure 5.6 PID effect in a PV module string: the negative side of the string is affected more by PID than the positive side.

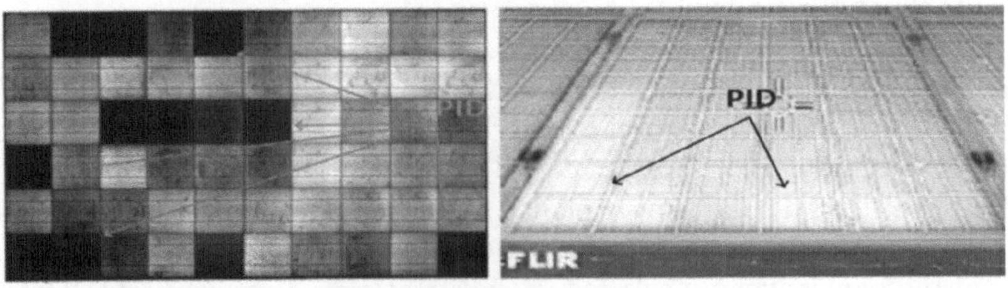

Figure 5.7 Dark cells/spots detected in PID-affected modules [20].

5.2.10 Damage in solar cell, PV panel back sheet, and junction box failure

These are not the type of degradation which occurs annually at a defined rate, but this event can impact the performance of the SPV. The cost of the PV module decreases as a result of the reduced material use, but it also increases the risk of a crack in the cell or glass. In the process of manufacturing a PV module, about 6% of solar cells develop breaking/cracking [4].

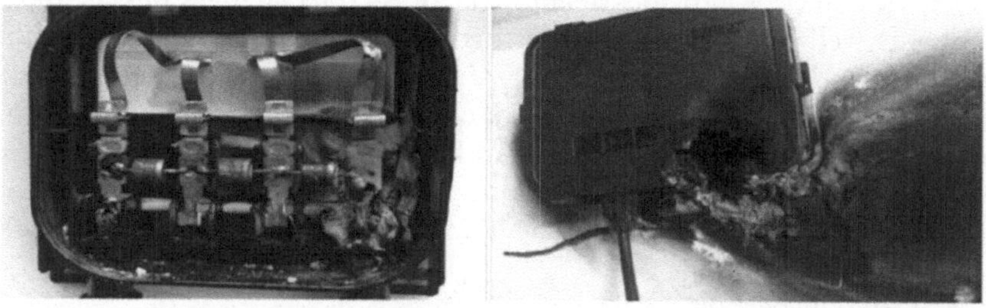

Figure 5.8 Burned junction boxes and bypass diodes [11].

Despite not always being evident to the unaided eye, a crack in a solar cell is an obvious flaw. For finding cracks in solar cells, EL is one of the most reliable technique [7]. The effects of a crack in a solar cell can range from not affecting the power to completely shutting down the module, depending on the size and position of the crack. It is probable that cracks enlarge over time, isolating individual cell components or obstructing current passage in the cells, causing faster deterioration and associated power loss. A PV module's power loss due to cell cracks is proportional to the area of disconnected cells and ranges for PV modules between 1 and 15% [7].

The power output of the module is normally unaffected by minor frame damage, but over time, any small imperfection might cause a severe safety or performance problem, such as chances of moisture or water penetration into the module. In a survey, for 39% of the instances, the module frame was rusted, deformed, or somehow damaged [10]. The EVA encapsulant swells or deforms as a result of the rear sheet's cracking, which makes it easier for moisture to get inside.

The junction box provides housing to the bypass diodes and the output terminals. Burned junction boxes and bypass diodes, as shown in Figure 5.8 are common failure mechanisms that, in the worst situations, increase the risk of fire. The system's severe stress, the junction box's inadequate attachment to the back sheet, the lid's inappropriate closure, moisture, and improper wiring are among the major offenders for junction box failure [11].

5.3 DEGRADATION ANALYSIS CHARACTERIZATION METHODS

A plethora of module inspection methods have recently undergone thorough evaluation in laboratories, and the majority of those methods have been implemented as on-site procedures. O&M companies are keenly interested in these techniques but frequently struggle to select the most effective one [21].

The next section provides a summary of each technique's on-site application as well as an explanation of the physical principles underlying the measurement's interpretation. Figure 5.9 shows the visual demonstration of the different modes of degradation characterization techniques. All the techniques, excluding measurements of the IV-characteristic curve, produce qualitative data. Quantitative information can only be obtained via the IV-characteristic curve approach [7, 9, 21].

- Visual inspection
- I-V characteristics
- Infrared thermography (IR imaging)
- Ultraviolet fluorescence
- Ultrasonic inspection
- Luminescence techniques
  - Electroluminescence (EL imaging)
  - Photoluminescence (PL imaging)
- Laser beam-induced current (LBIC)

5.3.1 Visual inspection

The simplest yet highly efficient method is a visual examination. Although many performance changes are imperceptible and require investigation using more complex characterization techniques, many other failures, such as hot spots, burn marks, delamination, yellowing, back sheet blistering, junction box failures, and many others, can be detected by visual inspection [21]. The visual inspection after PV module production has been made

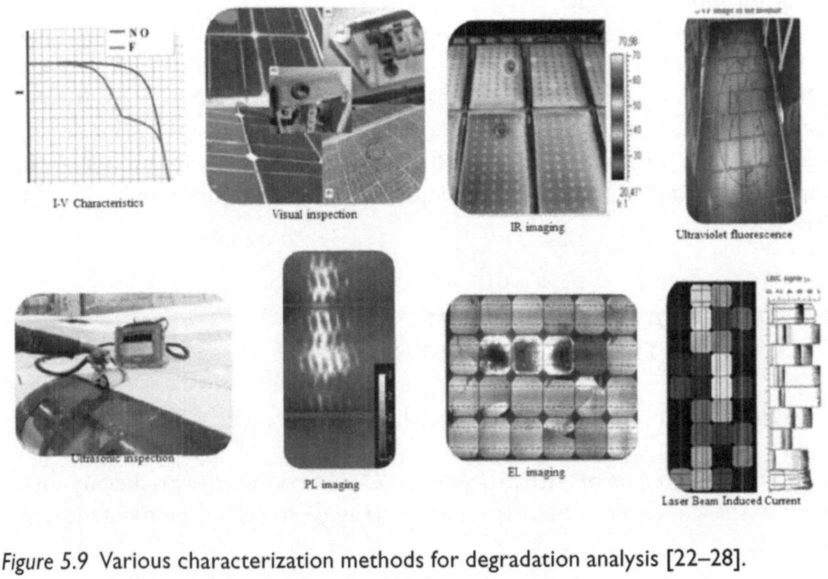

Figure 5.9 Various characterization methods for degradation analysis [22–28].

mandatory as per IEC 61215-1; however, it is also recommended for field inspections [7].

5.3.2 I-V characteristics

The IV characteristics (IV), as the name implies, is a quantitative evaluation approach that specifies certain parameters using the I-V curve of an illuminated or dark (applying external supply) PV module. The illuminated I-V curve shows the module's current and voltage output under different light levels. It helps determine the maximum power output at Pmax. The dark I-V curve shows output without light and is useful for identifying defects. The slope of the dark I-V curve at higher voltages can indicate any defects or irregularities in the module's construction, such as a damaged cell or a faulty connection. There are several circumstances in which an I-V curve can be measured inside standard test conditions (STC), i.e., AM1.5G spectrum in a laboratory [7] or outside in the field and then converting the measured values to the comparable values under STC. This process is specifically laid up to rate the PV module by the IEC 60904. The most crucial variables measured are short circuit current (I_{sc}), open circuit voltage (V_{oc}), and power at maximum power point (P_{max}). Based on the characteristic curve, the R_s, R_{sh}, and FF can be determined [21].

Investigations using dark IV (DIV) at PV plants are also helpful for preselection. Sometimes EL measurements are paired with the technique. With the assistance of a power supply and a multi-meter, DIV characteristic curves can be obtained [21].

5.3.3 Infrared thermography (IR imaging)

Infrared thermography, where the measurement is carried out under steady sunlight illumination, is the non-destructive technique, quick (real-time analysis), contactless, and most frequently employed for on-site analysis of PV-generating plants. IR imaging is majorly used to detect "hot spots" in solar cells. This technique makes use of the idea of localized heat creation caused by the joule heating effect. This joule heating effect is caused by faulty connections, short circuits, and shunted cells. Any series-connected cell that generates less current than the others and acts as a resistor or load to the others generates heat. When thermal imaging is done using an IR camera (spectral range of 8–14 μm), the dissipated heat creates a temperature gradient that appears as bright spots (commonly known as "hot spots") [9]. The IEC 62446-1:2016 specifies general standards for IR thermography. The measurements can be done using a Testo 890 high-resolution camera with a tiny bolometer detector. Many IR thermography methods, including forward looking infrared (FLIR), dark lock-in (DLIT), and illuminated lock-in thermography (ILIT), have been developed recently to detect failures in SPV modules [21].

5.3.4 Ultraviolet fluorescence

The goal of UV-fluorescence (UV-F) method is to detect the presence of luminophores, which are produced when the encapsulating material EVA breaks down when it is exposed to UV radiation. They stop becoming fluorescent when these luminophores interact with oxygen. Because they appear darker in the UV-F picture, this behavior can be exploited to identify cell fissures and other oxygen-leaking locations. The fluorophores reach a sufficient fluorescence signal after 80 kWh/m^2, which equates to 1.5 years of outdoor operation in the mid-latitudes [7]. The measurement comprises the UV-fluorescence unit of UV-LEDs, the remaining visible light being blocked by a filter glass [21].

5.3.5 Ultrasonic inspection

There are two subcategories for ultrasonic inspection: pulse-echo method and transmission method. By moving an ultrasonic transducer across the PV panel, the PV module is scanned using both techniques. By reflecting ultrasonic pulses from the defects and recording those pulses, defects are found using the pulse-echo method. When using a transmission method to find flaws, a PV module is subjected to an ultrasonic energy scan by an ultrasonic signal being passed through it. Some signals are attenuated during the scanning process because the module has defects, and these attenuated signals are then recorded. Therefore, the transmission method locates the defect's magnitude in this manner [9].

5.3.6 Luminescence techniques

Luminescence radiation is a form of light emitted by the radiative recombination of charge carriers in semiconductors. It can be initiated from the outside and the data gathered to produce a picture of the solar cells' conductive qualities. This technique is divided into EL and PL techniques, which are used to effuse the luminescence radiation emission in solar cells [7].

5.3.6.1 Electroluminescence (EL imaging)

EL imaging works on the same principle as a light emitting diode (LED). Radiative carrier recombination results in light emission when current is injected into a solar cell, which is effectively a big diode. Since silicon is an indirect bandgap semiconductor, the majority of recombination in it happens as a result of defects or auger recombination. Cell regions that emit more photons and those that emit fewer photons can be distinguished in the EL picture. Since there aren't many photons released, EL imaging must be carried out in complete darkness. If optical filters are used, a measurement at dusk (outdoor) or low ambient light (in a lab, below 100W/m^2) is also possible. EL is frequently carried out with I_{sc} and occasionally additionally with 1/10 I_{sc} to assess the severity of cell cracks [21]. An EL image enables the quantitative interpretation of solar cell characteristics such as series resistance, shunt resistance, minority carrier lifetime, and diffusion length [7]. In the IEC 60904-13, the EL technique is defined for use in lab settings and at nighttime ground inspections on single modules. For the measurements, a modified consumer camera (Canon EOS 700D) in which, the IR blocking filter has been removed can be used to detect the emitted 1,100 nm infrared light [21]. It is non-destructive and rather quick, with measurement periods of just one second.

5.3.6.2 Photoluminescence (PL imaging)

The process by which the luminescence signal is triggered differs between PL and EL. In traditional PL, which is often done during the manufacturing of solar cells, the solar cell or PV module is lit by outside, homogenous artificial light sources. Images at two separate operation points of the PV module are taken to retrieve the weak luminescence signal that is overlaid by the intense sunshine. One image, taken at open circuit voltage, retains the luminescence signal but another, taken at a high current state like full power, does not because the carriers either participate in the current or recombine. Sun radiation is removed from the photos by subtracting them, leaving only the luminescence signal [7]. A new methodology Day Light Luminescence System (DaySy) is developed in which measurement can be done under daylight conditions. DaySy system uses both electroluminescence and photoluminescence to give additional information [21].

5.3.7 Laser beam-induced current (LBIC)

The LBIC technique is employed to find any early defects that might have developed in the module during the manufacturing process. A thorough scan of the module surface with the aid of a concentrated HeNe laser beam with a wavelength between 638 and 850 nm is done. An electron-hole pair is formed in the semiconductor when a light beam scans across the surface of a photosensitive device, causing the direct current to flow. This current may then be measured using the appropriate tools [29], and photocurrent maps of the module are created. The module's photocurrent as a function of the laser beam's position creates an image that depicts the module's useful output current [9]. More variation in the current indicates that the cell is more defective [29].

5.4 THE OCCURRENCE FREQUENCY OF VARIOUS DEGRADATION MODES

The most frequent issues with modern PV modules are hot spots, internal circuitry discoloration, and glass breakage, whereas PID, glass breakage, a faulty back sheet, and delamination rank highest among PV module risks [7]. To discover several defects and failure scenarios for mono-c-Si solar modules over a 25-year period, a case study was conducted [4]. Figure 5.10 depicts various degradation and failure modules in percentage, with EVA discoloration occurring most frequently. Similar outcomes were observed in another study, as illustrated in Figure 5.11 [9].

In India, 57 c-Si modules installed at 26 different sites spread across 5 different climatic zones were examined for I-V data. As shown in Figure 5.12, discoloration and corrosion were present in nearly all age groups of modules older than 5 years, while delamination and back sheet degradation was only observed in a module older than 10 years. Higher-than-average degradation rates, i.e., a median rate of 1.22% per annum, were asserted. It was observed that the I_{sc} reduction and the degradation rate are closely connected. The principal cause of the decrease in I_{sc} and consequent acceleration of power deterioration was the discoloration of EVA caused by metal corrosion in hot and humid conditions [10]. In a visual inspection carried out for 177 c-Si PV modules installed in southern China's hot and humid location for 30 years, it was observed that 100% of the PV modules were affected by the EVA yellowing (discoloration), busbar corrosion, and glass soiling (Figure 5.13). The optical properties of the EVA degrade

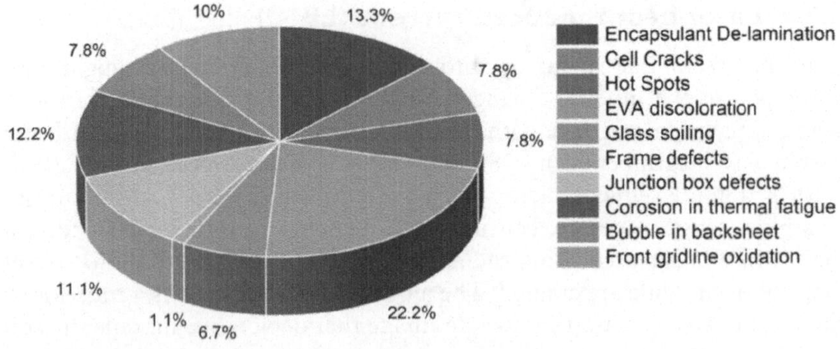

Figure 5.10 Various degradation and failure modules [4].

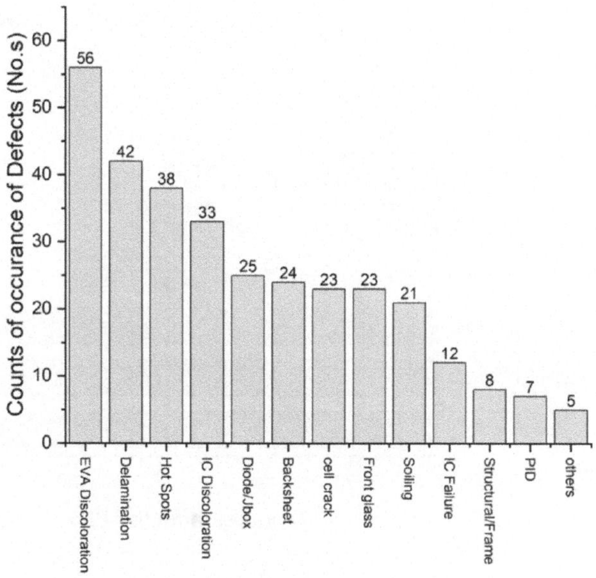

Figure 5.11 Count of occurrence of Defects 9.

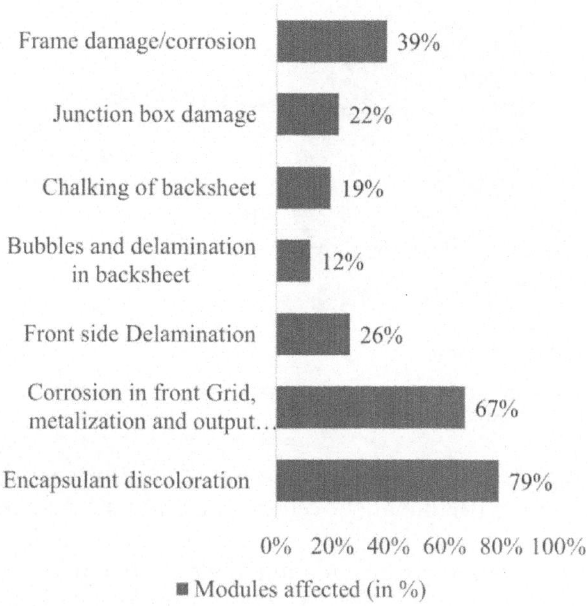

Figure 5.12 Frequency of occurrence of defects in PV [10].

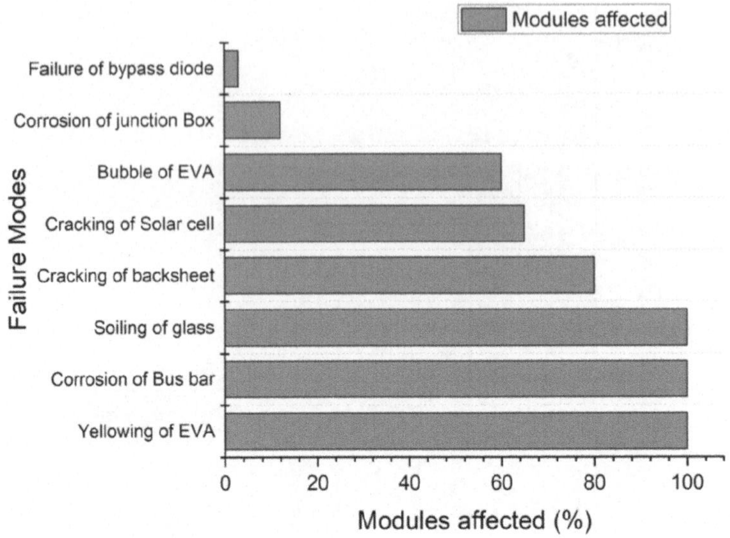

Figure 5.13 Modules affected by failure modes [12].

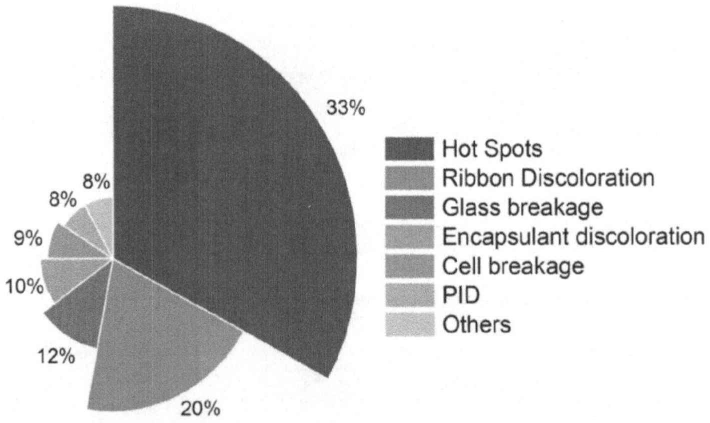

Figure 5.14 Occurrence of failures mode [30].

with yellowing, decreasing transmission by 12.6%. In more than 60% of the tested modules, delamination, cell cracks, and EVA bubbles were also found [12].

A much greater variety of degradation modes were reported for modules placed in hot and humid regions than for those in the desert and temperate conditions. In comparison to other climates, hot and humid regions are more likely to experience delamination and diode/J-box problems. Hot spots and internal circuitry discoloration seem to be the biggest degradation-related

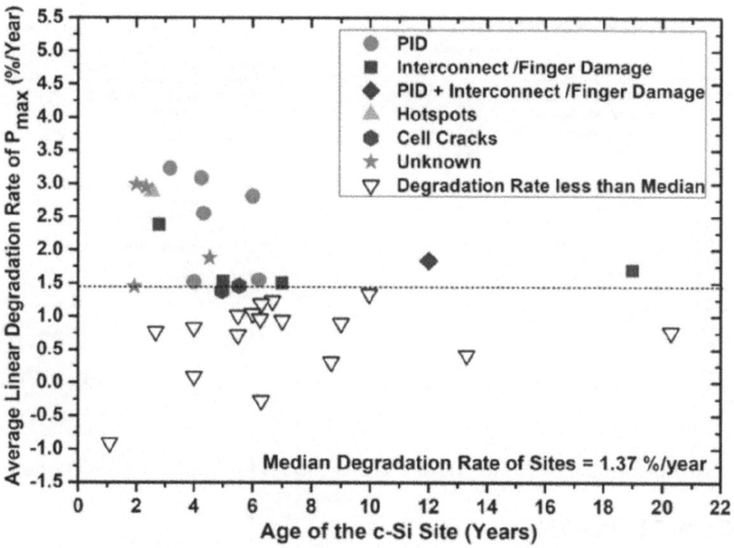

Figure 5.15 Average linear Pmax degradation rate as a function of age [32].

security risks for systems deployed in the last ten years (Figure 5.14). The most frequent form of degradation, especially in older systems, was encapsulant discoloration [30]. Also in another study, the predominant failure mode (83.5% of failures) of the 25-year-old c-Si PV modules operating in a mild climate was EVA discoloration [31].

The average linear P_{max} degradation rate across all c-Si sites is depicted in Figure 5.15 as a function of age. For locations where the rate of degradation is higher than the dataset's median (1.37% per annum), most young sites (those between 1 and 5 years old) are degrading more quickly than the old sites (older than 5 years). Surprisingly, the majority of the young sites decay quicker even after subtracting 2% for early rapid degradation brought on by LID. High degradation rates in younger sites are thought to be primarily caused by PID, which is then followed by interconnect/finger degradation [32].

The risk priority number (RPN) of various failure types is calculated in a study by examining the effects of performance and safety flaws. The performance RPN for various kinds of degradation for various climate zones is shown in Figure 5.16. Hot spots, internal circuitry discoloration, back sheet issues, and grounding wire corrosion were the main issues with older PV systems. Back sheet burn marks, back sheet peeling, and grounding wire corrosion were the main safety concerns in hot zones while frame grounding corrosion, back sheet difficulties, and hot spots were present in non-hot zones. These defects are observed in some young modules as well, but the degree of occurrence is less than in the older ones [33].

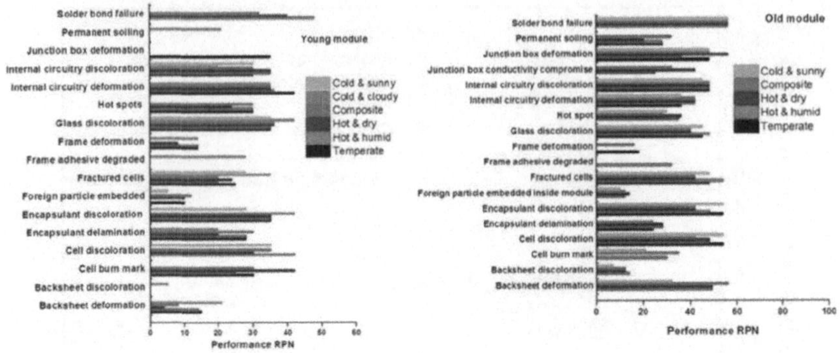

Performance RPN of the young module in the different climatic zone. Performance RPN of the old module in the different climatic zone.

(a) (b)

Figure 5.16 Performance RPN for various failure modes for Young and Old modules [33].

5.5 CONCLUSION

In summary, it has been observed that various factors impact the degradation and mode of failure for SPV, and among all, ambient weather conditions and the PV module packaging quality have a significant impact on the degradation. A comprehensive summary has been represented for the various degradation modes or faults for SPV cells and modules, including corrosion of front grid fingers, EVA discoloration, snail trails, hotspots, bubble formation, delamination, light-induced degradation, light and elevated temperature-induced degradation, potential-induced degradation, damage to solar cells, PV panel back sheets, and junction box failure. Different characterization techniques for degradation analysis are summarized briefly, including visual inspection, I-V characteristics, infrared thermography, ultraviolet fluorescence, ultrasonic inspection, luminescence techniques, electroluminescence, photoluminescence, and laser beam-induced current. It has been observed that EL imaging is the most preferred technique by the SPV O&M team as it is non-destructive, quick (real-time analysis), contactless, and most frequently employed for on-site analysis of PV-generating plants.

A concise summary of the occurrence frequency of various degradation modes for PV modules performing in outdoor conditions, as observed in many studies, is presented. It was discerned that, among all degradation modes, EVA discoloration shows the highest occurrence, with a value ranging between 18 and 20%, followed by delamination, i.e., 12 to 15%, cell hotspot, and corrosion. It was claimed that PV modules operating in hot and humid climates with age over 20 years were affected by EVA discoloration, i.e., around 80–100% of PV modules, followed by 60–100% by corrosion in the front grid, metallization, and output terminals. The variety of degradation mechanisms reported for modules deployed for extended periods in hot and humid environments was significantly greater than for modules deployed in the desert and mild climates. It has been observed that the PV modules installed in hot and dry conditions are more prone to EVA discoloration, while the hot and humid zone are most prone to the PID effect, corrosion in their metallization interconnects, and output terminals as well as back sheet degradation, while the modules placed in the cold zone experience the least degradation.

REFERENCES

1. "GOAL 7: Affordable and clean energy | UNEP - UN Environment Programme." https://www.unep.org/explore-topics/sustainable-development-goals/why-do-sustainable-development-goals-matter/goal-7 (accessed Nov. 22, 2022).
2. C. Ma and Z. Liu, "Water-surface photovoltaics: Performance, utilization, and interactions with water eco-environment," *Renew. Sustain. Energy Rev.*, vol. 167, p. 112823, Oct. 2022, doi: 10.1016/J.RSER.2022.112823.
3. A. Goswami and P. K. Sadhu, "Degradation analysis and the impacts on feasibility study of floating solar photovoltaic systems," *Sustain. Energy, Grids Netw.*, vol. 26, Jun. 2021, doi: 10.1016/j.segan.2020.100425.
4. P. Rajput, M. Malvoni, N. M. Kumar, O. S. Sastry, and G. N. Tiwari, "Risk priority number for understanding the severity of photovoltaic failure modes and their impacts on performance degradation," *Case Stud. Therm. Eng.*, vol. 16, Oct. 2019, doi: 10.1016/j.csite.2019.100563.
5. A. Abdelouahed, S. Elmamoun, A. Berrada, and A. Ameur, "Photovoltaic modules degradation assessment using different statistical techniques," *Int. J. Energy Res.*, Jul. 2022, doi: 10.1002/er.8320.
6. F. K. A. Nyarko, G. Takyi, and E. H. Amalu, "Robust crystalline silicon photovoltaic module (c-Si PVM) for the tropical climate: Future facing the technology," *Sci. African*, vol. 8, Jul. 2020, doi: 10.1016/j.sciaf.2020.e00359.
7. L. Koester, S. Lindig, A. Louwen, A. Astigarraga, G. Manzolini, and D. Moser, "Review of photovoltaic module degradation, field inspection techniques and techno-economic assessment," *Renew. Sustain. Energy Rev.*, vol. 165, p. 112616, Sep. 2022, doi: 10.1016/j.rser.2022.112616.
8. S. K. Singh and N. Chander, "Mid-life degradation evaluation of polycrystalline Si solar photovoltaic modules in a 100 kWp grid-tied system in east-central India," *Renew. Energy*, vol. 199, pp. 351–367, Nov. 2022, doi: 10.1016/j.renene.2022.09.013.

9. M. Kumar and A. Kumar, “Performance assessment and degradation analysis of solar photovoltaic technologies: A review,” *Renew. Sust. Energy Rev.*, vol. 78. Elsevier Ltd, pp. 554–587, Oct. 01, 2017, doi: 10.1016/j.rser.2017.04.083.
10. S. Chattopadhyay, Shashwata Dubey, Rajiv Kuthanazhi, Vivek Kuthanazhi, Jim Joseph John, Chetan Singh Solanki, Anil Kottantharayil, Brij Mohan Arora, K. L. Narasimhan, Vaman Kuber, Juzer Vasi, Arun Kumar, and O. S. Sastry “Visual degradation in field-aged crystalline silicon PV modules in India and correlation with electrical degradation,” *IEEE J. Photovoltaics*, vol. 4, no. 6, pp. 1470–1476, 2014, doi: 10.1109/JPHOTOV.2014.2356717.
11. J. Kim, M. Rabelo, S. P. Padi, H. Yousuf, E. C. Cho, and J. Yi, “A review of the degradation of photovoltaic modules for life expectancy,” *Energies*, vol. 14, no. 14. MDPI AG, Jul. 02, 2021, doi: 10.3390/en14144278.
12. Huili Hana, Xian Dong, Bingzhi Li, Huan Yan, Pierre J. Verlinden, Jiangfeng Liu, Jiapei Huang, Zongcun Lianga and Hui Shen “Degradation analysis of crystalline silicon photovoltaic modules exposed over 30 years in hot-humid climate in China,” *Sol. Energy*, vol. 170, May, pp. 510–519, 2018, doi: 10.1016/j.solener.2018.05.027.
13. Nikoleta Kyranaki, Alex Smith, Keith Yendall, David A. Hutt, David C. Whalley, Ralph Gottschalg and Thomas R. Betts “Damp-heat induced degradation in photovoltaic modules manufactured with passivated emitter and rear contact solar cells” *Prog. Photovoltaics Res. Appl.*, 2022, doi: 10.1002/pip.3556.
14. Wei Luo, Yong Sheng Khoo, Peter Hacke, Volker Naumann, Dominik Lausch, Steven P. Harvey, Jai Prakash Singh, Jing Chai, Yan Wang, Armin G. Aberle and Seeram Ramakrishna “Potential-induced degradation in photovoltaic modules: A critical review,” *Energy and Environ. Sci.*, vol. 10, no. 1. Royal Society of Chemistry, pp. 43–68, Jan. 01, 2017, doi: 10.1039/c6ee02271e.
15. Advanced Energy Industries, “Understanding potential induced degradation,” White Paper of Advanced Energy, Doc. ENG-PID-270-01, 8 p. Bend, Oregon, 2013.
16. J. Berghold, S. Koch, A. Ukar, M. Leers, and P. Grunow, “Potential-induced degradation (PID) and its correlation with experience in the field,” *Photovoltaics Int.*, pp. 85–92, Jan. 2013.
17. S. Mekhilef, R. Saidur, and M. Kamalisarvestani, “Effect of dust, humidity and air velocity on efficiency of photovoltaic cells,” *Renew. Sustain. Energy Rev.*, vol. 16, no. 5, pp. 2920–2925, 2012, doi: 10.1016/j.rser.2012.02.012.
18. C. Peike, S. Hoffmann, P. Hülsmann, B. Thaidigsmann, K. A. Weib, M. Koehl and P. Bentz, “Origin of damp-heat induced cell degradation,” *Sol. Energy Mater. Sol. Cells*, vol. 116, pp. 49–54, 2013, doi: 10.1016/j.solmat.2013.03.022.
19. S. Dietrich, J. Froebel, M. Ebert, and J. Bagdahn, “Experiences on pid testing of Pv modules in 2012,” vol. 49, p. 6120, 2012.
20. “PID & Regeneration - PV performance EN.” https://www.pv-performance.com/pid-regeneration-153.html (accessed Nov. 14, 2022).

21. W. Mühleisen, C. Hirschl, G. Brantegger, L. Neumaier, M. Spielberger, H. Sonnleitner, B. Kubicek, G. Ujvari, R. Ebner, M. Schwark, G. C. Eder, Y. Voronko, K. Knöbl and L. Stoicescu,"Scientific and economic comparison of outdoor characterisation methods for photovoltaic power plants," *Renew. Energy*, vol. 134, pp. 321–329, 2019, doi: 10.1016/j.renene.2018.11.044.
22. A. Pozza and T. Sample, "Crystalline silicon PV module degradation after 20 years of field exposure studied by electrical tests, electroluminescence, and LBIC," *Prog. Photovoltaics Res. Appl.*, vol. 24, p. n/a-n/a, Nov. 2015, doi: 10.1002/pip.2717.
23. "PHOTOVOLTAICS: Photoluminescence imaging speeds solar cell inspection | Laser Focus World." https://www.laserfocusworld.com/test-measurement/research/article/16552041/photovoltaics-photoluminescence-imaging-speeds-solar-cell-inspection (accessed Nov. 17, 2022).
24. Sumaira Ilyas, Muhammad Noman, Fazle Samad, Bushra Mahnoor, Sai Tatapudi, Fahad U. Zafar and Govindasamy Tamizhmani, "Design and development of outdoor daylight ultraviolet fluorescence imaging setup: an inexpensive tool to detect degradation of photovoltaic modules," *https://doi.org/10.1117/1.JPE.11.025501*, vol. 11, no. 2, p. 025501, Mar. 2021, doi: 10.1117/1.JPE.11.025501.
25. P. Mastny, J. Moravek, and J. Drapela, "Practical experience of operational diagnostics and defectoscopy on photovoltaic installations in the Czech Republic," *Energies*, vol. 8, no. 10, pp. 11234–11253, 2015, doi: 10.3390/EN81011234.
26. D. A. Quansah, M. S. Adaramola, and G. Takyi, "Degradation and longevity of solar photovoltaic modules—An analysis of recent field studies in Ghana," *Energy Sci. Eng.*, vol. 8, no. 6, pp. 2116–2128, Jun. 2020, doi: 10.1002/ESE3.651.
27. A. Haque, K. V. S. Bharath, M. A. Khan, I. Khan, and Z. A. Jaffery, "Fault diagnosis of Photovoltaic Modules," *Energy Sci. Eng.*, vol. 7, no. 3, pp. 622–644, Jun. 2019, doi: 10.1002/ESE3.255.
28. "(PDF) Task 13: Review on infrared and electroluminescence imaging for PV field applications." https://www.researchgate.net/publication/324703087_Task_13_Review_on_Infrared_and_Electroluminescence_Imaging_for_PV_Field_Applications (accessed Nov. 17, 2022).
29. M. Israil, S. A. Anwar, and M. Z. Abdullah, "Automatic detection of micro-crack in solar wafers and cells: A review," *Trans. Inst. Meas. Control*, vol. 35, no. 5, pp. 606–618, 2013, doi: 10.1177/0142331212457583.
30. D. C. Jordan, S. R. Kurtz, K. VanSant, and J. Newmiller, "Compendium of photovoltaic degradation rates," *Prog. Photovoltaics Res. Appl.*, vol. 24, no. 7, pp. 978–989, Jul. 2016, doi: 10.1002/pip.2744.
31. J.-S. Jeong, "Effect of EVA discoloration in 25-year-old single crystalline silicon photovoltaic modules operated under moderate climate," *Microelectron. Reliab.*, p. 114721, Nov. 2022, doi: 10.1016/j.microrel.2022.114721.

32. Yogeswara Rao Golive, Sachin Zachariah, Rajiv Dubey, Shashwata Chattopadhyay, Sonali Bhaduri, Hemant K. Singh, Birinchi Bora, Sanjay Kumar, A. K. Tripathi, Anil Kottantharayil, Juzer Vasi, and Narendra Shiradkar "Analysis of field degradation rates observed in all-india survey of photovoltaic module reliability 2018," *IEEE J. Photovoltaics*, vol. 10, no. 2, pp. 560–567, 2020, doi: 10.1109/JPHOTOV.2019.2954777.
33. Birinchi Bora, O. S. Sastry, Rajesh Kumar, Rajiv Dubey, Shashwata Chattopadhyay, Juzer Vasi, Som Mondal, and Basudev Prasad "Failure mode analysis of PV modules in different climatic conditions," *IEEE J. Photovoltaics*, vol. 11, no. 2, pp. 453–460, 2021, doi: 10.1109/JPHOTOV.2020.3043847.

Chapter 6

Performance analysis of a building-integrated solar PV/T system located in Gulmarg

Subhra Das

6.1 INTRODUCTION

India is a land having diverse geographical features ranging from snow-capped mountains to desert and lush green plain land to plateau. It has been reported by researchers (Ramachandra et al., 2011) that during February, a major part of India receives above 5 $kWh/m^2/day$ while the Western and Eastern Himalayan region receives insolation in the range of 3–4 $kWh/m^2/day$. The socioeconomic status of the people is also impacted by the geographical conditions as the hilly areas in the northern and eastern Himalayan region face scarcity of energy and food due to the extreme environmental conditions (Sati, 2015; Konwar, 2015), which is attributed to moderate-to-heavy snowfall with temperature ranging from –2°C to 22°C during winters in the hilly regions depending on the altitude of the place (Dikshit et al., 2014). Thus, technology needs to be devised to address these issues related to scarcity of energy supply. Research on solar photovoltaic-thermal systems provides evidence that these systems can to some extent provide solution to the energy problems in these regions.

A photovoltaic-thermal (PV/T) collector is a module in which the photovoltaic panel not only produces electricity but also serves as a thermal absorber. These collectors can produce heat and power simultaneously (Das, 2022). It consists of solar photovoltaic (PV) panel, heat transfer tubes, header pipes, thermal insulation, and casing as show in Figure 6.1.

Solar radiation incident on the solar PV panel is absorbed by it. It utilizes a part of the incident solar radiation to produce electricity and the rest is converted to heat energy. A part of the heat energy is transferred to the fluid flowing through the heat transfer tubes and the rest is lost to the ambient. This helps in lowering the temperature of the solar PV panels and thereby improving the efficiency of the panels.

The performance of the PV/T collector is affected by the design and control parameters such as glazing, mass flow rate, absorber, heat transfer fluid characteristics, and many more (Singh, 2009). Researchers have designed a concentrating solar PV/T system to increase the efficiency of the solar cell by concentrating the solar radiation on the solar cell and attaching

DOI: 10.1201/9781003623168-8

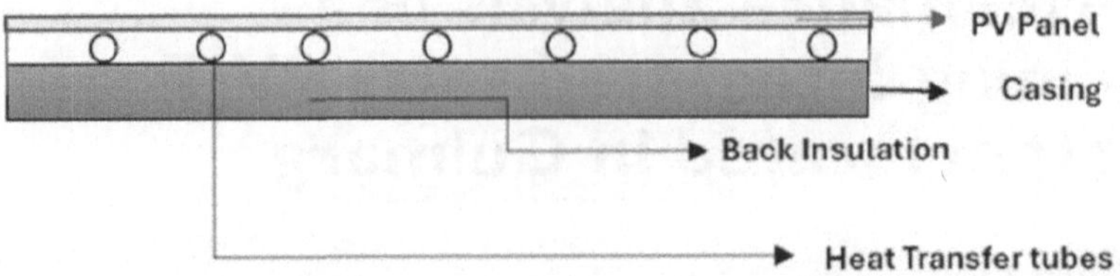

Figure 6.1 Solar PV/T collector.

it with fins to transfer heat to the heat transfer fluid flowing through its back (Othman et al., 2005). Kribus et al. (2006) designed a miniature concentrating solar PV/T system that is useful for high temperature applications, and it produces 140–180 W of electricity and 400–500 W of thermal energy. Researchers have shown that a solar concentrating PV/T collector performs better than a flat solar PV/T collector (Othman et al., 2005).

Solar PV/T is also classified based on the heat transfer fluid. It is observed that an air-based PV/T collector is preferred over the water-based PV/T collector because of the ease of designing, ease of flow of fluid underneath the absorber surface, and danger of electrical faults due to leakage of water in the circuit (Tonui and Tripanagnostopoulos, 2007). However, air type solar PV/T also has some disadvantages related to low density, low heat capacity, and low thermal conductivity, which leads to poor heat extraction by the circulating air (Tonui et al., 2007). To extract heat from the solar PV panel, air is forced through the channels having in-built fins. These limitations of air as a heat transfer fluid in solar PV/T systems inspired researchers to consider water as the coolant and observed that a water-based solar PV/T system has a better efficiency than an air-cooled solar PV/T system (He et al., 2006).

6.1.1 Building-integrated solar photovoltaic-thermal system

Building-integrated solar photovoltaic-thermal (BIPV/T) system is the integration of photovoltaic and thermal (PV/T) system into the building envelope (Misara, 2011). The solar PV/T modules are used to build the outer layer of the exterior walls of building thus replacing conventional building envelope materials. The solar panels generate power during the daytime which can be stored in batteries or can be fed into the grid. The thermal system at the back of the collector collects the heat from the panel using air or phase change materials to produce hot fluid, which can be stored in a thermal storage tank. The life cycle cost of these buildings is lower than a conventional building without BIPV/T or with rooftop PV, which require dedicated mounting structures.

A complete BIPV/T system (Das, 2022) consists of PV/T modules which may be thin-film or crystalline, transparent or opaque; heat transfer tubes at the back to collect heat from the solar PV panel; a charge controller to regulate the power into and out of the battery storage bank (in stand-alone systems); a power storage system which is generally a battery bank in case of off grid system and the utility grid in case of grid interactive systems; power conversion equipment including an inverter to convert the DC output to AC compatible with the utility grid; backup power supplies (optional) in case of stand-alone systems; and mounting structures, wiring, and safety devices to disconnect the system in case of any electrical faults. BIPV/T systems can be designed as stand-alone systems or can be connected to the grid. Power generation at the point of use reduces transmission and distribution losses thereby contributing to savings to utility. It reduces the electricity bills of the consumer because of peak shaving, which is like matching peak production with periods of peak demand. Buildings generating power and heat through renewable energy sources help in reducing the overall greenhouse gas emissions.

6.2 PERFORMANCE OF BUILDING-INTEGRATED SOLAR PV/T SYSTEM

The performance of a building-integrated solar PV/T system can be performed following the steps below:

Step 1: Specify the location of the building, day, and duration of the study.

Step 2: Estimate the incident solar radiation on the walls of the building having solar PV/T system using the Klucher model (Das, 2021):

$$I_T = I_B \frac{\cos\theta}{\cos\theta_z} + I_D \left(\frac{1+\cos\frac{\theta_t}{2}}{2} \right) \left[1 + F sin^3 \frac{\theta_t}{2} \right] \left[1 + F cos^2\theta \, sin^3 (\theta_z) \right]. \quad (6.1)$$

I_T is the total insolation received on the tilted surface, I_D is diffuse insolation received on horizontal surface, I_B is the direct insolation received on horizontal surface, θ_t is tilt angle. Incidence angle, θ, and zenith angle θ_z are computed for each wall having different orientations (Duffie and Beckmann, 1980).

The ASHRAE model can be used to estimate I_B and I_D on the horizontal surface using the following relations:

$$I_B = I_{bn}\cos\theta_z = A\exp\left[\frac{-B}{\cos\theta_z}\right]\times\cos\theta_z, \tag{6.2}$$

$$I_D = C I_{bn}, \tag{6.3}$$

where A, B, and C are constants whose values are determined month wise (Sukhatme et al., 2012).

Step 3: Estimate the temperature of the solar PV panel using the following relation (PV Performance Modeling collaborative):

$$T_p = I_T exp(a + bV) + T_a \tag{6.4}$$

Here the parameters a and b depend on the module construction, materials, and mounting configuration. In the present case, the module configuration is glass/cell/polymer sheet with insulation at the backside. Thus, the values of a = –2.81 and b = –0.0455 are considered.

The overall heat loss coefficient U_L is expressed as:

$$U_L = U_t + U_b + U_s \tag{6.5}$$

The top loss coefficient U_t can be expressed as:

$$U_t = \left[\frac{\delta_1}{K_1} + \frac{\delta_2}{K_2} + \frac{1}{h}\right]^{-1} \tag{6.6}$$

Where δ_1 and δ_2 are the thickness of the encapsulation material and the glass cover of the solar panel respectively and K_1 and K_2 are the thermal conductivity the encapsulation material and the glass cover of the solar panel respectively and h is the convective and radiative heat transfer coefficient from the glass and is expressed as (Duffie and Beckmann, 1980):

$$h = 5.7 + 3.8V \tag{6.7}$$

where V is the wind speed and h is measured in W/m^{2}K.

The back loss coefficient is computed using the following relation:

$$U_b = \left[\frac{\delta_3}{K_3} + \frac{\delta_4}{K_4}\right]^{-1} \tag{6.8}$$

Here (K_3, K_4) and (δ_3, δ_4) is the thermal conductivity and thickness of the brick and concrete material respectively used as the building material. The side loss coefficient is neglected in this calculation as it will be considered while computing the top loss coefficient for the adjacent wall.

Step 4: Assuming that all the solar PV/T panels are connected in series in a wall and receives uniform solar insolation at any time *t*. The electrical power output P_E from the solar PV panel can be estimated using the following relation [HOMER Pro 3.15]:

$$P_E = P_{r,E} f_{PV} \frac{I_T}{I_{T,STC}} \left[1 + \alpha_p \left(T_p - T_{p,STC}\right)\right] \tag{6.9}$$

Where $P_{r,E}$ is the rated capacity of the PV array [kW], f_{PV} is the PV derating factor [%], $I_{T,\,STC}$ is the incident radiation at standard test conditions [kW/m^2], α_p is the temperature coefficient of power [%/°C], and $T_{p,STC}$ is the PV cell temperature under standard test conditions.

Step 5: Estimate the thermal output P_T from the solar PV/T collector using the following relation:

$$P_T = \dot{m} c_p \left(T_o - T_a\right) \tag{6.10}$$

Step 6: Overall electrical and thermal performance of the solar PV/T wall can be obtained using the following relation:

$$\eta = \frac{P_E + P_T}{P_{in}} \tag{6.11}$$

Note that the output from each of the wall will be different so these arrays are not connected. The output from each of these arrays is fed separately to the battery bank or inverter.

6.3 CASE STUDY: PERFORMANCE OF A BUILDING-INTEGRATED SOLAR PV/T SYSTEM IN GULMARG

Consider a house made of integrated PV/T walls where a conventional air heater is fabricated at the back side of the solar PV panel which is fitted to the wall made of bricks. Consider that all the outer walls of the house are of same area Ac (10' × 12') facing east, west, north, and south. Three solar PV panels (having dimensions 1,655 mm × 990 mm × 34 mm) of 250 kWp are connected in series on each wall. The rated output from the array is 750k Wp.

The performance of the house with building-integrated solar PV/T system is analyzed for a particular day using MATLAB following the steps described in section 6.2. The assumptions that had been made regarding the specifications of the panel, material characteristics, site details, are tabulated in Table 6.1.

Table 6.1 Assumptions made for simulation

Parameters	*Values*
$I_{T,STC}$, W/m^2	1,000
α_p, per °C	−0.3% to −0.5%
$T_{p,STC}$, °C	25
δ_1, mm	0.5
δ_2, mm	3
k_1, W/mK	0.311
k_2, W/mK	2
Thickness of cement mortar plaster, m	0.0125
Thickness of brick wall, m	0.1778
Thermal conductivity of cement mortar plaster, W/mK	1.515
Thermal conductivity of brick wall, W/mK	0.8
τ_{glass}	0.9
α_{cell}	0.85
Day	June 11, 2021
Location	Gulmarg
Latitude	34.0484° N
Longitude	74.3805° E
Mass flow rate of air, kg/s	0.02

Firstly, the incident solar radiation on the four walls is estimated using Klucher's model, and the amount of radiation absorbed by the solar panels is computed. The absorbed radiation is partly converted to electrical power, partly transferred to the heat transfer fluid flowing through the channels below the PV panel, and the rest is lost to the ambient from the top surface of the panel and through the wall to inside the room. The temperature of the PV panel is computed from morning 5 am to 5 pm using equation (6.4) and is tabulated in Table 6.2.

The electrical and thermal outputs from the four walls are computed separately and are depicted in Figure 6.2. The eastern wall produces power in the morning half till 1 pm and attains a maximum around 7 am to 9 am. The eastern wall does not receive sunlight in the afternoon and consequently does not produce any electrical or thermal power output.

During this time, the inlet duct to the air heater needs to be closed to prevent heat loss from the room. During the morning, the eastern wall produces an average electrical output of 199 W and thermal output of 1157.4 W utilizing the incident solar radiation on the wall.

The western wall receives sunlight in the afternoon, and hence, generates electricity and heat from 1 pm to 5 pm as shown in Figure 6.3. The building-integrated solar PV/T wall produces an average electrical output

Table 6.2 Temperature of the building-integrated solar PV/T panels

S. No.	*Time, h*	*Ta, C*	*TpE, C*	*TpW, C*	*TpN, C*	*TpS, C*
1	5	12	12.7	12.0	12.2	12.0
2	6	13	42.7	13.0	25.5	13.0
3	7	15	57.8	15.0	28.4	15.0
4	8	18	61.1	18.0	27.7	21.8
5	9	20	58.4	20.0	25.5	29.1
6	10	22	51.9	22.0	23.8	35.4
7	11	24	43.1	24.0	24.0	39.9
8	12	24	31.6	27.7	24.0	40.6
9	13	24	24.0	38.8	24.0	39.5
10	14	24	24.0	48.6	24.0	36.7
11	15	25	25.0	57.6	25.0	33.6
12	16	25	25.0	62.3	34.3	28.7
13	17	25	25.0	61.2	37.5	25.0

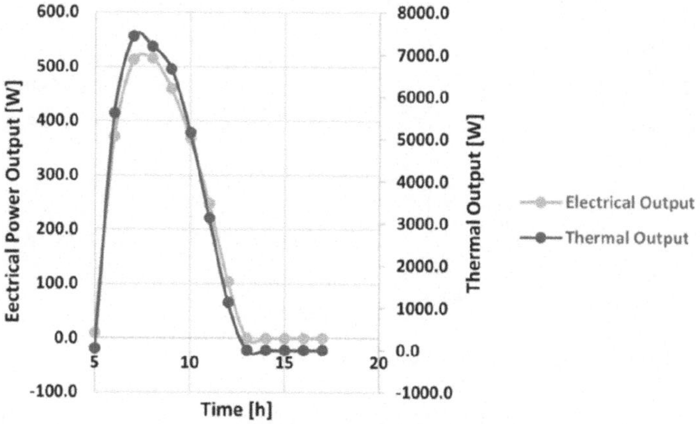

Figure 6.2 Electrical and thermal output of the eastern wall.

of 146.2 W and thermal output of 724.0 W using the average solar radiation of 222.1 W/m^2 over the day.

The northern wall receives solar radiation in the early morning hours and in the late evening. Thus, the output from this wall is low compared to the other walls as shown in Figure 6.4. It records an average electrical output of 68.6 W and thermal output of 307.5 W over the day. An average radiation of 93.4 W/m^2 is recorded over the day.

The southern wall contributes the most in terms of the number of hours of generation of electrical and thermal output as shown in Figure 6.5. It receives solar radiation from 7 am to 5 pm and generates an average electrical power output of 103.1 W and thermal output of 468.6 W over the day.

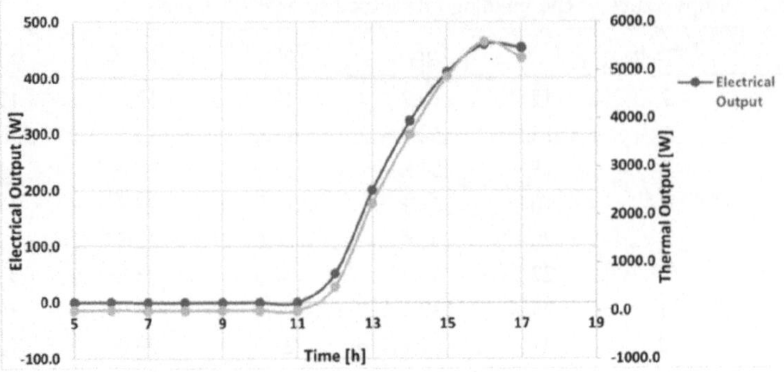

Figure 6.3 Electrical and thermal output of the western wall.

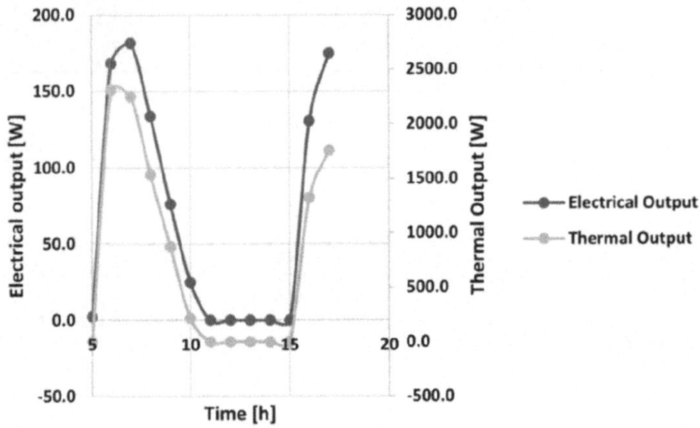

Figure 6.4 Electrical and thermal output of the northern wall.

The average electrical and thermal efficiency of each of the wall separately and the average efficiency of the combined solar PV/T system is tabulated in Table 6.3.

It is observed that individually the system performance is not very high but when we combine both solar PV and thermal system together, the system efficiency for the southern and eastern wall is above 50%. And for the other two walls, the overall efficiency is greater than 35%. This shows that building-integrated solar PV/T walls are more efficient than any conventional wall or a simple solar PV plant.

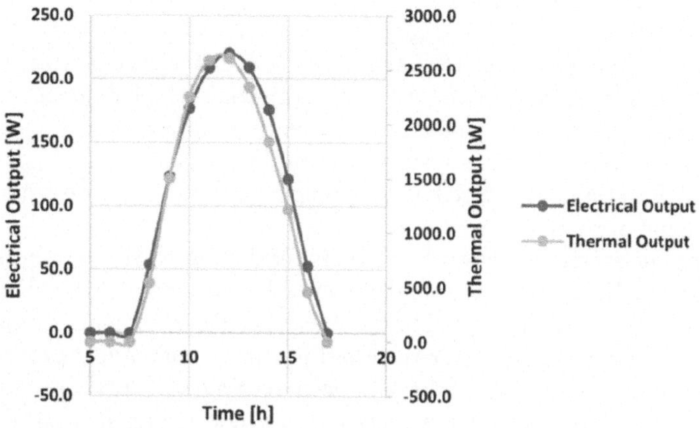

Figure 6.5 Electrical and thermal output of the southern wall.

Table 6.3 Average electrical and thermal efficiency of the four walls

Average Efficiency	*East*	*West*	*North*	*South*
Electrical	0.09	0.06	0.09	0.10
Thermal	0.41	0.29	0.33	0.43
Solar PV/T wall	0.50	0.36	0.42	0.54

6.4 CONCLUSIONS

In the present case study, solar panels which are available in the market have been used for the simulation. The solar panels only cover 44% of the total wall area. Better results can be achieved by increasing the area covered by solar PV/T panels. The analysis shows that there is a potential to generate both electrical and thermal output by the east, west, and southern walls but the northern wall does not contribute significantly to energy generation. Thus, it is advisable to build northern wall using conventional building materials having poor thermal conductivity. The output from the other three walls varies over the day. So, these need not be connected in series rather the input from each wall should be fed in the inverter separately.

The hybridization of solar PV and thermal system helps in the utilization of the same space to generate energy and thus results in improved energy efficiency of the overall system. These systems will be particularly useful for places where both electricity is scare and heating load is high like places in hilly areas.

ABOUT THE AUTHOR

Subhra Das has more than 16 years of research experience in areas of solar energy technologies and its energy and exergy analysis. Her current research interests are solar radiation estimation, thermal application of nanoparticles and nanofluids in solar systems and building material, and analysis of solar PV/T technologies. She has academic collaborations with Universities in India and abroad.

She has delivered consultancy projects funded by NISE, an autonomous institute of Ministry of New and Renewable Energy and DST, India and project funded by Ministry of Human Resource Development. She has published research papers in peer-reviewed journals, conference proceedings, book chapters, and books. She has been serving as a reviewer for eight international journals and has received an Outstanding Reviewer Award from *Solar Energy Journal.*

REFERENCES

Das, S. 2021. Short term forecasting of solar radiation and power output of 89.6kWp solar PV power plant. *Materials Today: Proceedings*, 39 (5), 1959–1969. ISSN: 2214-7853. https://doi.org/10.1016/j.matpr.2020.08.449.

Das, S. 2022. Thermal analysis of a house made of composite wall comprising of solar PV panel and epoxy/Al polymer located in Gulmarg. *International Journal of Ambient Energy*, 43 (1), 7207–7217. https://doi.org/10.1080/01430750.2022.2063179.

Dikshit, K. R., and Dikshit, J. K. 2014. Weather and climate of North-East India. In: *North-East India: Land, People and Economy: Advances in Asian Human-Environmental Research.* Springer. https://doi.org/10.1007/978-94-007-7055-3_6

Duffie, J. A., and Beckman, W. A. 1980. *Solar Engineering of Thermal Processes Solar Engineering.* 1st edition. John Wiley and Sons, Inc.

He, W., Chow, T. T., Ji, J., Lu, J., Pei, G., and Chan, L. S. 2006. Hybrid photovoltaic and thermal solar-collector designed for natural circulation of water. *Applied Energy*, 83, 199–210.

HOMER Pro 3.15. https://www.homerenergy.com/products/pro/docs/latest/how_homer_calculates_the_pv_array_power_output.html

Konwar, P. 2015. *Socio-economic conditions, inequality and deprivation in North East India.* Munich Personal RePEc Archive.

Kribus, A., Kaftori, D., Mittelman, G., Hirshfeld, A., Flitsanov, Y., and Dayan, A. 2006. A miniature concentrating photovoltaic and thermal system. *Energy Conservation and Management*, 47, 3582–3590.

Misara, S., Henze, N., and Sidelev, A. 2011. *Thermal Characteristics of BIPV (U Value and g Value).* ISES Proceedings for SWC. http://proceedings.ises.org/paper/swc2011/swc2011-0192-Misara.pdf

Othman, M. Y. Hj., Yatim, B., Sopian, K., and Abu Bakar, M. N. 2005. Performance analysis of a double-pass photovoltaic/thermal (PV/T) solar collector with CPC and fins. *Renewable Energy*, 30, 2005–2017.

Ramachandra, T. V., Jain, R., and Krishnadas, G. 2011. Hotspots of solar potential in India. *Renewable and Sustainable Energy Reviews*, 15, 3178–3186.

Sandia Module Temperature Model. PV performance modelling collaborative. https://pvpmc.sandia.gov/modeling-steps/2-dc-module-iv/module-temperature/sandia-module-temperature-model/

Sati, V. P. 2015. Natural resources potentials and socio-economic status in the Indian Himalayan region. *Nature & Environment*, 20 (2), 48–58.

Singh, B., and Othman, M. Y. 2009. A review on photovoltaic thermal collectors. *Journal of Renewable and Sustainable Energy*, 1, 062702. https://doi.org/10.1063/1.3266963

Sukhatme, S. P., and Nayak, J. K. 2012. *Solar Energy, Principles of Collection and Storage*. 3rd edition. Tata McGraw Hill Education Pvt Ltd.

Tonui, J. K., and Tripanagnostopoulos, Y. 2007. Air-cooled PV/T solar collectors with low-cost performance improvements. *Solar Energy*, 81, 498–511.

Tonui, J. K., and Tripanagnostopoulos, Y. 2007. Improved PV/T solar collectors with heat extraction by forced or natural air circulation. *Renewable Energy*, 32 (4), 623–637.

Section III

Control and management of solar photovoltaics

Chapter 7

Multi-functional prosumer converters design and control for solar photovoltaic energy conversion systems

T. Sathiyanarayanan

7.1 INTRODUCTION

The electrical network along with the core building components have been constantly changing and upgrading in order to meet the ever-growing demand, to include the modern technological advancements, and to rectify the socioeconomic challenges concerned with the electrical energy sources. The fast-depleting fossil fuels together with their growing environmental concerns added to the issues in transmission and distribution systems, and this has led to the widespread adoption of distributed renewable energy resources (DRER) across all parts of the globe [1]. The high price related with the DRER, which has been a major roadblock in the past, has gradually come down over the recent years [2], and it is now a viable and potent alternative to replace the fossil fuel–based power generation. The bigger installations are constantly planned, designed, and implemented [3, 4], and of the renewable sources, solar photovoltaic energy conversion system (SPECS) and wind energy conversion system (WECS) are more prominent.

The race for fully DRER-based grids has started, and SPECS has a major contribution due to the supple nature of photovoltaic (PV) coming in different ranges of power, varying from small-sized cells of few Watts to big-sized farms of Million/Mega Watt range; the adept nature of it to be connected in series and parallel to increase the voltage, current, and power ratings give lot of flexibility and freedom for the users. There are many standards framed, reformed, and re-framed over the course of the installations in order to connect and operate SPECS with the conventional electrical grids [5–8]. The countries, the states within, and the regions within delineate the operation codes of SPECS (and DRER, in general) [9].

Sunlight, the primary energy source for SPECS, isn't available all-around the day and can be erratic in nature/level due to cloud movements; this uncertain nature and the generated power from PV panels being direct current (DC) begets power electronic converters for interfacing and controlling the PV [10–12] and battery energy storage system (BESS) to smoothen the fluctuations and provide power in the absence of sunlight. The power electronic converters are consisting power electronic switches, which in turn

DOI: 10.1201/9781003623168-10

requires to be controlled at high speed, and this leads to modification of the performance of the SPECS as per the requirements [13–16].

The conventional grid generators had slow dynamic response due to the widespread usage of big synchronous machines and only the loads had faster dynamics (switching, variations); the modern electric grids with DRER not only have faster dynamics on load side, but also on the source side due to the usage of small-sized power sources, demanding faster control for source side. Among the various controllers, model predictive control (MPC) has been found to be an alternate for traditional proportional integral (PI) control and having very good dynamic response [17–21]. With the ability of converting system operation into objective function, the MPC can be easily applied for control of SPECS in futuristic grids [22–26] where the user is becoming an active Prosumer = producer + consumer as against the existing passive consumer.

7.2 SMART GRIDS, PROSUMERS, AND DISTRIBUTED RENEWABLE ENERGY RESOURCES

The electric power grid as of now consists of big centralized generating stations, coal-fired, nuclear-powered, hydroelectric, which are usually located near the fuel source, and load centers are fed through transmission and distribution networks as shown in Figure 7.1. The safety, environmental, and depletion of sources are major concerns, which added with aggregated technical and commercial losses in generation, transmission, and distribution call forth the imminent change.

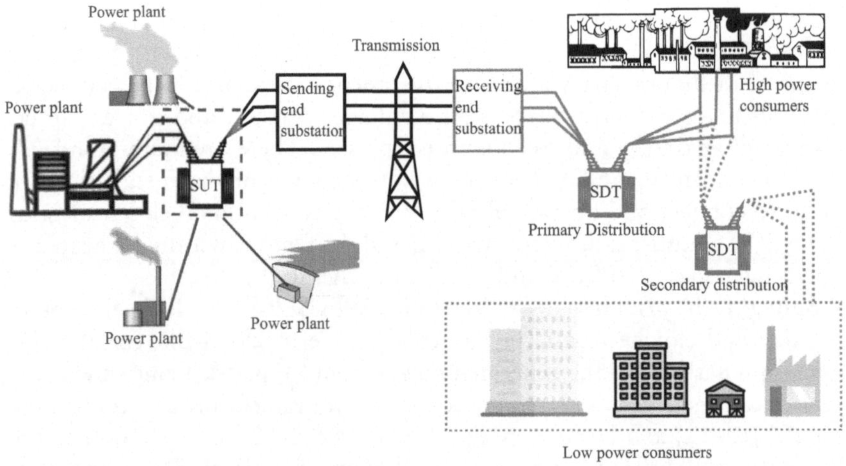

Figure 7.1 Conventional electrical power network.

Aforementioned points are generally overlooked in the past where the primary aim has been developing the economy and due to high cost of alternate sources of energy. This conventional power grid has unidirectional power flow, from the generating stations to the load centers and the end user is a customer buying the electricity with little-to-zero involvement in the structure. In smart grids, the prosumer can buy as well as sell power making it more active, and based on the buying/selling, it can be beneficial to the prosumer or the network or both; this means that there is bidirectional power flow and the overall losses can be reduced. A simple smart grid with an addition of high and low power prosumers is shown in Figure 7.2. This has made the technical committees and the national/state/regional grid authorities across the globe to revamp the standards set for DRER.

The prosumers form clusters of generators and loads establishing a localized/decentralized market structure where they can sell/buy active power, reactive power, and ancillary services for which the MPC is suited [27, 28] with different control frames. The changing standards and grid codes dictates the technical specifications, transforming the normal inverters to smart converters [29–31], of which low-voltage ride through (LVRT), frequency support, and reactive power support are notable[32–35].

7.3 SOLAR PHOTOVOLTAIC POWER CONVERSION SYSTEM CONTROL

The initial years of SPECS inclusion to the electric grid didn't call for advanced control strategies as the amount of solar PV installed wasn't high

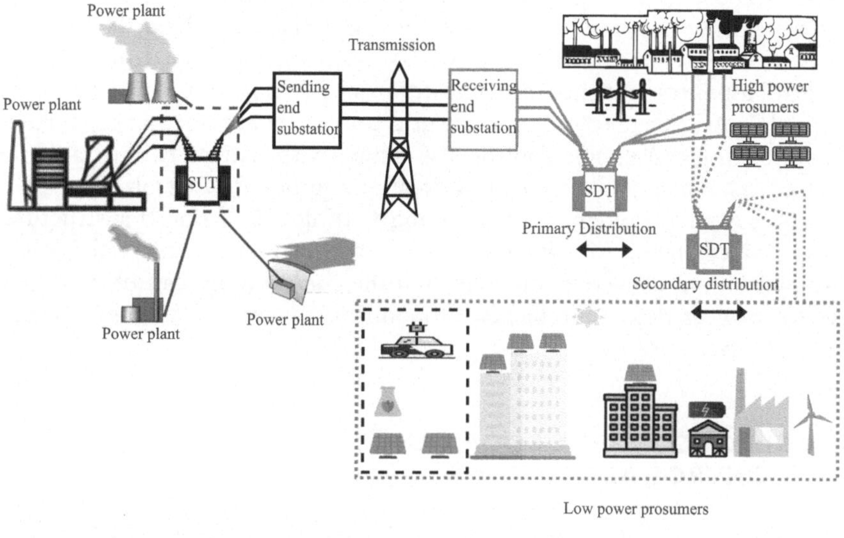

Figure 7.2 Electrical power network with prosumers.

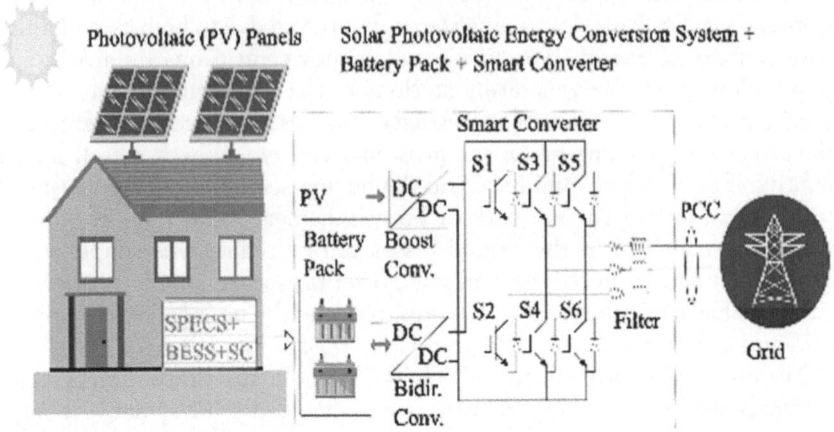

Figure 7.3 A domestic prosumer SPECS.

and simple power injection to grid was the goal. The penetration of DRER as a whole and SPECS in particular has been constantly increasing to the extent where it can be said like *"A* ***SPECS*** *for every rooftop/every home."* The changeover of centralized power system to decentralized system means that electricity has become a commodity which a prosumer can buy and/or sell from home, making power trading a possible part-time or even a full-time job further enhancing the economy while catering to the growing energy demand.

The solar cells are put in series parallel combination to get solar panels, and a BESS is taken in parallel to smoothen the erratic PV output and store power during daytime to use for night time. The control can be broadly classified as DC side controls and AC side controls, with the DC side having control for extracting maximum power from PV and maintaining DC voltage while AC side taking care of the grid connection, codes, and supportive functionalities. A simple rooftop PV system with battery is illustrated in Figure 7.3, which is connected to grid after appropriate controls.

The control topology of each section with detailed converter structure is explained in the following sections and note that the controller, control algorithm, converter topology, and the control loops can be modified so as to meet the requirements, economical constraints, and the market availability.

7.4 SPECS CONTROL LOOPS AND CONTROL ALGORITHMS

The most basic and the most researched control loop for the SPECS is the maximum power point tracking (MPPT) control. The cost of the solar panels is high, and in order to efficiently operate it, the maximum possible power needs to be extracted. Older iterations of the tracking have been toward mechanical tracking with moving panels physically to receive sunlight at one particular angle and the newer iterations of MPPT have been toward electrical tracking. There are many possible methods for MPPT like perturb observe, incremental conductance, and AI based, and from the MPPT algorithm employed, the gate signal for the boost converter can be obtained as shown in Figure 7.4.

The power from solar PV is erratic, and even though in-depth research has given many MPPT algorithms to track properly, the variations in the MPP can cause Vdc to fluctuate which makes the BESS an essential support; this in turn helps in operating the SPECS during night time with no sunlight available, and the stored power in the BESS can be utilized for enabling the prosumer.

A bidirectional converter is used for BESS, and its control is depicted in Figure 7.5. The sizing of PV panels for BESS needs to be done with the consideration of load ratings, economics, and space constraints.

The control of PV with MPPT and BESS makes the DC voltage constant as long as the assumption of BESS capacity not hit; now the system can be treated as a voltage source converter as the dynamics on DC side are fast. For the AC side, it is not as simple as DC side as the parameters to be controlled increases. In order to have a grid connection, it is required to have 3-phase output; for low load, domestic SPECS 1 phase is sufficient, but for generalized case, domestic/industrial/commercial loads 3 phase is suitable. For the three-phase system, let a, b, c be the three phases being equal magnitude and spatially displaced by 120°. For effective control operation and calculations, two transformations are shown in Figure 7.6, which are abc to d-q *(direct axis, quadrature axis)* and abc to $\alpha - \beta$ (alpha-beta).

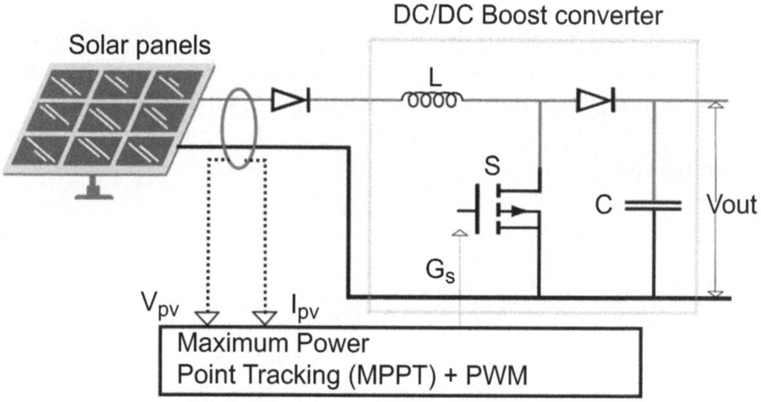

Figure 7.4 Solar PV MPPT control.

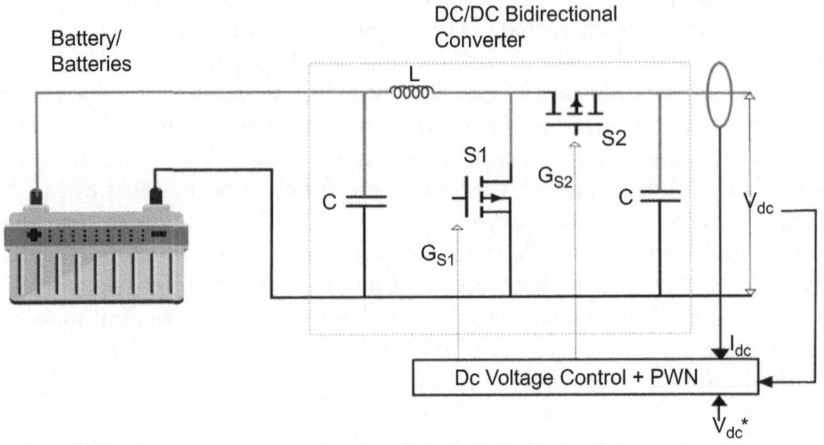

Figure 7.5 Bidirectional control for Vdc control.

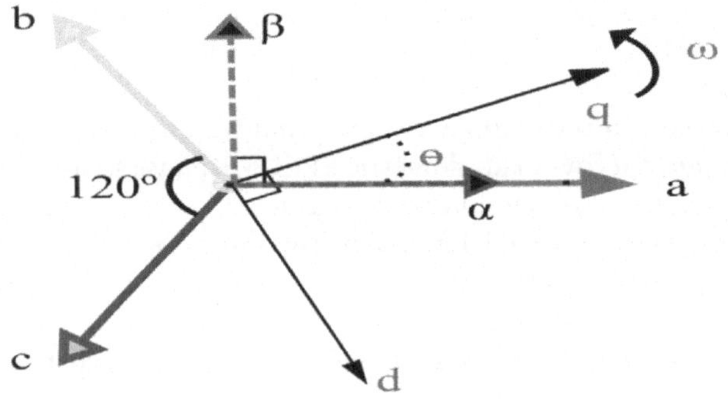

Figure 7.6 Different frames of control.

The basic principle behind these transformations is to reduce the number of equations to be solved by the controller, from 3 phase to 2 quantities; in alpha-beta frame, these two quantities are sinusoidal and in direct-quadrature frame, these two quantities are constants. The transformation matrices are given below:

$$I_{\alpha\beta} = \frac{2}{3}\begin{bmatrix} 1 & \frac{1}{2} & -\frac{1}{2} \\ 0 & \frac{\sqrt{3}}{2} & -\frac{\sqrt{3}}{2} \end{bmatrix}\begin{bmatrix} I_a \\ I_b \\ I_c \end{bmatrix}$$

$$I_{dq} = \frac{2}{3}\begin{bmatrix} \cos(\theta) & \cos(\theta-120^{o}) & \cos(\theta+120^{o}) \\ \sin(\theta) & \sin(\theta-120^{o}) & \sin(\theta+120^{o}) \end{bmatrix}\begin{bmatrix} I_a \\ I_b \\ I_c \end{bmatrix}$$

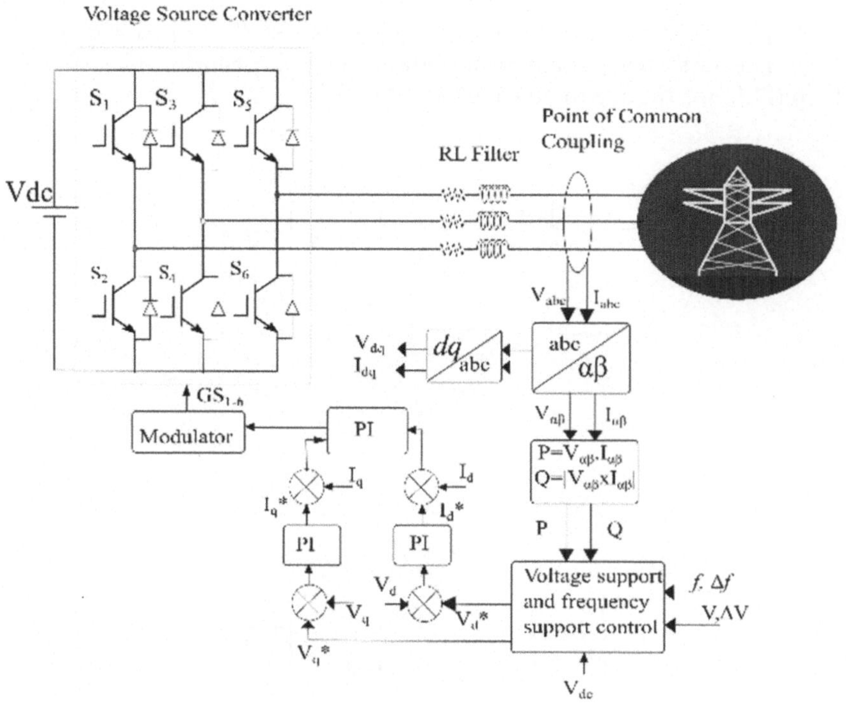

Figure 7.7 DC-AC voltage source converter control.

Under balanced condition assumption, the third quantity in $I\alpha\beta$ and Idq is zero. The point to be noted here is that, there will be unbalanced conditions temporarily due to faults and permanently due to unbalanced loads, which requires the Io component and a special 4-leg converter for balancing through neutral. The concepts explained for the 3-leg converter shown in Figure 7.7 can be extended to 4-leg converters too.

7.5 SIGNIFICANCE OF SPECS CONTROLLERS

The vast literature on the controllers can be viewed as a constant search for a better controller and although PI controllers have been deployed across all industries, the scope for improvement has always pushed researchers from industries and academia to try out different controllers [15, 16]. The control frame needs to be dq (or constant reference) for PI controller to work well and the tuning is a tedious job. This is the place where advanced controllers like model predictive control (MPC) is coming in and due to the inherent nature of fast dynamic response matching to the fast-changing smart grid environment it works better for power electronic systems [19].

As shown in Figure 7.8, there is no issue of tuning of controllers and no constraints on the frame of control, which can be viewed as a potential candidate for system not requiring additional components like phase locked loop (PLL) for the adc to dq transformation.

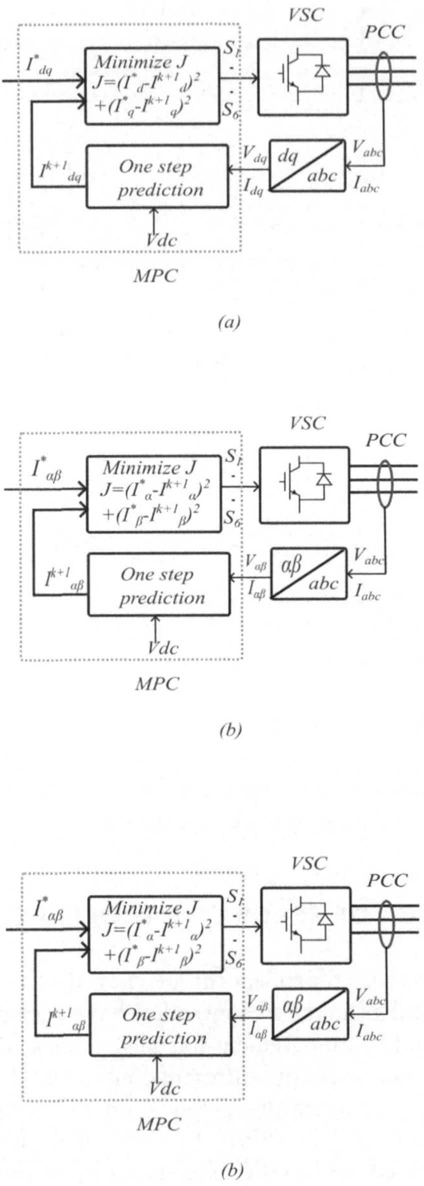

Figure 7.8 MPC in different control frames: (A) dq, (B) alpha-beta, (C) abc.

The fast response of the current controller determines the system response, and with the newer grid requirements like fast frequency response (FFR), having the faster controller is highly beneficial to the stability of the system.

In Figure 7.9 the active-reactive power capability curve of a voltage source converter is given for producer and consumer zones, considering active power *P* as the generation/consumption criteria.

The actual apparent power rating of the converter (Snameplate) is slightly higher than the maximum possible apparent power in producer and consumer zone (Spro and Scon) considering future expansion and SPECS momentary overloading. The difference is determined by economics mostly, as SPECS can be included module by module.

The producer and the consumer ratings usually are different, as the maximum active or reactive power in producer zone (Ppro, Qpro) comes from both SPV and BESS while in consumer zone Pcon, Qcon comes only from BESS. Assuming the safety considerations, the limits of P and Q for both

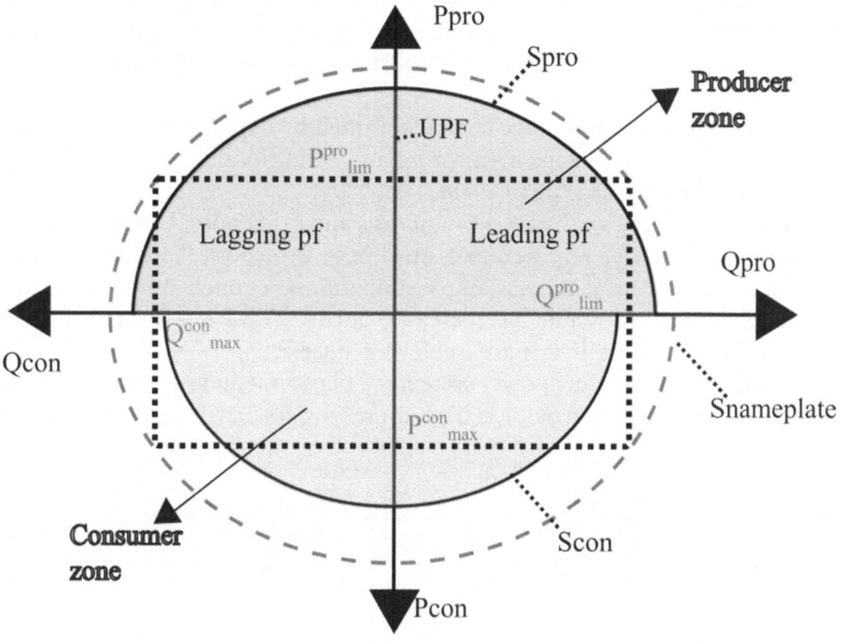

Figure 7.9 Active-reactive power capability of converter for producer and consumer zones (producer = P generated, consumer = P consumed).

zones can be calculated. The operation of SPECS ultimately lies in these two zones and with newer standards and grid codes, the need for different power factor operation can also be effectively solved by fast and proper control.

7.6 MULTI-FUNCTIONAL CONVERTER CONTROL

The SPECS with modified prosumer controls has to meet some of the grid requirements; as of now, the grid codes for bigger DRER is taken and is generalized for prosumer control with the futuristic opinion that every prosumer will eventually contribute to grid support in the race for full DRER-based smart grids for both economic and technical reasons.

The multiple functions of the converter can be broadly split as voltage support and frequency support or reactive power support and active power support.

Figure 7.10 depicts a general voltage ride through curve and till the voltage recovers to a certain point the SPECS need to stay connected to give support to the grid. If correct support isn't provided, the voltage may deteriorate leading to a possible brownout which might end up in a black-out. The property of smart grids being self-healing lies in their components and SPECS need to take part in grid support during events like fault or load change or generation change.

The grid support for voltage can be obtained by deploying a control loop like Figure 7.11; the voltage monitoring and conditioning needs to be done in order to classify the type of event.

The voltage limits are based on standards and grid codes; based on nature of the event, it can be required to control only Q or both P and Q. Point to be noted here is that, if there is ample sunlight and enough storage capacity P can be maintained while Q is increased during LVRT but due to the need of fault limiting, usually it is advisable to reduce P.

The major threat to any power system is a black-out and smart grid SPECS need to do its share to avoid it. The main reason for black-out is generators

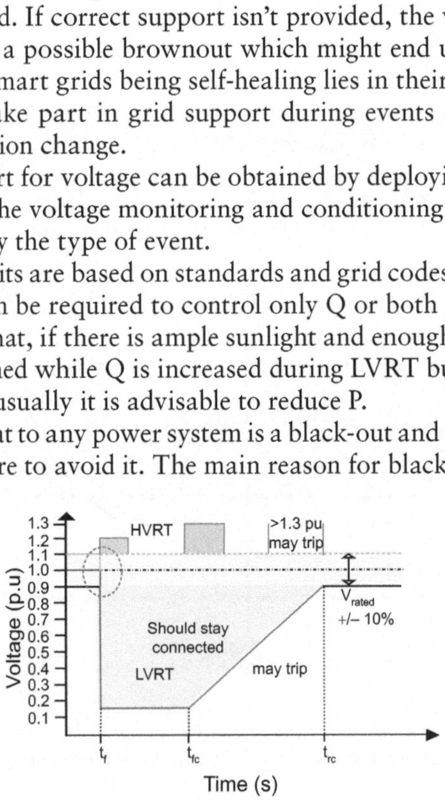

Figure 7.10 Voltage ride through operations.

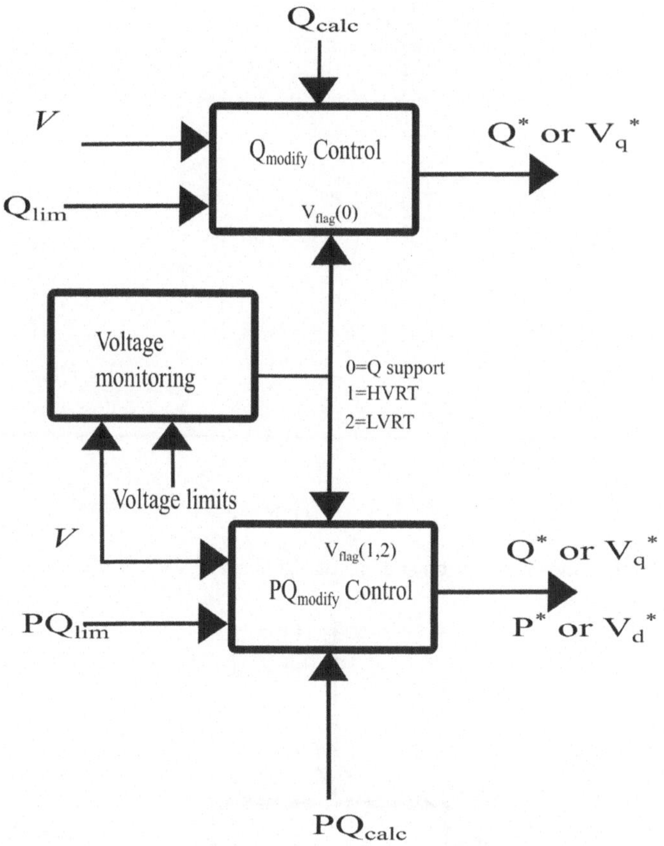

Figure 7.11 Voltage-reactive support loop for SPECS.

falling out of synchronism, which is caused by power imbalance between the generation and the load; the parameter that can be measured locally to get the information about power imbalance globally is frequency and by contributing to frequency restoration every SPECS can try to heal and restore the system in case of major events. The lack of inertia from SPECS due to no-rotating parts and so no kinetic energy need to be addressed for this, and the impact of inertia over frequency restoration can be seen in Figure 7.12.

The guidelines for frequency support are also available from various standards and grid codes; for the frequency range of operation, the normal control is followed and when it deviates from rated frequency, the active power *P* is modified in order to restore the frequency. This is depicted in Figure 7.13 and based on *f-P* modifications, and different types of frequency responses can be obtained.

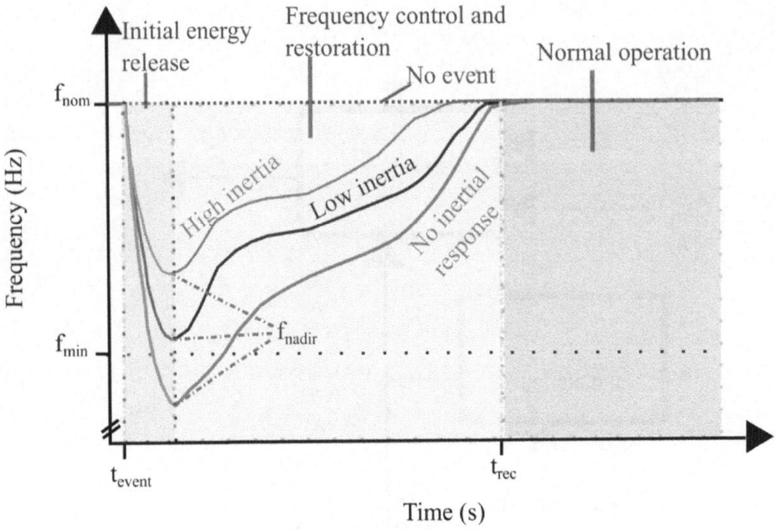

Figure 7.12 Frequency restoration and inertia significance.

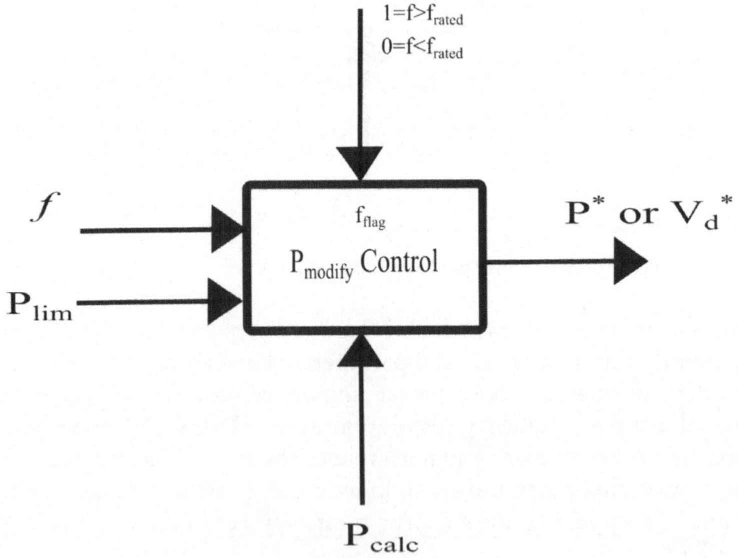

Figure 7.13 Frequency-active power support loop for SPECS.

7.7 PROSUMER SPECS SMART CONVERTERS

The prosumer SPECS smart converters control loops are summarized in Figure 7.14, and the existing control frame of dq frame is used for evaluating the performance of the multiple grid-supportive functionalities. The measurements for the phase locked loop (PLL) are taken from the point of common coupling (PCC). The measured quantities are converted to per-unit bases so as to present the general response of SPECS irrespective of the system size.

7.7.1 CaseI: Major fault at PCC, t = 5 s and cleared by t = 5.3 s (Indian grid code for LVRT)

The power system is very dynamic, and the faults are occurring despite the best efforts of safe operation across the countries. Due to a major fault, the voltage dips to 0.15 p.u. from 1 p.u. and then recovers. This is shown in Figure 7.15 below and the change in active, reactive powers is shown in Figure 7.16:

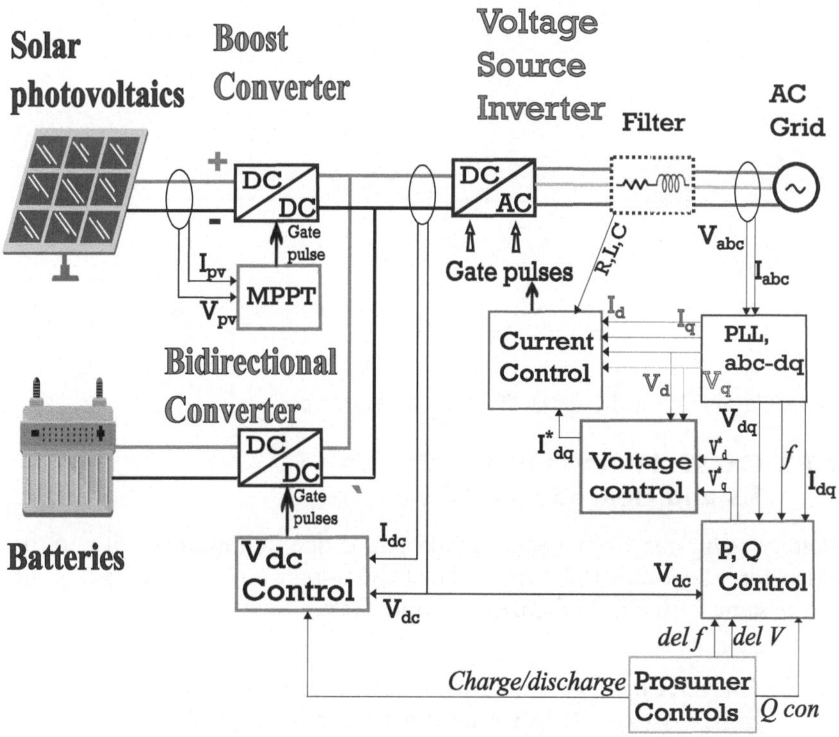

Figure 7.14 Multi-functional prosumer smart converter for SPECS.

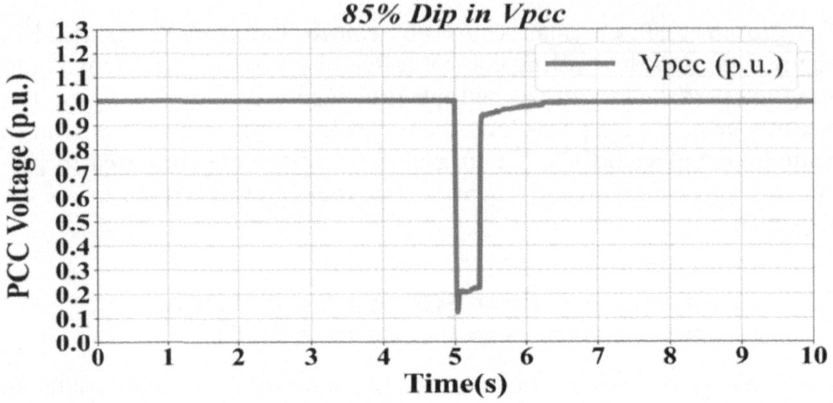

Figure 7.15 LVRT voltage at PCC: 1 p.u. to 0.15 p.u.

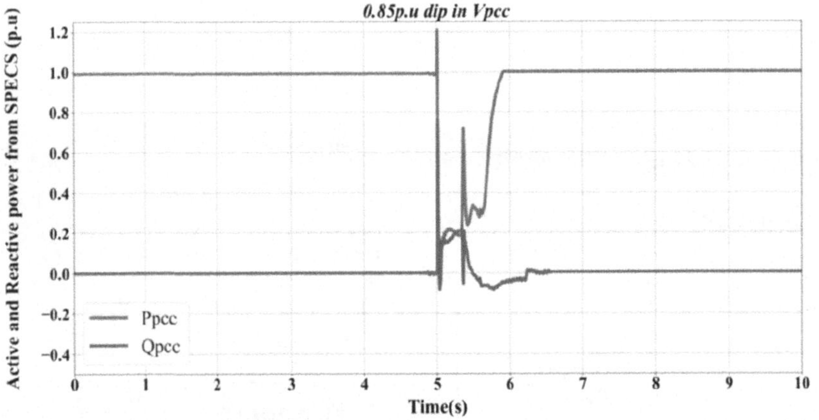

Figure 7.16 SPECS *P* and *Q* for LVRT: −0.85 p.u. dip in Vpcc.

7.7.2 Case2: Minor fault at PCC, t = 5 s and cleared by t = 8 s

Voltage reduction from 1 p.u. to 0.85 p.u. is due to a minor fault and that exists for longer time than case 1. The PCC voltage and the SPECS response to the same is shown in Figures 7.17 and 7.18.

7.7.3 Case3:Load/generation change leading to high voltage (high voltage ride through)

Voltage shoot-up from 1 p.u. to 1.2 p.u. The PCC voltage and the SPECS response to the same is shown in Figures 7.19 and 7.20.

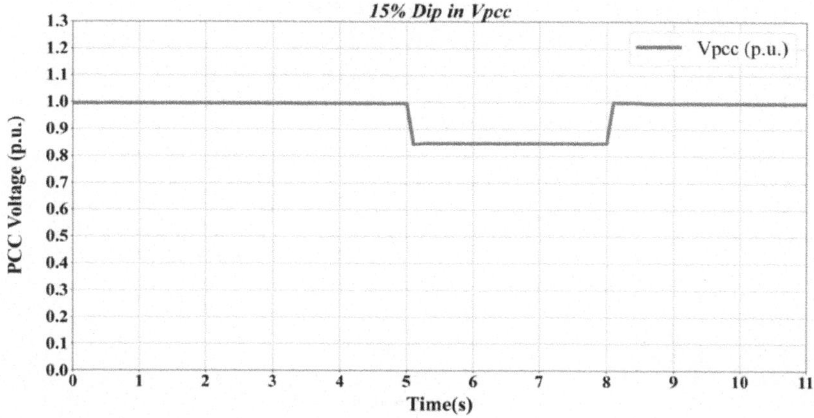

Figure 7.17 LVRT voltage at PCC: 1 p.u. to 0.85 p.u.

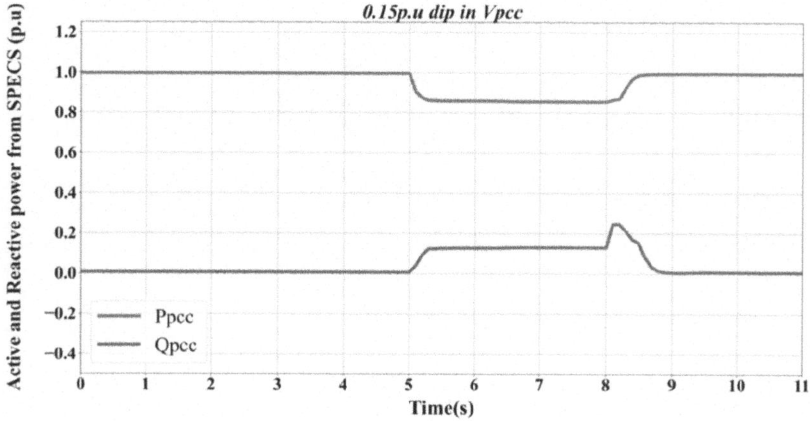

Figure 7.18 SPECS *P* and *Q* for LVRT: −0.15 p.u. dip in Vpcc

7.7.4 Case 4: Load/generation change leading to voltage variations

Voltage shoot-up from 1 p.u. to 1.2 p.u. The PCC voltage and the SPECS response to the same is shown in Figures 7.21 and 7.22.

7.7.5 Case 5: Load/generation change leading to frequency variations (*f* is lesser than *f*rated)

The case study for overall load is higher than generation and so the frequency of system is reducing. The active power support delivered by SPECS

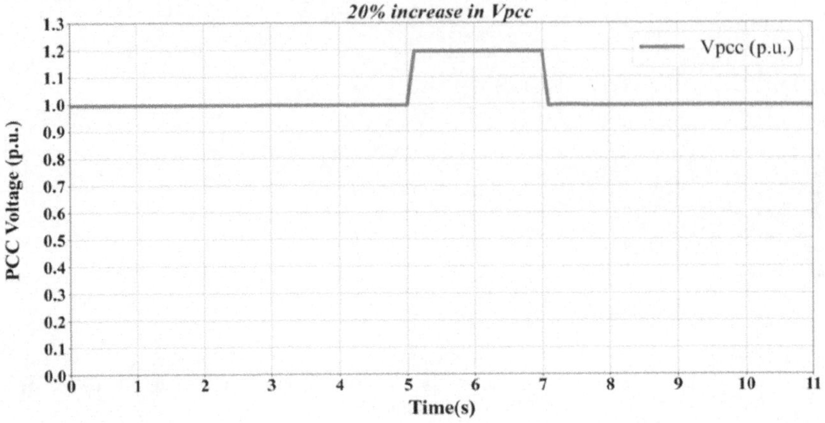

Figure 7.19 HVRT voltage at PCC: 1 p.u. to 1.2 p.u.

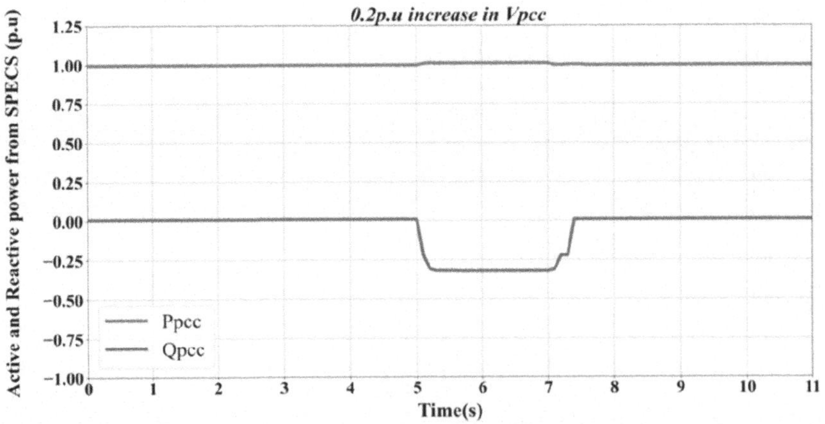

Figure 7.20 SPECS *P* and *Q* for HVRT: −0.2 p.u. shoot-up in Vpcc.

is limited by their *Pmax* in producer and consumer zone. The PCC frequency and the SPECS response to the same is shown in Figures 7.23 and 7.24.

7.7.6 Case 6: Load/generation change leading to frequency variations (*f* is greater than *f*rated)

The case study for overall load is lesser than generation and so the frequency of system is increasing. The active power support delivered by SPECS is

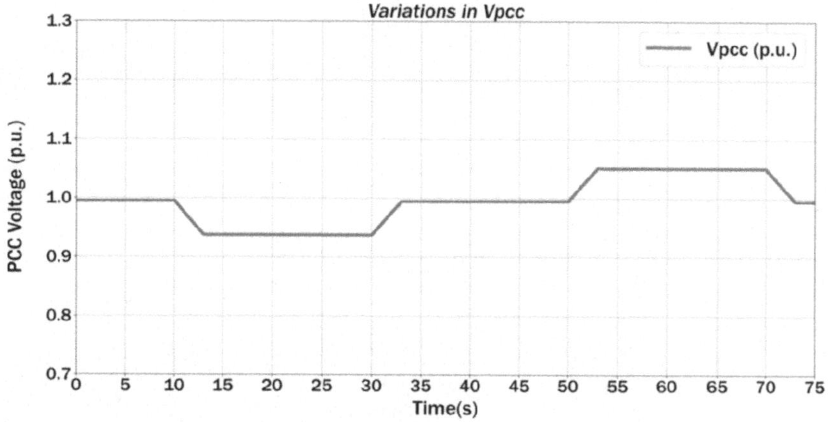

Figure 7.21 PCC voltage variations due to loads/generation changes.

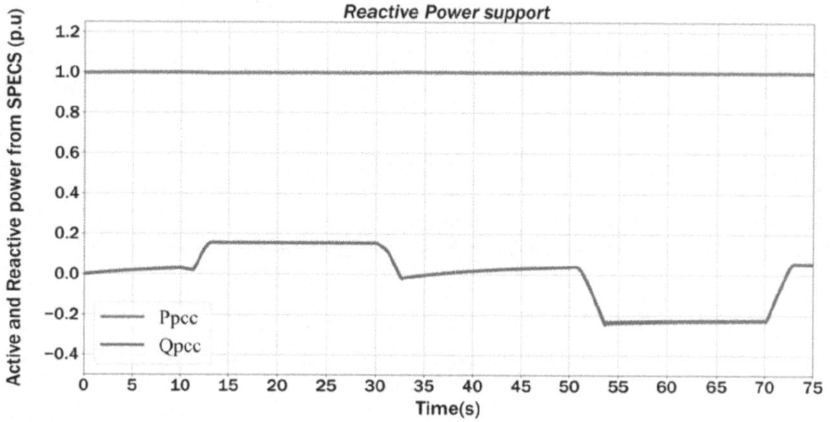

Figure 7.22 SPECS *P* and *Q* for variations in Vpcc.

limited by their *Pmax* in producer and consumer zone. The PCC frequency and the SPECS response to the same is shown in Figures 7.25 and 7.26.

These results visualize the multi-functional smart prosumer converter operation for various cases and following grid connectivity standards and codes for smart control. The scope of this can be expanded in multiple domains like the following:

1. Controller performance evaluation (PI, proportional resonant, hysteresis, sliding mode control, deadbeat, MPC, H-∞, LQR),
2. Converter topology (4-leg converter, multilevel converter),

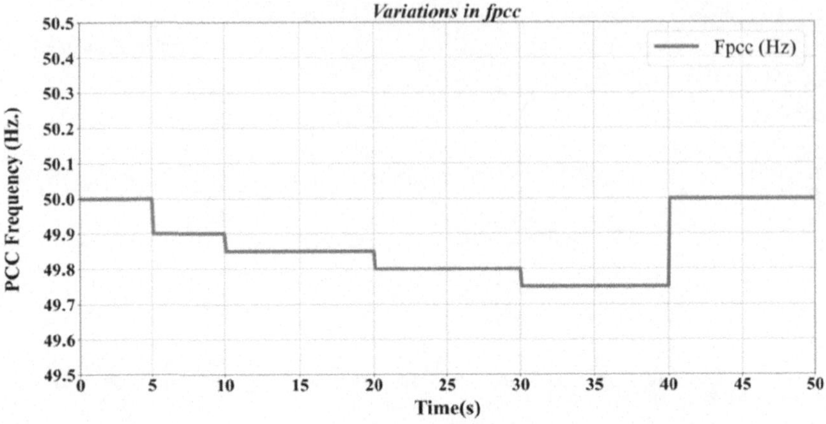

Figure 7.23 PCC frequency variations due to loads/generation changes.

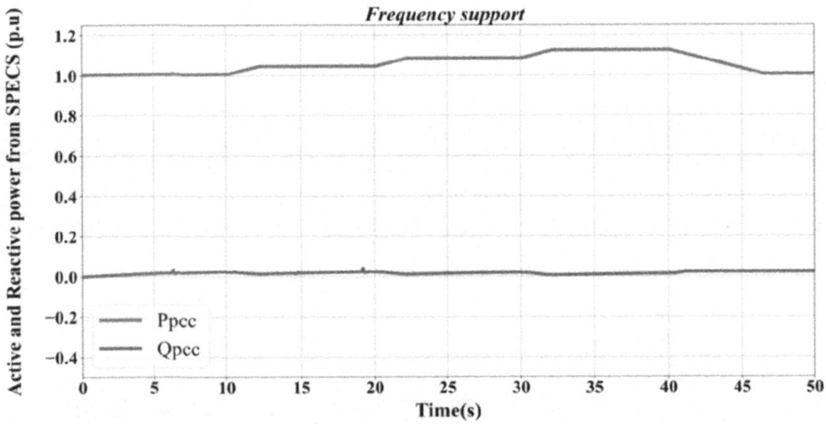

Figure 7.24 SPECS *P* and *Q* for variations in fpcc.

3. Fast frequency response (rate of change of frequency-based control, Pinj within 2s for events),
4. Weak grid operation, islanded/synchronising operation,
5. PLL-less control algorithms for highly dynamic systems,
6. Fully DRER-based smart grids.

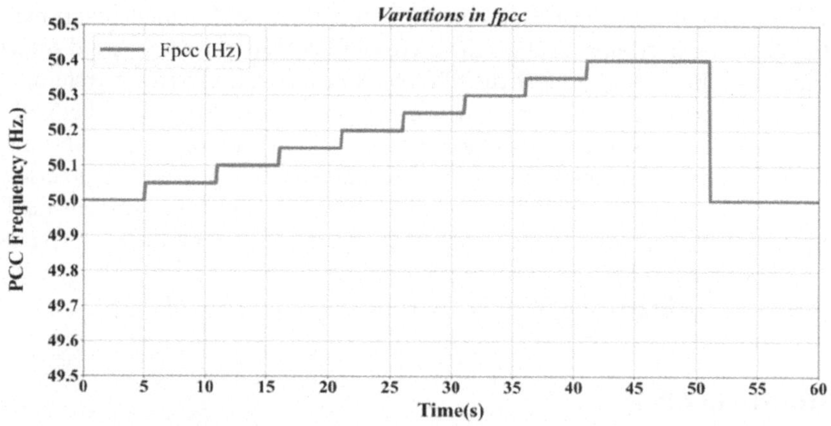

Figure 7.25 PCC frequency variations due to loads/generation changes.

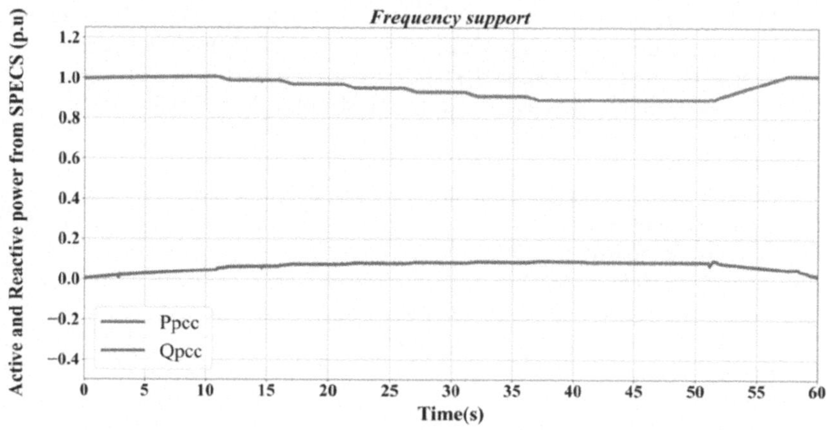

Figure 7.26 SPECS *P* and *Q* for variations in fpcc.

7.8 CONCLUSION

The SPECS have become a valuable solution for many of the current problems in the electrical sector, and the growth of SPECS is going to be ramping up due to its overall cost reducing and the impressive flexibility. The ubiquitous SPECS will require a good control for its satisfactory operation, and with the present electrical grids transforming into smart grids, the SPECS will also need a multi-functional converter control for the prosumers. The new guidelines from the worldwide standards and the country codes depict the various changes needed for facilitating a myriad of modifications that are beneficial to both the prosumers and the grid. These are implemented in the various control loops, and scope for improvement of the current controllers used is also discussed. The SPECS control with added functionalities is a pragmatic approach to the increasing power converter-based smart grids, and this would bolster the SPECS inclusion in the existing system.

The performance of the prosumer SPECS smart converter control corroborates the benefits in the fast-changing generation and loading conditions. With the use of artificial intelligence techniques, it is possible to precisely forecast and control the SPECS as well as the loads. The futuristic self-healing smart grid requires multi-functional smart converters, and it converts the challenges in fully renewable powered grids into opportunities for the prosumers, indirectly motivating the widespread usage of SPECS.

BIBLIOGRAPHY

1. Renewables 2022: Global Status Report (GSR), Jun. 2022, [online] Available: http://www.ren21.net/.J. Clerk Maxwell, *A Treatise on Electricity and Magnetism*, 3rd ed., vol. 2. Oxford: Clarendon, 1892, pp. 68–73.
2. IRENA, *Renewable Power Generation Costs in 2021.* International Renewable Energy Agency, 2022, [online] Available: https://irena.org/.
3. D. Palchak et al., "Greening the Grid: Pathways to Integrate 175 Gigawatts of Renewable Energy into India's Electric Grid, vol. 1, National Study," National Renewable Energy Laboratory, Technical Report, 2017.
4. Khare Saxena, A.; Saxena, S.; and Sudhakar, K. Solar energy policy of India: An overview," *CSEE J. Power Energy Syst.*, 2020, 1, 1–32.
5. IEEE Recommended Practice for Utility Interface of Photovoltaic (PV) Systems, Standard 929-2000, 2000.
6. International Electrotechnical Commission (IEC), "Characteristics of the Utility Interface for Photovoltaic (PV) Systems," *IEC*, vol. 61727, 2004.
7. "IEEE Standard for Interconnection and Interoperability of Distributed Energy Resources with Associated Electric Power Systems Interfaces," IEEE Standard, 1547–2018 (Revision of IEEE Std 1547-2003), 2018.
8. "IEEE Standard for Harmonic Control in Electric Power Systems," IEEE Standard 519–2022 (Revision of IEEE Std 519-2014), pp. 1–31, 5 Aug. 2022.
9. IRENA, "Grid Codes for Renewable Powered Systems," *International Renewable Energy Agency*, 2022, [online] Available: https://irena.org/.
10. F. Blaabjerg, Z. Chen and S. B. Kjaer, "Power Electronics as Efficient Interface in Dispersed Power Generation Systems," *IEEE Transactions on Power Electronics*, 2001.
11. M. P. Kazmierkowski, R. Krishnan and F. Blaabjerg, *Control in Power Electronics: Selected Problems.* Amsterdam, The Netherlands: Elsevier, 2002.
12. R. Teodorescu, M. Liserre and P. Rodriguez, *Grid Converters for Photovoltaic and Wind Power Systems.* New York: Wiley, 2011.
13. J. Holtz, "Pulsewidth Modulation-a Survey," *IEEE Transactions on Industrial Electronics*, vol. 39, no. 5, pp. 410–420, Oct. 1992.

14. L. Malesani and P. Tomasin, "PWM Current Control Techniques of Voltage Source Converters-a Survey," *International Conference on Industrial Electronics,Control, and Instrumentation*, vol. 2, pp. 670–675, 15–19 Nov. 1993.
15. M. P. Kazmierkowski and L. Malesani, "Current Control Techniques for Threephase Voltage-Source PWM Converters: A Survey," *IEEE Transactions on Industrial Electronics*, vol. 45, no. 5, pp. 691–703, Oct. 1998.
16. Timbus, M. Liserre, R. Teodorescu, P. Rodriguez and F. Blaabjerg, "Evaluation of Current Controllers for Distributed Power Generation Systems," *IEEE Transactions on Power Electronics*, vol. 24, no. 3, pp. 654–664, Mar. 2009.
17. Linder and R. Kennel, "Model predictive control for electrical drives," 2005 IEEE 36th Power Electronics Specialists Conference, pp. 1793–1799, Jun. 2005.
18. P. Cortes, M. P. Kazmierkowski, R. M. Kennel, D. E. Quevedo and J. Rodriguez, "Predictive control in power electronics and drives," *IEEE Transactions on Industrial Electronics*, vol. 55, no. 12, pp. 4312–4324, Dec. 2008.
19. Jose Rodriguez and Patricio Cortes, *Predictive Control of Power Converters and Electrical Drives*. Wiley and IEEE Press, 2012.
20. S. Vazquez et al., "Model Predictive Control: A Review of Its Applications in Power Electronics," *IEEE Industrial Electronics Magazine*, vol. 8, no. 1, pp. 16–31, Mar. 2014.
21. S. Kouro, M. A. Perez, J. Rodriguez, A. M. Llor and H. A. Young, "Model Predictive Control: MPC's Role in the Evolution of Power Electronics," *IEEE Industrial Electronics Magazine*, vol. 9, no. 4, pp. 8–21, Dec. 2015.
22. J. Holtz, "Advanced PWM and Predictive Control—An Overview," *IEEE Transactions on Industrial Electronics*, vol. 63, no. 6, pp. 3837–3844, June 2016.
23. R. Sreejith, Naran M. Pindoriya and Babji Srinivasan, "A Novel Control algorithm based DSTATCOM for Load Compensation," Journal of Physics: Conference Series, IOP, 2015.
24. T. Geyer, D. E. Quevedo, "Multistep Finite Control Set Model Predictive Control for Power Electronics," *IEEE Transactions on Power Electronics*, vol. 29, no. 12, pp. 6836–6846, Dec. 2014.
25. T. Sathiyanarayanan and S. Mishra, "Model Predictive Control of multi functional inverter for grid tied photovoltaic generators," 2016 IEEE 6th International Conference on Power Systems (ICPS), 2016.
26. Bouzid, A.M., Guerrero, J.M., Cheriti, A., Bouhamida, M., Sicard, P. and Benghanem, M., 2015. A survey on control of electric power distributed generation systems for microgrid applications. *Renewable and Sustainable Energy Reviews*, 44, 751–766.
27. Bevrani, H., François, B. and Ise, T., 2017. Microgrid dynamics and control. John Wiley & Sons.
28. T. Sathiyanarayanan and S. Mishra, "Performance of FCS-MPC based current controller for VSC based DG sources," 2018 20th National Power Systems Conference (NPSC), 2018.

29. "Grid Integration with High PV Penetration: Smart Inverter Standards and Uses, Coordination with Voltage Regulation Resources, and Operation in Low Inertia Power Systems," EPRI, Palo Alto, CA: 2018. 3002014260.
30. P. Michael and P. Joe, "Impact of IEEE 1547 Standard on Smart Inverters", IEEE PES Industry Technical Support Task Force Technical Report, 2018.
31. IEA-PVPS, "PV as an Ancillary Service Provider," 2021, [Online]. Available: www.iea-pvps.org.
32. C.-Y. Tang, Y.-T. Chen and Y.-M. Chen, "PV Power System with Multimode Operation and Low-Voltage Ride-through Capability," *IEEE Transactions on Industrial Electronics*, vol. 62, no. 12, pp. 7524–7533, Dec. 2015.
33. M. Talha, S. R. S. Raihan, N. A. Rahim, M. N. Akhtar, O. M. Butt and M. M. Hussain, "Multi-Functional PV Inverter With Low Voltage Ride-Through and Constant Power Output," *IEEE Access*, vol. 10, pp. 29567–29588, 2022.
34. Aldbaiat, B., Nour, M., Radwan, E. and Awada, E., 2022. Grid-connected PV system with reactive power management and an optimized SRF-PLL using genetic algorithm. *Energies*, 15(6), p.2177.
35. Parvez, M., Elias, M.F.M., Rahim, N.A. and Osman, N., 2016. Current control techniques for three-phase grid interconnection of renewable power generation systems: A review. *Solar Energy*, 135, 29–42.

Chapter 8

Inverters for photovoltaic generation

Novel features

Alexandre Hugo da Silveira, Fausto Bastos Líbano, Roberto Chouhy Leborgne, Maicon Jaderson Silveira Ramos, and Eric Lacerda Silva

In recent years, the world energy matrix, traditionally based on fossil fuels (nearly 80%), has experienced significant changes toward a cleaner source where renewable energy is becoming more prominent. Both thermal and hydroelectric power generation, which have always accounted for most of the electric power generated in the planet, have been losing space for wind and solar power generation, which need power electronic devices to be connected to the power grid.

An example is found in countries such as Germany, which intends to stop power generation from mineral coal until 2038, gradually replacing this type of power generation with sources such as wind power and photovoltaic power (PV), mainly. Another important player in this transition is Australia, which in 2010 had one of the largest carbon-based matrices in the world, and today has over one-third of its power generation based on renewable energy sources (RES). In that country, 20% of end users already have installed PV generation systems [1]. In Brazil, from 2019 to 2020 there was an increase in electric power generation from all sources (with greater representativity of solar and wind power), except for those produced with coal and fossil fuels [2].

Power loads have changed in the last decades. They are not linear and passive as they traditionally used to be. For instance, electric vehicle chargers and LED lighting generate great harmonic distortion and phase imbalance.

In face of this, a new challenge has arisen in power systems all over the world. How to add auxiliary functions to frequency inverters, so that they perform roles that are usually performed by synchronous generators? For example, delivery of reactive power, not only in regular operation but also under sustained or transient fault, work and operation, even with loads with non-sinusoidal currents, feeding heavily imbalanced three-phase loads, and more.

In this chapter, multifunctional inverters (MIs), developed by several researchers, will be presented. These inverters inject power from renewable sources and also improve the power quality (PQ) by mitigating harmonic distortions, correcting phase imbalances and power factor, among others.

DOI: 10.1201/9781003623168-11

8.1 STATE OF THE ART

In this section, some recent research regarding the development of MIs is discussed. Aiming at a faster response of the inverter control, in regard of network disturbances and changes in the system parameters due to the climate, Ray et al, in [3] used a series hybrid active filter (SEHAF) to a PV array and connected them to the system via an isolating transformer. This was characterized by having a passive filter for the 5th and 7th harmonics and for those above the 25th harmonic, and another active filter controlled by the Robust Extended Complex Kalman Filter (RECKF) technique [4] in series with the network and the load.

Then a dc-link was connected to the three-phase voltage source inverter (VSI). The RECKF technique aimed to generate reference voltages from the fundamental components of estimated voltage and current. Thus, the author compared the traditional PI and a Fuzzi Logic controller (FLC) to control the RECKF, showing that the proposed system is superior. Figure 8.1 shows the basic configuration of this system.

In [5], a control using the theory of instantaneous real and reactive powers *pq* was implemented in a three-phase four-wire system (Figure 8.2). Further details on the *pq* theory and the equations used can be seen in [5, 6]. Consequently, the system acts to correct harmonic distortions and current imbalances caused by unbalanced and non-linear loads. An Adaptive Hysteresis Band Controller (AHBC) is used to generate the inverter pulses by comparing the reference currents and measurements on the load. The choice for adaptive control was that the common hysteresis band controller, although simple, generates a non-constant and harmonic switching

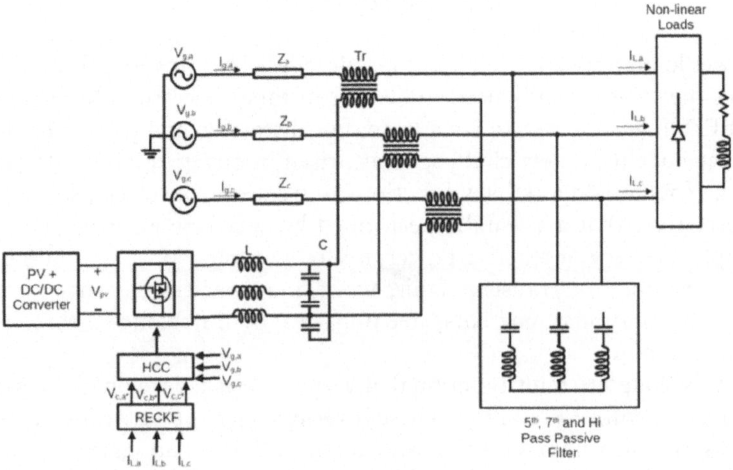

Figure 8.1 Diagram of the system proposed by [3].

frequency, which can be problematic for the network. The AHBC used in the work is presented in [7].

The system used by the cited author is similar to the one used in the present chapter. The use of AHBC makes the design of passive grid connection filters simpler. To make the MI control structure proposed in this chapter simpler, it was decided to use a conventional hysteresis controller.

In [8], the *pq* and *pqr* theories were assessed and compared in a PV (Figure 8.3). The traditional *pq* theory, proposed by [6], was developed for a three-phase system with three conductors, and later modified, including the zero sequence for an unbalanced system with four conductors. However, according to [8], the modified *pq* theory does not provide an independent control for the real, reactive, and zero sequence powers. In [9], the so-called *pqr* theory based on a three-dimensional rotational reference for a three-phase

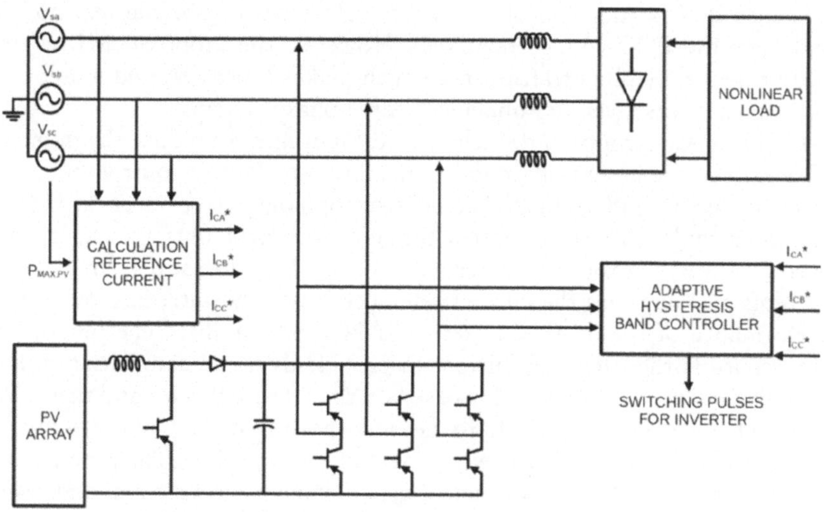

Figure 8.2 Diagram of the MI system proposed by [5].

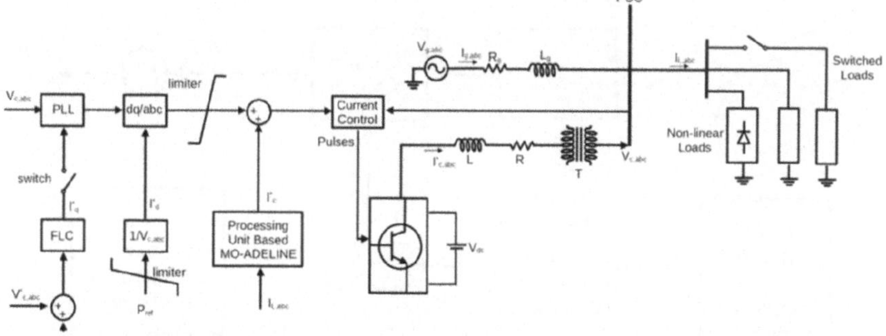

Figure 8.3 Diagram of the MI system proposed by [8].

system with four conductors, makes the three power components linearly independent from each other and allows the control of each one of them.

In the work [10], was proposed a new processing unit to estimate symmetrical components, which are used to improve PQ. This unit is based on a new structure called multi-output adaptive linear neuron (MO-ADELINE) with greater capacity to manage multi-output (MO) systems than the original (ADELINE) [11], since the latter is only suitable for managing single output systems. The study in question, a Fuzzi Logic Controller (FLC) was used in place of the conventional PI controller so that the system can follow the voltage reference at the point of common coupling (PCC). The performance of the proposed system was tested using the PSCAD/EMTDC simulator. The system topology can be seen in Figure 8.4. The control system based on MO-ADELINE receives the load current, composed by linear, non-linear, and susceptible to switching elements and generates a current reference that is added to other current references to provide real power and keep the PCC voltage regulated. However, the proposed MI by the author uses a coupling transformer which makes it heavier, and moreover, no emphasis was given to unbalanced three-phase systems.

In [12] it was proposed to include a Kalman filter to reduce the number of sensors in the design of the inverter. Thus, an adaptive control was used to estimate the grid voltage. The control topology can be seen in Figure 8.5 and includes three loops: capacitor current control, output current, and output power control. The outermost control loop measures three-phase currents and provides the current reference in steady state, improving the performance of the inverter in terms of PQ and stability. The innermost loop controls the capacitor current to provide dynamic compensation for grid disturbances. The control is based on the SRF system *dq* and therefore uses a phase-locked loop (PLL) to measure the ω frequency of the network which is also used for the current control of the capacitor. The power loop uses information from the estimated grid voltage to provide real and reactive power to the system.

In the work of [13], the improved linear sinusoidal tracer (ILST) [14] algorithm was used, where the reference currents were extracted from the load current measurements. The voltages were measured in the network,

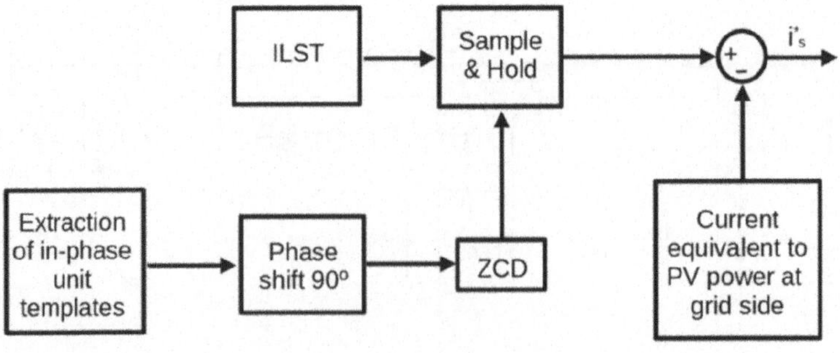

Figure 8.4 System and control proposed by [10].

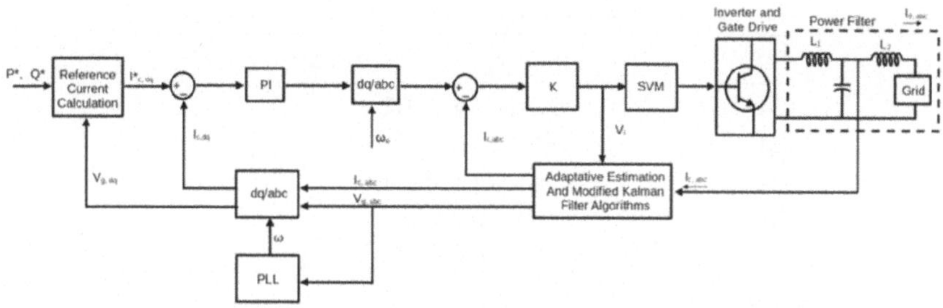

Figure 8.5 Control system proposed by [12].

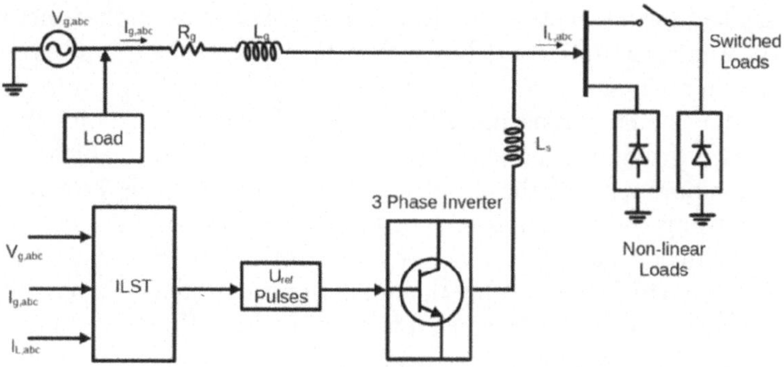

Figure 8.6 Diagram of the systems proposed by [13].

and the phase components were calculated in p.u. After that, the components were obtained. In this way, the system is able to compensate harmonic distortion and reactive power. Figure 8.6 shows the system used in the article. However, in this article, a simple L filter was used for grid connection and [13] also does not emphasize the case of unbalanced grids.

In [15] was presented a three-phase PV, connected to the grid. The modeling of the PWM converter is detailed, and the control strategy used is the instantaneous powers theory (*pq* theory) using the *dq*0 transform (SRF). The system runs as an active filter compensating harmonic components and reactive power generated by the local loads. An input voltage clamping technique, making the system operate close to the maximum power point (MPP) and delivering this energy from the PV system to the grid. The proposed system uses a Texas Instruments DSP model TMS320F2812 in its assembly. Tests on the prototype were performed with distinct types of loads. Simulation and experimental results are presented.

Table 8.1 shows the main models and control strategies found in the references described above.

It should be emphasized that all the works cited above have analyzed the behavior of PCC voltages and currents that used MIs. This behavior has not

Table 8.1 Summary of PV models and control strategies

Ref.	*Strat. inv. ctrl.*
[5]	*pq* theory AHBC
[3]	RECKF
[8]	*pq* and *pqr* theory
[10]	MO-ADELINE
[12]	Modif. Kalman F.
[13]	ILST
[15]	*pq* theory SRF

been studied for the external distribution network or when these networks are not running at an infinite bus where there is no voltage and frequency variation in the grid.

Regarding smart microgrids, a literature review conducted on important databases showed an exponential increase in publications, which indicate the contemporaneity of the problem. In addition, there is growing concern regarding the PQ improvement in islanded microgrid operations. A poor PQ affects the performance of a network either running in a bulk power system (BPS), concerning the quality of the product delivered by the power distribution utility or when it is running through islanded microgrid operation, regarding control systems, especially when energy resources are predominantly renewable and frequency inverters are used. These control systems can be affected when voltages and currents are highly harmonic distorted [16–18]. Thus, the use of equipment that integrates active filters with inverters for electricity generation (MI), presents itself as an intelligent solution to improve the performance of these networks. In [19] an active shunt filter based on the theory of instantaneous powers was used, also called the *pq* theory proposed by [6].

However, the active filter does not inject active power from an energy source such as PV for example, and there is also the fact that the microgrid model used in the article is not detailed. In [20] an active filter is also used for the mitigation of harmonics. Although this study uses a real microgrid model, it is a three-wire MV grid, and it lacks the proper focus on the integrated filter usage in a power source. On the other hand, [21] presents the use of a MI acting on a microgrid to improve PQ, same as [22] that brings computational analysis, as well as experimental results.

Although these papers detail the MI controls for microgrids operation, they do not present the microgrid model, and use three-phase balanced systems models, lacking to assess the imbalance loads and neutral currents, common in LV grids. On the other hand, in the study [23], the test model chosen was a three-phase 220 V line-to-line and 60 Hz LV distribution network. Several single-phase MI are connected along the network. Through master/slave communication, these inverters are controlled to minimize the imbalance between the phases. Operating modes were analyzed both with the LV network connected to the MV transformer as well as disconnected, functioning as an islanded microgrid and with asymmetric and distorted voltages.

Thus, there are rare or non-existent studies that analyze the performance of a PV running on a four-wire LV microgrid. Hence, this study represents an important contribution to the development of alternatives to reduce the impacts caused by low PQ in unbalanced four-wire LV networks. To simulate the LV distribution system, the four-wire commercial CIGRÉs network was implemented in MATLAB/Simulink software, where it was possible to observe the neutral current. The simulation model for a LV network capable of operating with unbalanced and non-linear loads is an important contribution of this chapter.

The models of power and control circuits used are seen in detail in Section 8.2. This system allows to assess the typical behavior of a LV network, which is still little explored in the literature, since most studies usually consider balanced and sinusoidal systems. In addition, few studies show the MI operating on a more complete network and its benefits across the grid. Currently, standardized indexes of PQ have been used to compare if results are in consonance with the limits considered. A multifunctional PV system requires the development of several components and control strategies. Such system is comprised of PV modules, dc-dc converters, frequency inverters with dc-link, and passive filter. Regarding a control system, it is necessary to develop strategies to find the maximum power point tracking (MPPT) through control of the dc-dc converter, and it is necessary to control the real power injection that can be made by the dc-link control voltage and injecting the maximum power available from PV modules. Moreover, regarding control, to mitigate harmonic contents, it is necessary to control the injection of equalization currents to balance the phases required by non-linear loads. The main components of such system are presented below.

8.2 BACKGROUND

This section will present the models of the main components, as well as some of the control techniques used in MIs.

8.2.1 Photovoltaic modules

The PV modules are built from the serial or parallel association of several units called PV cells. Each small PV cell can generate approximately between 1 and 2 Watts, and a commercial module has an average generation capability of up to 600 W. There are several types of modules that are

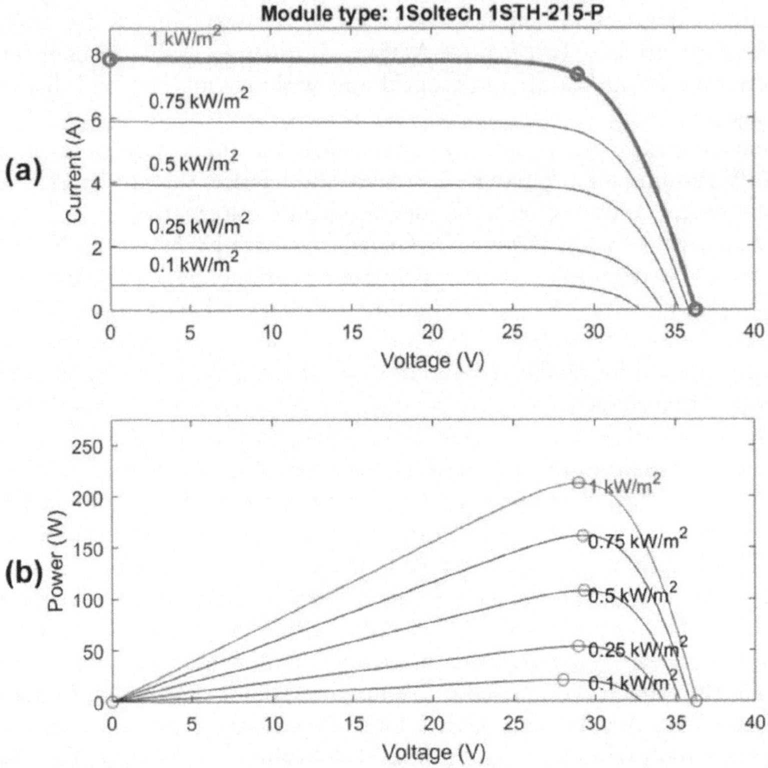

Figure 8.7 (a) Current versus voltage and (b) power versus voltage according to solar radiation for a PV module.

different in the manufacturing process and the physical aspect and energy conversion efficiency. The most used are those of single crystal Si and multicrystalline (also called polycrystalline), which account for 90% of the market. With much easier manufacturing process, the thin-film panels have lower cost, but also less efficiency. It is also worth mentioning that the amorphous silicon (a-Si), which typically has lower efficiency compared to single crystal and polycrystalline.

The output voltage of a PV module is usually around 30 V, with a current of some amperes and that is the reason modules are usually associated in arrays. These arrays may be connected in series or parallel, depending on the intended voltage and power in the dc-dc converter input. Figure 8.7(a) shows how current and voltage vary, depending on solar radiation, in a 215 Wp polycrystalline solar panel (model Soltech 1STH-215-P). With a fixed temperature of 25°C, it is noticeable that the current remains constant until it gets close to 30 V, and then it suffers a fast reduction. Figure 8.7b shows how the power available in the module and the voltage vary for the same value range of solar radiation. Here, it is possible to notice that, for a given voltage, there is a maximum power value for each range of solar radiation. Such power value is known as MPP. A detailed presentation of the various cell models and PV modules is presented in [24].

8.2.1.1 Mathematical model of photovoltaic modules

There has been abundant research on electric circuit models that represent the behavior of PV panels. Some authors, such as [25], used a simplified model composed of an ideal source and a diode connected in parallel with the source as well as a series resistor. Therefore, the output current of the cell will be represented by (8.1). The most used model is quite more complex, since it has a resistance in series and another in parallel [26] as shown in the Figure 8.8. In this case, the current that passes by the resistance in parallel I_p (8.2) must be considered and (8.1) gets one more term referring to this current, (8.3).

$$I = I_{pv} - I_d = I_{SC} - I_0 \cdot \left[e^{\frac{V+IR_s}{nV_t}} - 1 \right] \tag{8.1}$$

$$I = I_{SC} - I_d - I_p \tag{8.2}$$

$$I = I_{SC} - I_d = I_{SC} - I_0 \cdot \left[e^{\frac{V+IR_s}{nV_t}} - 1 \right] - \frac{V + IR_s}{R_p} \tag{8.3}$$

Where I_0 is the saturation current of the diode [A]; V is the terminal voltage of the module [V]; R_s and R_p are the resistances in series and in parallel, respectively, [Ω]; n is the ideality factor of the diode; V_t is the terminal voltage of the module [V], ($V_t = m(kT/q)$); k is the Boltzmann constant ($1.38 \cdot 10^{-23}$J/K); T is the temperature of the cell [K]; q is the Coulomb constant ($1.6 \cdot 10^{-19}$ C); I_d is the current of the diode. The current of the PV

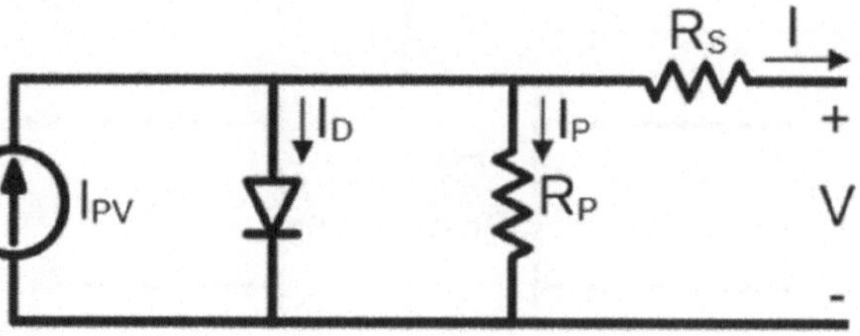

Figure 8.8 Equivalent circuit of PV module with R_s and R_p.

module I_{pv} depends on the irradiance S and on the temperature of the cell, as may be seen in (8.4) [27].

$$I_{pv} = I_{SC,ref}\left[\left(\frac{S}{1000}\right) + \frac{J}{100}\left(T - T_{ref}\right)\right] \tag{8.4}$$

Where $I_{SC,ref}$ is the current of reference short-circuit at the temperature (T_{ref}) and radiation (S_{ref}) reference.

8.2.2 DC-DC converter

Aiming at adjusting voltage and current in the PV array in the connection with the dc-link of the frequency inverter, a dc-dc converter is commonly used. Such converter may be defined as a system comprised of power semiconductor switches, as Mosfets and IGBTs, in addition to the association of inductors and capacitors aimed at controlling the power flow between an input and an output [28] (Figure 8.9), where V_s is the source voltage and V_o is the output voltage of the dc-dc converter. The control variable is known as duty-cycle (D), and when its value is between 0 and 1, it activates the power switches through a control loop. The most common types of dc-dc converters are buck converter(step-down), boost converter (step-up), buck-boost (step-down/step-up) converter, Cuk, and SEPIC. The buck converter is used to raise the input voltage, while the boost is used to lower such voltage. Buck-boost, Cuk, and SEPIC converters may be used for both raising and lowering such voltage in relation to the output. Table 8.2 shows the voltage gains between input and output in relation to the duty-cycle for the converter operating in continuous mode ([28]).

8.2.3 DC-AC converter

The DC-AC conversion stage of a photovoltaic inverter has a simple topology, with a low number of components, low cost, and high efficiency. Inverters can be configured as half-bridge or full-bridge (or H-bridge), with

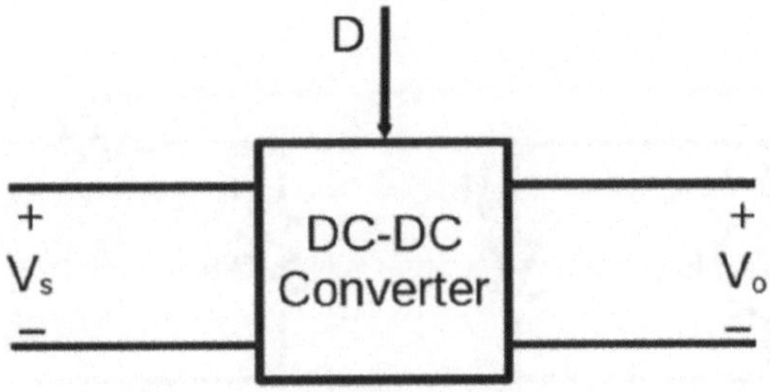

Figure 8.9 Simplified representation of a dc-dc converter.

Table 8.2 Voltage gain with duty-cycle relation

Converter	*Voltage gain*
Buck	D
Boost	1/(1 - D)
Buck-Boost	−D/(1 - D)
Cuk	−D/(1 - D)
SEPIC	D/(1 - D)

Source: [28].

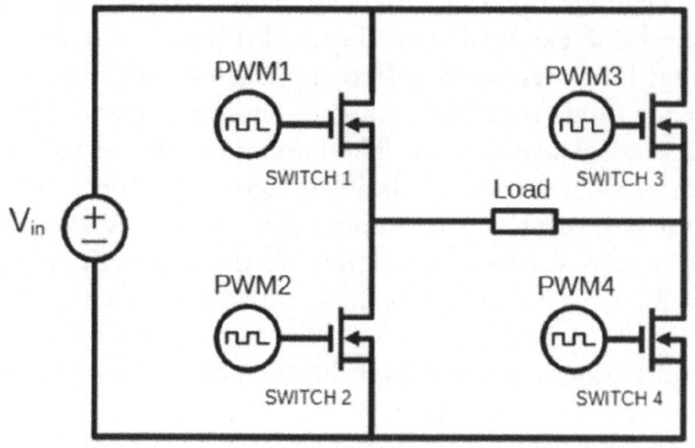

Figure 8.10 Single-phase H-bridge.

the latter being more commonly used due to its simplicity of implementation and high efficiency. This is also because in a half-bridge circuit, the input voltage needs to be twice the required amplitude, requiring a capacitive divider that serves as an intermediate voltage point between the capacitors and provides the reference for zero voltage, to enable the return path for load current [29].

Figure 8.10 shows an H-bridge for a single-phase inverter, where each semiconductor is controlled using pulse width modulation (PWM), and the effective output voltage value in one switching cycle is directly proportional to the modulation index and the amplitude of the DC bus voltage. Positive voltage (positive half-cycle) is applied to the load when switches 1 and 4 are turned on simultaneously, and reverse voltage (negative half-cycle) is applied to the load when switches 2 and 3 are turned on.

When it comes to three-phase inverters, they have a similar topology as shown in Figure 8.10, with the difference that they have three arms, one for each phase of the system.

Nowadays, many applications have been using multilevel inverters, which are called so because they provide multiple levels of output voltage, instead of the two levels (on and off) found in traditional inverters. One of their main advantages is the improvement in power quality, as the output voltage with multiple levels results in a smoother waveform closer to a pure sine wave, which reduces harmonic distortion.

Its disadvantage is the increased complexity of design, higher initial cost, and higher maintenance requirements due to the use of more advanced power electronic components, so the overall cost-benefit must be evaluated considering the specific needs of the application.

8.2.3.1 Multilevel inverters

There are three main topologies that can be used to implement a multilevel inverter. The diode clamped type (Figure 8.11) uses diodes and provides various voltage levels across the different phases to series-connected capacitor banks. Each diode transfers a limited amount of voltage, thus reducing stress on other electrical devices. The maximum output voltage is half of the DC input voltage, which is a disadvantage of this topology.

This type of inverter has main applications in static VAR compensation, variable speed drive for motors, interconnections in high voltage systems, and DC and AC transmission lines. Its main advantages are [30]:

- Entire phase share a common dc bus, which minimize the capacitance requirement of the converter;
- Capacitance of the capacitor is low;
- High efficiency at fundamental frequency.

And disadvantages:

- Practical limit on the number of levels, because it requires clamping diode;
- If the control and monitoring are not precise, DC level will discharge.

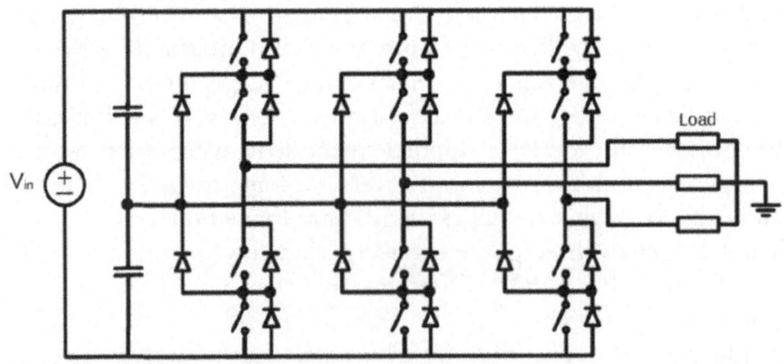

Figure 8.11 Diode clamped multilevel inverter.

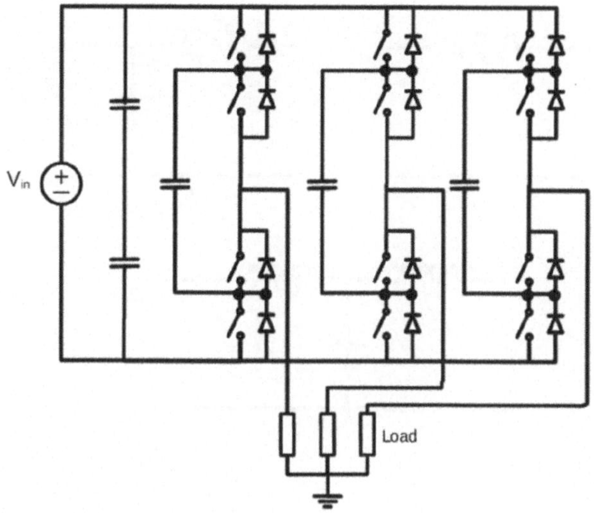

Figure 8.12 Flying capacitor multilevel inverter.

The flying capacitor model (Figure 8.12) is made through a series connection of switching cells clamped by capacitors. In this case, instead of diodes, capacitors limit the voltage across the circuit components. In this type of inverter, the switching states occur in the same way as in the inverter with fixed diodes, and as mentioned before, the output voltage is half of the DC input voltage. Its main applications are related to induction motor torque control and sinusoidal current rectifiers. Its main advantages are:

- Eliminates the clamping diode problems;
- Phase redundancies are available for balancing the voltage levels of the capacitors, real and reactive power flow controlled;
- The large number of capacitor enables the inverter to save from short duration outage and deep voltage sags.

And disadvantages:

- Complex start-up;
- Lower switching efficiency;
- Capacitors are expensive than diodes;
- Voltage control across all the capacitors is difficult.

8.2.4 Passive filter for grid-connected inverters

The main objective of a passive filter is filtering the wave signals of voltage and current that flow from the inverter toward the load. This filter must

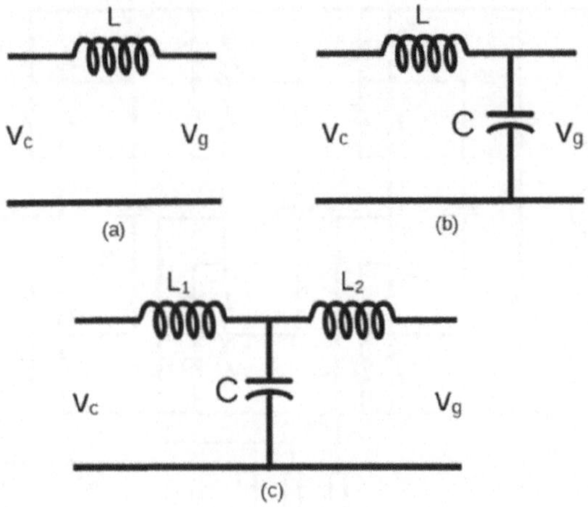

Figure 8.13 Circuits configuration for passive filter (a) L, (b) LC, and (c)LCL.

let the low frequency components pass and block those of high frequency, which derive mainly from the switching of the power semiconductors. According to [31], there are three main types of passive filters for frequency inverters, the L, LC, ad LCL types. The L type filter on Figure 8.13a is of first order, and it is capable of attenuating up to –20 dB/decade on all frequency ranges. This filter is commonly used in CCVSI-type converters.

The LC type filter (Figure 8.13b) is of the second order, and it attenuates -40db/decade, and it is used in VCVSI-type converters.

The third type of filter (third order), LCL (Figure 8.13c), has greater attenuation to switching frequencies of the inverter. It has an attenuation of -60 dB/decade and lower production of reactive power compared to the LC filter. Thus, like the L type, LCL type is commonly used in CCVSI-type inverters.

Due to having better attenuation, the LCL filter was chosen for this study and shall be better approached. The remaining types of filters may be known with more detail in [31]. In Figure 8.13, V_c and V_g are the inverter and the grid voltages.

8.2.4.1 LCL passive filter analysis

The design and analysis of the LCL filter, for the three-phase inverter, used in this study was based in the methodology proposed by [32]. The three-phase LCL filter with wye-connected capacitors and with a damping resistor in serial connection with the capacitor has the single-phase simplified model presented in Figure 8.14. The conductor's inductor resistance is not considered in this model. Where L_1 and L_2 are respectively the inductances of the input and output inductors, C is the capacitance of the capacitor, R

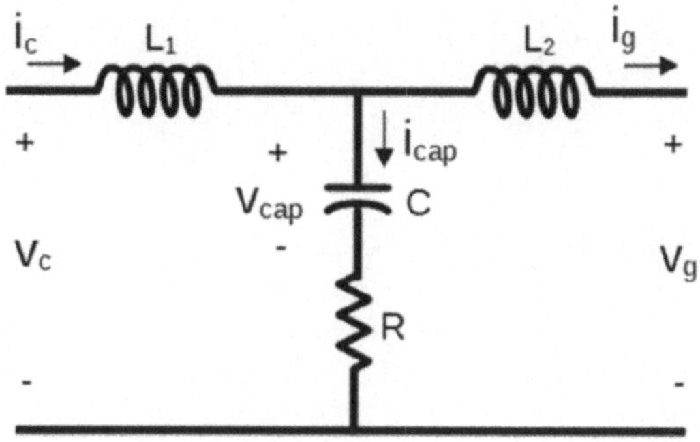

Figure 8.14 LCL filter per-phase model.

is the damping resistance of the LCL filter. The variables V_c and I_c are the converter voltages and currents, V_g and I_g are the grid voltages and currents, and V_{cap} and I_{cap} are the LCL filter capacitor voltage and current.

The equations of voltages and currents are obtained by applying the Kirchhoff laws and are represented by (8.5), (8.6), and (8.7) below.

$$\frac{di_c}{dt} = \frac{1}{L_1}\left[v_c - v_{cap} - \left(i_c - i_g\right)\right] \quad (8.5)$$

$$\frac{di_g}{dt} = \frac{1}{L_2}\left[-v_g - R\left(i_g - i_c\right) + v_{cap}\right] \quad (8.6)$$

$$\frac{dv_{cap}}{dt} = \frac{1}{C}\left(i_c - i_g\right) \quad (8.7)$$

By organizing the equations and rewriting in the form of state space, the result is (8.8).

$$\dot{x} = Ax + Bu \quad (8.8)$$

Where:

$$\dot{x} = \begin{bmatrix} \dot{i}_c \\ \dot{i}_g \\ \dot{v}_{cap} \end{bmatrix} \quad (8.9)$$

$$A = \begin{bmatrix} -\frac{R}{L_1} & \frac{R}{L_1} & \frac{1}{L_1} \\ -\frac{R}{L_1} & \frac{R}{L_1} & \frac{1}{L_1} \\ \frac{1}{C} & \frac{1}{C} & 0 \end{bmatrix} \tag{8.10}$$

$$x = \begin{bmatrix} i_c \\ i_g \\ v_{cap} \end{bmatrix} \tag{8.11}$$

$$B = \begin{bmatrix} \frac{1}{L_1} & 0 \\ 0 & -\frac{1}{L_2} \\ 0 & 0 \end{bmatrix} \tag{8.12}$$

$$u = \begin{bmatrix} v_c \\ v_g \\ 0 \end{bmatrix} \tag{8.13}$$

According to [32, 33], considering that the filter is connected to a grid with ideal sinusoidal voltage, it may be considered that, for high frequencies, the LCL filter output is in short-circuit, therefore $v_g = 0$. And the transfer function $G(s) = i_g / v_i$, shall remain as (8.14) below.

$$G(s) = \frac{i_g}{v_c} = \frac{(CR)s + 1}{(CL_1L_2)s^3 + (L_1 + L_2)(CR)s^2 + (L_1 + L_2)} \tag{8.14}$$

As an example, the following parameters are considered for the LCL filter: $L_1 = 1\,mH$, $L_2 = 0.1\,mH$, and $C = 4.973\,\mu H$. For such values, a Bode diagram was traced with the damping resistance varying from zero to $R = 10\,\Omega$, which can be seen in Figure 8.15. A proper value for this resistor is the key to combine the damping in the resonance frequency and the effectiveness of the LCL filter. It is noticeable that as the damping resistance value R increases, there is also attenuation in the signal caused in the resonance frequency. However, the effectiveness of the filter decreases as this resistance increases, because there is lower attenuation in high frequencies.

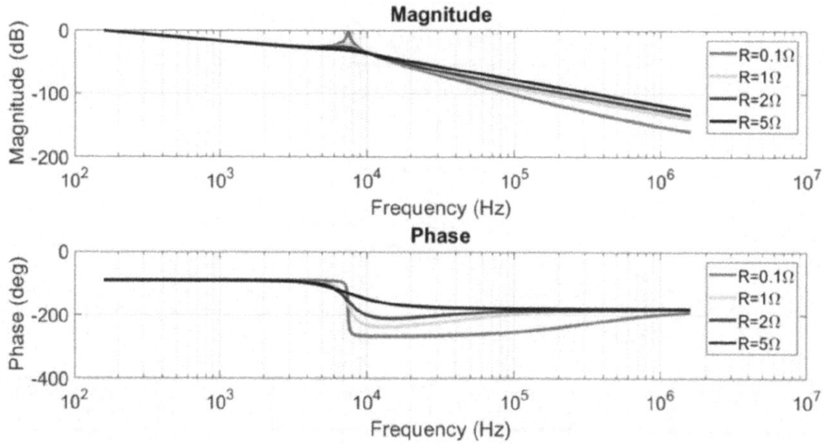

Figure 8.15 Bode diagram for damping resistance $R = 0.1\ \Omega$ (red), $R = 1\ \Omega$ (green), $R = 2\ \Omega$ (blue), and $R = 5\ \Omega$ (black).

8.2.5 Maximum power point tracker – MPPT

The maximum power point, as shown in Figure 8.7b, is a point in the voltage graphic *versus* power where the PV array supplies the highest electric power. And there are several control techniques aimed at varying the duty-cycle in the dc-dc converter (thus varying the terminal voltage of the panels), and, therefore, maximizing the generated power. Such methods are known as maximum power point tracker (MPPT). Among the most known techniques, it is worth mentioning incremental conductance (IncCond), perturb and observe (P&O), neural networks, and Fuzzy logic. The description and comparison of several of these techniques can be seen in [34]. As it is a technique that is widely applied and easy to implement, the P&O method shall be shown and used in this study.

8.2.5.1 The P&O technique applied to the boost converter

The P&O method seeks the maximum power based on the disturbance of the duty-cycle D and on the observation of the value obtained. Figure 8.16 shows the algorithm for implementation of this technique. At first, initial values are estimated for voltage $V(0)$ and current $I(0)$ and the actual values $V(k)$ and $I(k)$ are measured, where k is the iterations number. Power is calculated as the product of such values. With the product obtained, ΔP and ΔV are calculated subtracting the new calculated values from the old ones. In case the value of ΔP is positive, it means that there was a power increase, and it indicates that the algorithm is moving the voltage towards the MPP.

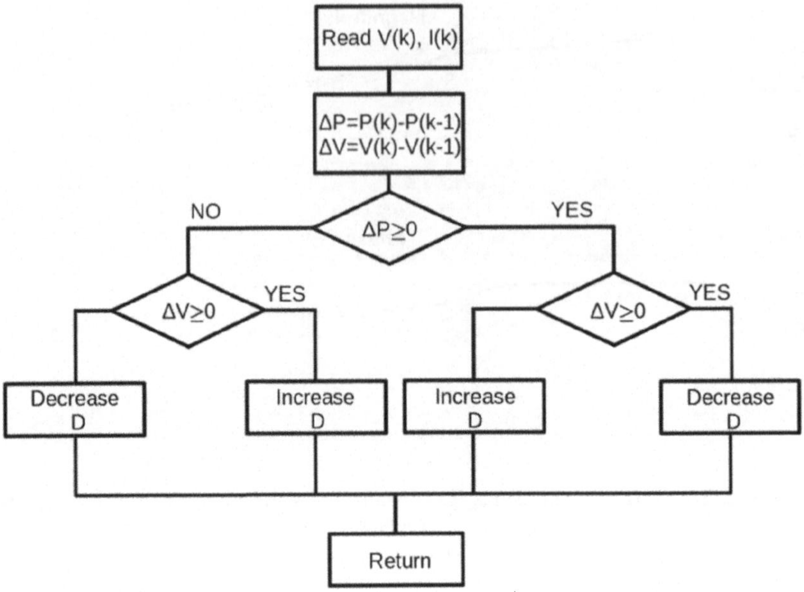

Figure 8.16 Algorithm for the P&T technique.

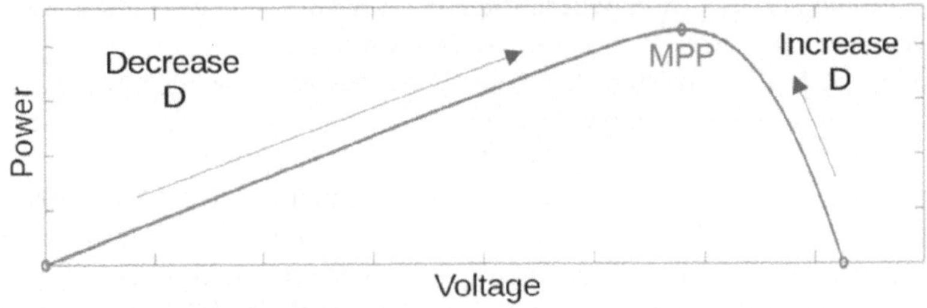

Figure 8.17 Duty-cycle increment and decrement directions for the MPP.

When ΔP is negative, it is a sign that the voltage is moving in the opposite direction of the MPP. As seen, in CCM, the relation between the output and input voltage in the boost converter is given by $V_o / V_i = 1/(1-D)$. Thus, when D increases, there is a decrease in the input voltage V_i.

Figure 8.17 shows that on the left side of MPP, it is necessary to increase the voltage in the terminals of the PV array by decreasing D and on the right side of MPP, it is necessary to do the opposite. After this process, new values of $V(k)$ and $I(k)$ will be obtained and the process is restarted. When the value of ΔP is within a tolerable margin of error, the MPP was reached, and the systems enters a steady state.

8.2.6 Phase-locked loop – PLL

The phase-locked loop (PLL) is a fundamental part of technologies such as frequency inverters, radio and wireless transmissions, and systems that require synchronization of a signal based on a reference. It is a negative feedback system that can be implemented either analogically or digitally, with the basic difference between the two modes being their components. While digital PLLs operate in discrete time and use digital filters, analog PLLs usually consist of a phase detector, a low-pass filter, and a voltage-controlled oscillator (VCO) via electronic circuits. The objective of this section is to present its operating principle and list the main techniques for application in frequency inverters [35].

The basic operating principle consists of continuously correcting the phase and/or frequency difference between the input and output signals of the loop. This occurs through the interaction among three main blocks: phase and frequency detector (PD/PEDD), low-pass filter (LPF), and voltage-controlled oscillator (VCO), as shown in Figure 8.18 [36].

The phase and frequency detector block provides an output voltage whose DC component is proportional to the phase and frequency difference between the input signal and the VCO signal, generating an error signal. The signal generated by the PD is then forwarded to the filter, where its DC component is extracted for use as a control signal for the oscillator. The filter is responsible for filtering the output of the PD and converting the current pulses into a continuous voltage value, which is used to control the VCO, i.e., it performs conversion while filtering. This block is also related to the stability of the feedback, attenuation of unwanted noises, and determination of the loop bandwidth, which influences the total noise of the PLL [35].

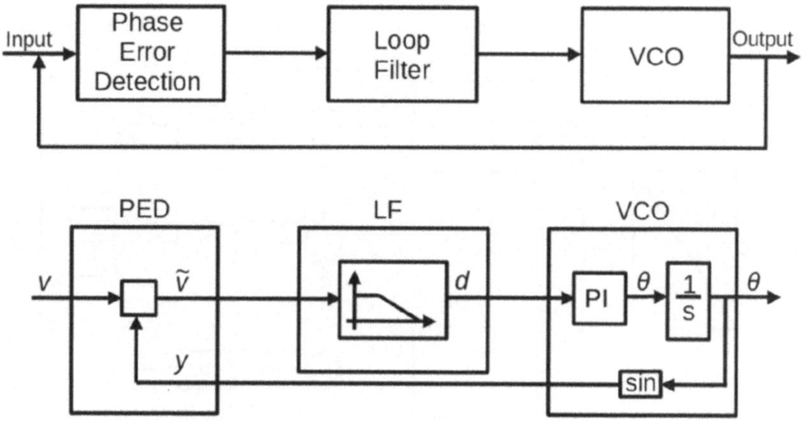

Figure 8.18 Control loop of a PLL.

The VCO is responsible for generating a signal whose frequency is dependent on the control voltage, and this signal is fed back into the loop as the input of the PD. As a result, it is observed that the frequency of the VCO output signal is a multiple of the frequency of the reference signal. In a situation where there is no input signal, the frequency of the output signal is determined solely by the VCO, based on its characteristics, and remains at a central value.

In summary, when an input is applied to the system, the PD compares the input with generated signals by the VCO. If they are different, an error proportional to this difference is generated, which passes through the filter and serves as the control voltage for the VCO, adjusting the frequency of the output signal from the VCO and bringing it closer to the frequency of the input signal. Once they are equal, the VCO locks or "captures" the input signal and the output of the PLL is a frequency-multiplied signal by a factor. A change in the frequency of the input signal will result in a new difference signal generated at the output of the PD, causing a change in the output voltage of the filter, which will then adapt the VCO to the new frequency.

Generally, these circuits are implemented digitally, where there are several possible topologies according to the requirements. When it comes to applications focused on multifunctional frequency inverters, two types stand out. The $\alpha\beta$-PLL has the main characteristic of using the Clarke transform for angular frequency detection, transforming the natural abc reference frame into the stationary $\alpha\beta$ reference frame. The block diagram shown in Figure 8.19 depicts the closed-loop diagram used to control the variation in θ in such a way that its value is zero, where θ is the angular velocity obtained by the PLL, and θ is the measured reference angular velocity in the power grid [37].

The dq-PLL (Figure 8.20) is very common in three-phase applications, and the main difference from the previous one is that the frequency detection is done using the Park transform, where the three-phase voltage vector is transformed from the natural abc reference frame to the rotating dq

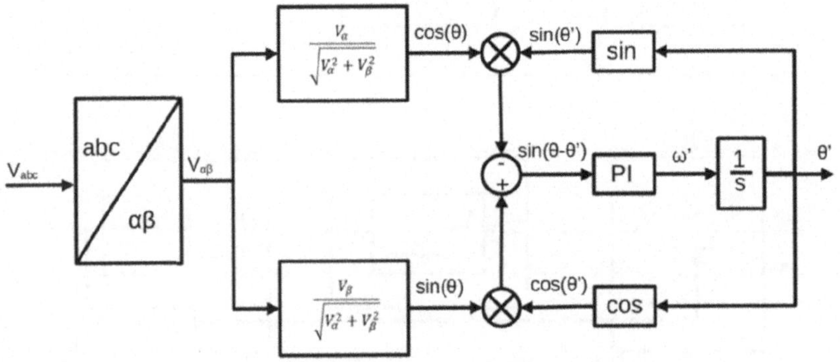

Figure 8.19 Block diagram of a $\alpha\beta$-PLL.

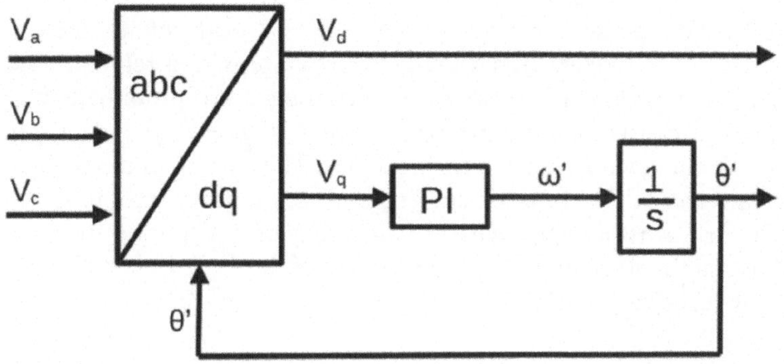

Figure 8.20 Block diagram of a *dq*-PLL.

reference frame. A feedback loop is used to regulate the *q* component to zero, controlling the angular velocity of the *dq* reference frame. The amplitude of the voltage vector is given by the *d* component, and the output of the feedback loop provides the grid angular velocity.

In balanced systems, this model is capable of eliminating steady-state error in phase and frequency tracking and achieving high bandwidth, which offers fast and accurate tracking performance. However, it is very sensitive to harmonics or voltage imbalances, and therefore, more sophisticated control schemes are needed in such situations.

8.2.7 CCVSI and VCVSI

The frequency inverter is the connection link between the solar source, which generates power in DC and the power grid, which usually operates in AC. When connected to the grid, the inverter has its voltage and frequency synchronized with the grid. It must operate supplying the grid with the maximum power that may be generated by the PV panels, under certain conditions of solar radiation and temperature. The inverters may be single-phase or three-phase.

According to [38], in a wider manner, inverters may be classified as voltage source inverters (VSI) or current source inverters (CSI). Regarding the control system, they may be classified as current controlled (CC) or voltage controlled (VC). The VSI is designed to supply power as a voltage source, keeping the voltage values fixed while the current changes. The CSI operates as a current source, keeping the current fixed and changing the voltage.

Although the VSI and CSI converters are remarkably similar constructively, there is a crucial difference in the connection between the DC side and the inverter bridge. Regarding VSI, the connection between the DC side and the inverter bridge is made by connecting the capacitors in parallel, while with CSI there is a serial connection of an inductor. Most inverters for use in PV power generation are the voltage source type. The voltage-controlled voltage source inverter (VCVSI) use grid voltage feedback in their control loop where it is intended to operate with a reference voltage. Nowadays, the VCVSI inverters are becoming more prominent with the development of islanded microgrids, where it is necessary to supply a voltage and frequency reference to the grid [39]. The current-controlled voltage source inverter (CCVSI) needs current feedback in its control loop.

Such current is adjusted with the injection of real power from the PV, which is usually done by the voltage control of the dc-link. It is also possible to supply reactive power to the grid through the same control loop.

8.2.8 Active filter for power quality enhancement

In recent years, with the advance of technological development of devices based in power electronics, active filters to correct harmonic distortions, especially for loads located in the same PCC have gained much space in the industry. When compared to conventional passive filters that traditionally are used for the mitigation of harmonics, the active filters (AFs) have better control and compensation adjustment of a wide range of the frequency spectrum, being automatically adjusted for load variations. Moreover, passive filters, due to their electric features, may get in resonance with the grid, which does not happen with the AFs. In case of passive filters, they are designed for certain load conditions. This conditions can be, changeling loads, both linear and non-linear, during the operation of this grid; for some situations, compensation may not be the best, letting pass certain harmonic components. In addition, the AFs may improve the PF in a dynamic manner, like steps, as in the case of modular capacitor banks. Furthermore, three-phase AFs may balance unbalanced loads, making an improved current distribution in phase conductors.

According to [40], AFs may be classified regarding type, topology, and phase number. They may be VSI or CSI type, series, or shunt, or they may be single-phase or three-phase. Figure 8.21 shows the topology of a shunt-type filter. Due to the simplicity and for not needing the use of transformers, this topology is usually the most applied. The active filter injects harmonic currents in opposite phase to those requested by non-linear loads, to reduce distortions in the current wave format. Figure 8.22 shows a series-type filter that uses a transformer to make the coupling between the AF and the grid, thus regulating the voltage in the connection. It is normally used when it is intended to decrease the negative sequence voltage [41] and regulate the voltage in three-phase systems [42]. Figure 8.23 presents a filter known as unified power quality conditioner (UPQC), which combines the connection of two types, series, and shunt. Both units share the same dc-link capacitor. As it combines both the function of current and voltage harmonic correction, it is considered an ideal AF [40]. It is used mainly in loads that are

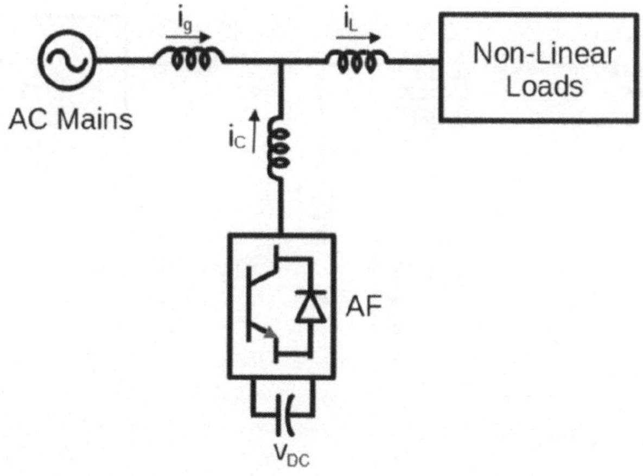

Figure 8.21 Shunt-type AF.

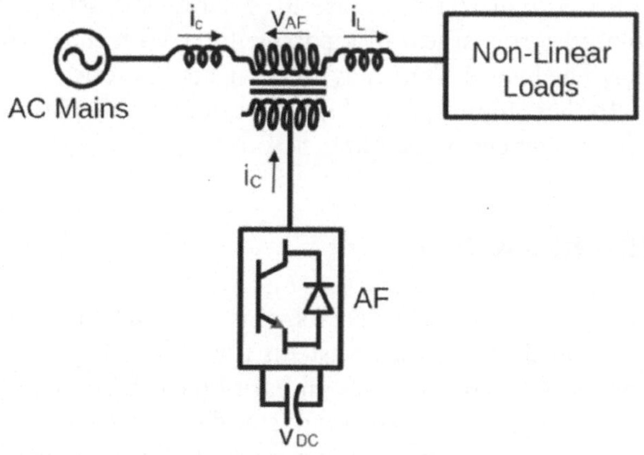

Figure 8.22 Series-type AF.

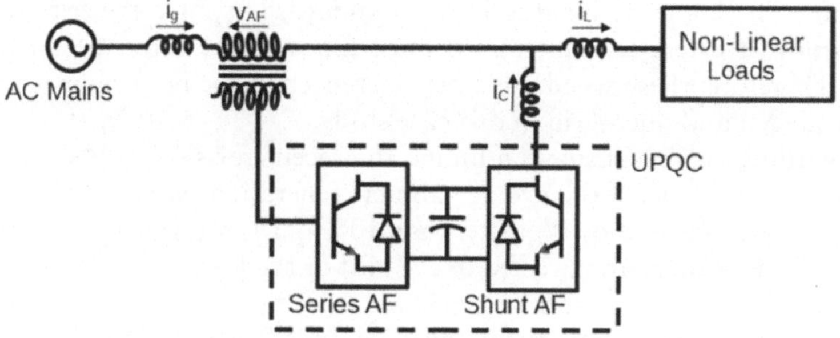

Figure 8.23 Unified power quality conditioner AF.

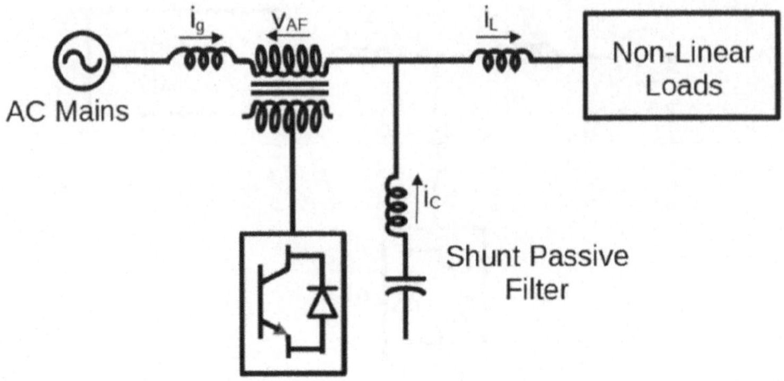

Figure 8.24 Hybrid-type AF.

overly sensitive to harmonic distortions, such as medical equipment. Due to the vast number of keys and greater complexity in control systems, it usually also has a higher cost. Figure 8.24 shows the setting of a hybrid filter that combines robustness of a passive filter with the performance of an active filter. The hybrid filters may be connected in series or parallel with the grid [43]. In these figures, i_g and i_L are both grid and load currents, and i_c is the capacitor current of the filter.

8.3 STANDARDIZATION

With the increasing adoption of DERs, it is essential to analyze and use PQ standards to ensure that these systems function effectively. Among the key standards is [44], which focuses on harmonic control in electric power systems. Additionally, recent revisions of standards, such as [45], have been introduced to establish compliance requirements for equipment that interconnects the network. Furthermore, [46] focuses on the interconnection and interoperability of inverter-based resources (IBRs) in electrical power systems. These international technical standards lay down the criteria that frequency inverters and microgrids must follow for a better performance [47]. Therefore, these standards were referenced in the performance analysis of the MI and microgrid in this case study.

Regarding current distortion limits, the recent revision of [44] presents a tree decision flow. If the average annual generation based on DERs and inverter-based generation is greater than 10% of the total energy demand in the grid, the readers are directed to the [45] or the [46].

Among the primary international standards regulating the connection of DERs in an electric power system, the [45] is worth mentioning. Where technical specifications of interconnection and interoperability tests between the electric power systems and the DERs are in focus. It includes general requirements, such as response to abnormal conditions, PQ, islanding, and other vital topics for the safe operation of DERs, such as FRT requirements.

The performance requirements of DERs under abnormal operating conditions is divided in three categories: namely I, II, and III.

According to [45], about abnormal operating performance:

- Category I is based on essential BPS stability and reliability needs and is achievable by all DER technologies that are in common use today.
- Category II covers all BPS stability and reliability needs and is coordinated with existing reliability standards to avoid tripping for a wider range of disturbances that concern BPS stability.
- Category III is based on both BPS stability and reliability and distribution system reliability and PQ needs and is coordinated with existing interconnection requirements for extremely high DER penetration.

Category I is less restrictive, on the other hand, category III is the most restrictive. Under abnormal conditions, the DER must keep its operation for extended periods to ensure system stability and reliability [45].

The parameters for cat. I systems are used in this study due to the characteristics of the LV circuit of the case study. According to the standard, the DERs must be designed to provide FRT capability during voltage sags, not exceeding their operational capabilities. The DER must operate indefinitely when the voltage remains between 0.88 and 1.10 p.u. It must keep operation for the time (T_{FRT}) given by (8.15) when the rms voltage sag (v_{sag}) is between 0.70 and 0.88 p.u. For voltage sags below 0.50 p.u., the DER may be disconnected instantaneously. The standard establishes minimum operation time during voltage swells, in such cases, the DER may disconnect in up to one cycle.

$$T_{FRT} = 0.7\ s + \frac{4\ s}{1\ p.u.}\left(V_{sag} - 0.7\ p.u.\right) \tag{8.15}$$

8.3.1 Limits of harmonic distortions

The [44] limits to 5% individual voltage harmonics and to 8% the THD (total harmonic distortion) for $V \leq 1.0$ kV. Referred here as THDv. In the case of currents, for rated voltages between 120 V and 69 kV, the standard limits harmonics according to the ratio between the short-circuit current (I_{SC}) and demand current (I_L) in the PCC. Table 8.3 shows the limits of harmonic distortions recommended by the standard for the case of $100 < I_{SC} / I_L < 1000$; values that were considered for this study [27].

Where TDD is the demand distortion rate that indicates harmonic distortions in percentage of the current of the maximum I_L demand. The individual current harmonic components (I_h) up to the 50th order, excluding

Table 8.3 Limits of harmonic distortions of total currents in % of the maximum demand current fundamental according to [44]

Order of harmonic currents of the current	$100 < \frac{I_{SC}}{I_L} < 1000$
$3 \leq h < 11$	12.0%
$11 \leq h < 17$	5.5%
$17 \leq h < 23$	5.0%
$23 \leq h < 35$	2.0%
$35 \leq h < 50$	1.0%
TDD	15.0%

the interharmonics, were considered for calculation purposes. TDD can be found using (8.16).

$$TDD = \frac{\sqrt{\sum_{h=2}^{50} I_h^2}}{I_L} \tag{8.16}$$

Measurement is obtained through statistic methods and may be conducted (very short time harmonic measurements – 3 s) with readings during the period of one day, considering a 99% percentile, or one week (short time harmonic measurements – 10 min), considering a 95% percentile. In other words, the magnitude is in conformity when 99% or 95% of the samples are below the limits set. Criteria to measure and obtain the samples used are based on [48], which defines methods for measurement and interpretation of results of quality parameters of electric power for alternate current systems at 50/60 Hz.

8.4 CHAPTER REMARKS

With the actual and irreversible transition from power sources mostly based on synchronous generators, which primary sources traditionally were coal and hydro power plants to more decentralized and cleaner power sources, with their indispensable power electronics devices, PQ related issues have arisen.

Likewise, electric loads have no longer been predominantly linear and passive, and today, loads such as LED lighting and electric vehicle chargers contribute to harmonics emission.

The purpose of this chapter was to show that by adding some auxiliary services to the frequency inverter (especially the PV), it may become an ally to improve the PQ.

This chapter presented the state of the art in the development of MIs, as well as some models of inverters with the function of active filters. All system components, such as PV arrays, DC-DC converter, inverter bridge, and control methods were shown in detail.

REFERENCES

1. N. Lal, T. Price, K. L., C. Wilson, F. Billimoria, T. Morrow, M. Garbutt, and D. Sharafi, "Essential system services reform," *IEEE, Ed.*, vol. 19, no. 5, pp. 16–19, 2021.
2. EPE, "Balanco energético nacional - BEN - Brazil," *Empresa de Pesquisas Energéticas*, pp. 1–268, 2021.
3. P. K. Ray, S. R. Das, and A. Mohanty, "Fuzzy-controller-designed-pv-based custom power device for power quality enhancement," *IEEE Transactions on Energy Conversion*, vol. 34, no. 1, pp. 405–414, 2019.
4. R. Panigrahi, P. C. Panda, and B. Subudhi, "A robust extended complex kalman filter and sliding-mode control based shunt active power filter," *Electric Power Components and Systems*, vol. 42, no. 5, pp. 520–532, 2014. https://doi.org/10.1080/15325008.2013.871609
5. S. Y. Mousazadeh, M. Savaghebi, A. Beirami, A. Jalilian, J. M. Guerrero, and C. Li, "Control of a multi-functional inverter for grid integration of pv and battery energy storage system," *2015 IEEE 10th International Symposium on Diagnostics for Electrical Machines, Power Electronics and Drives (SDEMPED)*, pp. 474–480, 2015.
6. H. Akagi, Y. Kanazawa, and A. Nabae, "Instantaneous reactive power compensators comprising switching devices without energy storage components," *IEEE Transactions on Industry Applications*, vol. IA-20, no. 3, pp. 625–630, 1984.
7. S. Y. Mosazadeh, S. H. Fathi, M. Hajizadeh, and A. R. Sheykholeslami, "Adaptive hysteresis band controlled grid connected pv system with active filter function," *2012 International Conference on Power Engineering and Renewable Energy (ICPERE)*, pp. 1–6, 2012.
8. R. R. Sawant and M. Chandorkar, "Methods for multi-functional converter control in three-phase four-wire systems," *Iet Power Electronics*, vol. 2, pp. 52–66, 2009.
9. K. Hyosung and H. Akagi, "The instantaneous power theory on the rotating p-q-r reference frames," *Proceedings of the IEEE 1999 International Conference on Power Electronics and Drive Systems. PEDS'99 (Cat. No.99TH8475)*, vol. 1, pp. 422–427, 1999.
10. M. I. Marei, E. F. El-Saadany, and M. M. A. Salamav, "A novel control algorithm for the dg interface to mitigate power quality problems," *IEEE Transactions on Power Delivery*, vol. 19, no. 3, pp. 1384–1392, 2004.

11. B. Widrow and M. A. Lehr, "30 years of adaptive neural networks: Perceptron, madaline, and backpropagation," *Proceedings of the IEEE*, vol. 78, no. 9, pp. 1415–1442, 1990.
12. K. H. Ahmed, A. M. Massoud, S. J. Finney, and B. W. Williams, "Sensorless current control of three-phase inverter-based distributed generation," *IEEE Transactions on Power Delivery*, vol. 24, no. 2, pp. 919–929, 2009.
13. L. S. Stanly, R. Divya, and M. G. Nair, "Grid connected solar photovoltaic system with shunt active filtering capability under transient load conditions," *2015 International Conference on Technological Advancements in Power and Energy (TAP Energy)*, pp. 345–350, 2015.
14. S. wu Du and J. ming Su, "Analysis of an improved harmonic currents detection method based on lst," *2011 2nd International Conference on Artificial Intelligence, Management Science and Electronic Commerce (AIMSEC)*, pp. 3755–3759, 2011.
15. M. F. Schonardie, "Sistema fotovoltaico trifásico conectado à rede com controle de potência ativa e reativa utilizando transformada dq0," *Universidade Federal de Santa Catarina - UFSC*, 2012.
16. V. Shalukho, A. I. Lipuzhin, and A. A. Voroshilov, "Power quality in microgrids with distributed generation," *2019 International Ural Conference on Electrical Power Engineering (UralCon)*, pp. 54–58, 2019.
17. K. DE BRANBANDERE, B. Bolsens, J. Van den Keybus, A. Woyte, J. Driesen, and R. Belmans, "A voltage and frequency droop control method for parallel inverters," *IEEE Transactions on Power Electronics*, vol. 22, no. 4, pp. 1107–1115, 2007.
18. K. Tang and C. Shen, "Harmonic emission level assessment method based on parameter identification analysis," *IET Generation, Transmission & Distribution*, vol. 13, no. 7, pp. 976–983, 2018.
19. M. A. Noroozi, T. M. Zarei, S. H. Fathi, and J. Milimonfared, "Enhanced shunt active power filter based on instantaneous power theory used in microgrids under non-ideal grid voltages," *2018 IEEE 7th International Conference on Power and Energy (PECon)*, pp. 360–365, 2018.
20. Z. Liu, X. Xu, and E. Makram, "Harmonics assessment and mitigation: A case study on an unbalanced stand-alone microgrid integrated with pv," *2015 North American Power Symposium (NAPS)*, pp. 1–6, 2015.
21. S. Varghese, A. Sebastian, and L. S. Lalu, "Power quality compensation in microgrid using mfgti," *2018 International Conference on Control, Power, Communication and Computing Technologies (ICCPCCT)*, pp. 478–482, 2018.
22. Z. Zeng, H. Yang, S. Tang, and R. Zhao, "Objective-oriented power quality compensation of multifunctional grid-tied inverters and its application in microgrids," *IEEE Transactions on Power Electronics*, vol. 30, no. 3, pp. 1255–1265, 2015.
23. D. I. BRANDÃO, T. Caldognetto, F. P. Marafão, M. G. Simões, J. A. T. Pomilio, and P., "Centralized control of distributed single-phase inverters arbitrarily connected to three-phase four-wire microgrids," *IEEE Transactions on Smart Grid*, vol. 8, no. 1, pp. 437–446, 2017.

24. Luque and S. Hegedus, *Handbook of Photovoltaic Science and Engineering.* Wiley, 2011. [Online]. Available: https://books.google.com.br/books?id=sLMkCsde1u4C
25. S.-K. Kim, J.-H. Jeon, C.-H. Cho, E.-S. Kim, and J.-B. Ahn, "Modeling and simulation of a grid-connected pv generation system for electromagnetic transient analysis," *Solar Energy*, vol. 83, no. 5, pp. 664–678, 2009. [Online]. Available: https://www.sciencedirect.com/science/article/pii/S0038092X08002843
26. F. CUPERTINO, R. M. Domingos, H. A. Pereira, W. U. Costa, and S. R. Silva, "Inversores multifuncionais para sistemas fotovoltaicos conectados a rede elétrica," *V Congresso Brasileiro de Energia Solar*, 2014.
27. H. Silveira, F. Libano, R. Leborgne, and M. Ramos, "Aplicacão de inversores multifuncionais usados em recursos energéticos distribuídos para melhoria da qualidade de energia em rede de distribuição secundária," *2020 Congresso Basileiro de Automática (CBA)*, p. 12, 2020.
28. D. Martins and I. Barbi, *Eletronica de potência Conversores CC-CC básicos não isolados.* Edição dos Autores, 2006. [Online]. Available: https://books.google.com.br/books?id=v6WVOgAACAAJ
29. D. Hart, *Eletrônica de Potência: Análise e Projetos de Circuitos.* McGraw Hill Brasil, 2016. [Online]. Available: https://books.google.com.br/books?id=-b5TDQAAQBAJ
30. Singh and R. Singh, *A Comparative Study of Multilevel Inverter Topologies.* IRJET, 2018. [Online]. Available: https://www.irjet.net/archives/V5/i3/IRJET-V5I3230.pdf
31. K. Ahmed, S. Finney, and B. Williams, "Passive filter design for three-phase inverter interfacing in distributed generation," *2007 Compatibility in Power Electronics*, pp. 1–9, 2007.
32. Reznik, M. G. Simões, A. Al-Durra, and S. M. Muyeen, "Lcl filter design and performance analysis for grid-interconnected systems," *IEEE Transactions on Industry Applications*, vol. 50, no. 2, pp. 1225–1232, 2014.
33. M. Liserre, F. Blaabjerg, and S. Hansen, "Design and control of an lcl-filter-based three-phase active rectifier," *IEEE Transactions on Industry Applications*, vol. 41, no. 5, pp. 1281–1291, 2005.
34. T. Esram and P. L. Chapman, "Comparison of photovoltaic array maximum power point tracking techniques," *IEEE Transactions on Energy Conversion*, vol. 22, no. 2, pp. 439–449, 2007.
35. T. Hornik and Q. Zhong, *Control of Power Inverters in Renewable Energy and Smart Grid Integration*, ser. IEEE Press. Wiley, 2012. [Online]. Available: https://books.google.com.br/books?id=m5kWmDIiuxQC
36. S. Nithya, C. R. Balamurugan, and S. P. Natarajan, "Design and analysis of new multilevel inverter topology with induction motor load," in *International Conference on Information Communication and Embedded Systems (ICICES2014)*, 2014, pp. 1–5.
37. S. Sumathi, L. Kumar, and P. Surekha, *Solar PV and Wind Energy Conversion Systems: An Introduction to Theory, Modeling with MATLAB/SIMULINK, and the Role of Soft Computing Techniques*, ser. Green Energy and Technology. Springer International Publishing, 2016. [Online]. Available: https://books.google.com.br/books?id=GY5QvgAACAAJ

38. M. Rashid, *Power Electronics Handbook*. Elsevier Science, 2011. [Online]. Available: https://books.google.com.br/books?id=eS1z95mzi28C
39. H. Silveira, F. B. Libano, R. C. Leborgne, and M. J. S. Ramos, "A novel application of multifunctional inverters to enhance power quality of smart microgrids: an analysis on a low voltage and four-wire grid," *Energy Systems*, March 2021. https://doi.org/10.1007/s12667-021-00434-y
40. Singh, K. Al-Haddad, and A. Chandra, "A review of active filters for power quality improvement," *IEEE Transactions on Industrial Electronics*, vol. 46, no. 5, pp. 960–971, 1999.
41. Campos, G. Joos, P. Ziogas, and J. Lindsay, "Analysis and design of a series voltage unbalance compensator based on a three-phase vsi operating with unbalanced switching functions," *IEEE Transactions on Power Electronics*, vol. 9, no. 3, pp. 269–274, 1994.
42. F. B. Libano, S. L. Muller, R. A. Marques Braga, J. V. Rossoni Nunes, O. S. Mano, and I. A. Paranhos, "Simplified control of the series active power filter for voltage conditioning," *2006 IEEE International Symposium on Industrial Electronics*, vol. 3, pp. 1706–1711, 2006.
43. F. Libano, J. Cobos, and J. Uceda, "Simplified control strategy for hybrid active filters," *PESC97. Record 28th Annual IEEE Power Electronics Specialists Conference.*, vol. 2, pp. 1102–1108, 1997.
44. IEEE-Std519, "Ieee standard for harmonic control in electric power systems," *IEEE Std 519–2022 (Revision of IEEE Std 519–2014)*, pp. 1–31, 2022.
45. IEEE-Std1547, "Ieee standard for interconnection and interoperability of distributed energy resources with associated electric power systems interfaces," *IEEE Std 1547–2018 (Revision of IEEE Std 1547–2003)*, pp. 1–138, 2018.
46. IEEE-Std2800, "Ieee standard for interconnection and interoperability of inverter-based resources (ibrs) interconnecting with associated transmission electric power systems," *IEEE Std 2800–2022*, pp. 1–180, 2022.
47. H. Silveira, F. B. Libano, R. C. Leborgne, and M. J. S. Ramos, "Evaluation of multifunctional pv inverters operation during faults," *2022 IEEE Electrical Power and Energy Conference (EPEC)*, pp. 296–301, 2022.
48. IEC 61000-4-30, "Electromagnetic compatibility (emc) - part 4-30: Testing and measurement techniques - power quality measurement methods," *IEC 61000-4-30:2015*, pp. 1–232, 2015.

Chapter 9

Multifunctional inverters connected to LV microgrids

Cases study

Alexandre Hugo da Silveira, Fausto Bastos Líbano, Roberto Chouhy Leborgne, Maicon Jaderson Silveira Ramos, and Eric Lacerda Silva

Practical applications on microgrids have evolved a lot in the recent years. With a significant amount of research, new and revised standards have been published, mainly in relation to networks with high penetration of distributed energy resources (DERs).

Although isolated grids have been used for several decades in remote communities – typically using fossil fuel resources – with the high penetration of DERs, microgrid operation has provided essential benefits for grids. One advantage is that microgrids can increase the reliability and supply of energy in distribution systems: in this case, the microgrid can operate both synchronized and isolated when part of the feeder fails.

However, along with this technological advancement, several new challenges, both in network protection and operation, are obstacles to the wide application of such systems. For example, load sharing, power quality (PQ), and renewable energy intermittency. Some researchers have developed proposals for multifunctional frequency inverters (MI) using instantaneous power theory and other harmonic mitigation techniques [1], but few works have evaluated the behavior of their MIs in more complete models of distribution networks.

Thus, new functionalities become important, such as reactive power control and the incorporation of active harmonic filters, for example. The main difference between an MI and an active power filter is that the MI is capable of not only improving PQ (power quality) but also generating energy through a primary source connected to its dc-link. Therefore, the main functionalities of an MI are to contribute to the improvement of PQ in an electrical network and also to generate energy through a DC source. There is no significant difference regarding the constructive aspects of an MI and a conventional frequency inverter. The main difference is in its control system that combines more advanced techniques to generate reference signals capable of compensating harmonic distortions, current imbalance, and low power factor.

In the previous chapter, the main components of a frequency inverter were presented, both for traditional generation and for generation that offers auxiliary services. In this chapter, a case study is performed, with the

DOI: 10.1201/9781003623168-12

application of more than one inverter of this type, in a microgrid that operates both connected to the main system and temporarily isolated.

A model of a multifunctional inverter (MI) based on PV, which, in addition to generating electric power, works as an active filter of current harmonics for local non-linear loads, is presented in this chapter. Such MI also has the capacity to correct phase imbalance and power factor.

Aiming at carrying out a detailed assessment of the behavior of MIs and of its contribution when they are connected to grids with major penetration of RES based in power electronics, a case study is presented, simulating the connection of two MIs in one LV grid.

This study shall present both the operation of this grid connected to the MV grid of the utility as well as when it is operating as an island microgrid. This grid has linear and non-linear loads, balanced and imbalanced loads, and a frequency inverter that operates sometimes as the main source during island microgrid operation.

9.1 NOVEL FUNCTIONALITIES OF A FREQUENCY INVERTER

Conventionally, a frequency inverter for use in PV, wind power, battery, fuel cell generation, etc., transforms a direct current (DC) voltage into an alternating current (AC) voltage of adjustable frequency and magnitude. Its main function is to change the voltage signal generated from a primary source, into voltages and frequencies that are usual in distribution and consumption. Thus, the inverter can deliver electric power to the system or to the load. Considering the new challenges of power systems, which migrate from synchronous machine central generation and linear passive loads to converter-based DER and non-linear loads, adding new functionalities to power converters is relevant. This study proposes a PV model that, in addition to generating power jointly with a PV array, also contributes to improve the PQ by compensating harmonic components of the current, correction of PF of local loads and line current balance.

9.1.1 Instantaneous power theory or *PQ* theory

In this study, the instantaneous power theory, also known as the *pq* theory, was employed to determine and generate the harmonic components at the load, which are generated by the inverter connected in parallel with the load. From the current readings of the non-linear load, it is possible to compensate for these distortions and keep the current from the grid sinusoidal. This method is usually applied for active filter control, and it enables the calculation, through Clarke transformation, of the instantaneous active and reactive powers. Based on such voltages (9.1) and currents (9.2) from the original system abc (angles in radians), it is possible to find the voltages and currents in the system $0\alpha\beta$, applying (9.3); and the instantaneous active (p) and reactive (q) powers applying (9.4), (9.5), and (9.6). Note that the presented system includes the zero sequence components (p_0, i_0 and v_0), which also allows its use for imbalanced systems.

$$\begin{aligned} v_a &= V_a \cdot cos(\omega t) \\ v_b &= V_b \cdot cos(\omega t - 2\pi / 3) \\ v_c &= V_c \cdot cos(\omega t + 2\pi / 3) \end{aligned} \tag{9.1}$$

$$\begin{aligned} i_a &= I_a \cdot cos(\omega t + \phi_a) \\ i_b &= I_b \cdot cos(\omega t - 2\pi / 3 + \phi_b) \\ i_c &= I_c \cdot cos(\omega t + 2\pi / 3 + \phi_c) \end{aligned} \tag{9.2}$$

$$\begin{bmatrix} x_0 \\ x_\alpha \\ x_\beta \end{bmatrix} = \sqrt{\frac{2}{3}} \cdot \begin{bmatrix} \frac{1}{\sqrt{2}} & \frac{1}{\sqrt{2}} & \frac{1}{\sqrt{2}} \\ 1 & -\frac{1}{2} & -\frac{1}{2} \\ 0 & \frac{\sqrt{3}}{2} & -\frac{\sqrt{3}}{2} \end{bmatrix} \begin{bmatrix} x_a \\ x_b \\ x_c \end{bmatrix} \tag{9.3}$$

$$p_0 = v_0 \cdot i_0 \tag{9.4}$$

$$p = v_\alpha \cdot i_\alpha + v_\beta \cdot i_\beta \tag{9.5}$$

$$q = v_\alpha \cdot i_\beta - v_\beta \cdot i_\alpha \tag{9.6}$$

Where x_0, x_α, and x_β can be both voltages ($v_{0\alpha\beta}$) and currents ($i_{0\alpha\beta}$) on the 0, α and β axis and x_a, x_b and x_c can be the voltages (v_{abc} in volts) or currents (i_{abc} in amperes) on the axes a, b and c. According to the pq theory, instantaneous active and reactive powers can be decomposed into two parts: an average real power ($\bar{p}$ in Watts) and an average reactive power ($\bar{q}$ in var), and an oscillating real power ($\tilde{p}$) and an oscillating reactive power ($\tilde{q}$), according to (9.7) and (9.8).

$$p = \bar{p} + \tilde{p} \tag{9.7}$$

$$q = \bar{q} + \tilde{q} \tag{9.8}$$

Knowing the p and q powers of the system, you can select the oscillating powers and take them as a reference by calculating the reference currents using (9.9) [2]:

$$\begin{bmatrix} \tilde{I}_{\alpha}^{*} \\ \tilde{I}_{\beta}^{*} \end{bmatrix} = \frac{1}{v_{\alpha}^{2} + v_{\beta}^{2}} \cdot \begin{bmatrix} v_{\alpha} & v_{\beta} \\ v_{\beta} & -v_{\alpha} \end{bmatrix} \begin{bmatrix} \tilde{p}^{*} \\ \tilde{q}^{*} \end{bmatrix} \tag{9.9}$$

9.1.2 Control system of the synchronous referential CCVSI for injection of active power

The main functionality of a frequency inverter for PV solar generation is to deliver to the grid the maximum active power generated by the PV array. In possession of the concepts of pq theory and Park transformation [3], it is possible, using the synchronous reference frame (SRF) to control the active power injected by PV based on the MPPT reference by adjusting the dc-link voltage of the inverter. The generation system, starting from the DC source, is designed to inject balanced currents into the grid. Any unbalanced current components are handled separately by the active filter control. Therefore, the zero-sequence component can be disregarded, which simplifies the system of equations. The transformation of the abc system to the synchronous system is given by (9.10) and (9.11) below [3]:

$$K = \frac{2}{3} \cdot \begin{bmatrix} cos(\omega t) & cos\left(\omega t - \frac{2\pi}{3}\right) & cos\left(\omega t + \frac{2\pi}{3}\right) \\ sin(\omega t) & sin\left(\omega t - \frac{2\pi}{3}\right) & sin\left(\omega t + \frac{2\pi}{3}\right) \end{bmatrix} \tag{9.10}$$

$$\begin{bmatrix} x_q \\ x_d \end{bmatrix} = K \cdot \begin{bmatrix} x_a \\ x_b \\ x_c \end{bmatrix} \tag{9.11}$$

The angular velocity can be found by $\omega = d\theta / dt$. For implementations, it is more common to use the variable θ that can be obtained using a PLL. Where θ is the lag angle of the reference axis that rotates synchronously with the angular velocity of the ω grid. In this case, simply replace ωt with θ in (9.11) [4]. It is now possible to calculate instantaneous powers in the dq system by (9.12) and (9.13), which defines the p active power and q reactive.

$$p = \frac{3}{2} \cdot \left(v_q \cdot i_q + v_d \cdot i_d \right) \tag{9.12}$$

$$q = \frac{3}{2} \cdot \left(v_q \cdot i_d - v_d \cdot i_q \right) \tag{9.13}$$

In a balanced feeding system, if V_a is taken as reference $V_d = 0$, isolating the currents and taking them as a reference, (9.14) and (9.15) are taken.

$$i_q^* = \frac{2}{3} \frac{P^*}{v_q} \tag{9.14}$$

$$i_d^* = \frac{2}{3} \frac{Q^*}{v_q} \tag{9.15}$$

9.1.3 Active filtering, PF increase, and current imbalance compensation

Based on the *pq* theory, it is possible to use VSI to compensate harmonic distortions caused by non-linear loads. As seen in the previous subsection, the real, and reactive powers may be decomposed in two components, a continuous one and an oscillating one (9.7) and (9.8). Therefore, the presence of harmonics in the current makes the signal of instantaneous real power more oscillating. As an example, Figure 9.1 shows the real power absorbed by a non-linear load. It is possible to perceive, in blue, an oscillating behavior of the instantaneous real power (p), which is the sum of the oscillating component, plus the average real power. The red curve shows

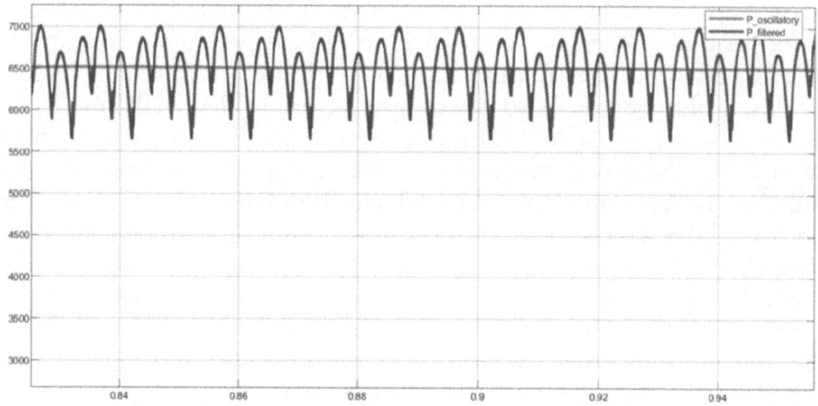

Figure 9.1 Waveforms of the oscillatory instantaneous power (before passing through the filter) and average (after the filter).

the average real power ($\bar{p}$) after the passage by the low-pass filter of the Figure 9.1.

The MI can also correct the PF of the load by injection of reactive power. In this case, the MI will generate the same power consumed by the load. Looking at the Figure 9.5, it is possible to see that the instantaneous reactive power, which is calculated based on the reading of the load current $I_{L,abc}$ is sent directly to the inverse transformation block. Furthermore, another great contribution of MI is current balancing in the phases. With the calculation of zero sequence currents, and, subsequently, insertion of this information in the inverse transformation, the active filter can compensate such currents, so that the system is balanced in the common connection point, between the source, the load, and the MI.

9.2 METHODOLOGY

This work proposes a methodology aimed at modeling, simulating, and building a prototype and a test microgrid for multifunctional photovoltaic inverters, in order to enable the evaluation of the performance of these inverters in power microgrids. Using simulation software, the various components of the photovoltaic system, such as panels, inverters, and batteries, as well as the low-voltage electrical network and linear and non-linear loads, will be modeled and evaluated. A case study will be conducted with simulations in two different scenarios. Performance evaluation will be based on indicators of electrical PQ, according to international standards.

To do so, a thorough research on theoretical fundamentals and the state of the art regarding MIs and power supply and quality control in microgrids is initially conducted.

Subsequently, the MI, the battery-powered VCVSI inverter, and a LV grid are modeled. The control of active filter and power injection from the PV is conducted from the MI model. The operating mode connected to the grid, where the grid references are read with a PLL, is programmed from the VCVSI. This system can run in parallel with the grid or in standby mode. For this, it must change its controls for an occasion when bulk power system (BPS) is not available. In this new control model, active and reactive power is calculated with methods of voltage and frequency drops. In grid modeling, model simulations were used which consider mutual impedances and that enable analysis at four lines. Subsequently, all these methods are grouped, forming a power microgrid that is more complex and closer to the practical one, with generation from the conventional grid and frequency inverters, as well as linear and non-linear loads, balanced or not.

In the case study, three different simulation scenarios are created, one with the grid connected to the BPS, other considering the grid operating as an island microgrid. At the end, some PQ indicators described in international standards are reviewed. The p.u. values shown in the graphs below are based on the power of the transformer, that is, 75kVA and rms network voltages, 400V.

9.3 PROPOSED MODELS

In this chapter, the modeling of the components of the inverters (MI-CCVSI and VCVSI) is presented.

9.3.1 Model of the multifunction inverter (MI-CCVSI) and its components

The proposed MI model is presented in Figure 9.3 where, from a PV array that is connected to the dc-link through a boost converter and, later, to an LCL filter, can deliver power to the grid and correct the harmonic distortions caused by the non-linear load connected to the same PCC. The boost converter is controlled from an MPPT that reads the voltage (V_{PV}) and the current (I_{PV}) of the PV array. A PLL extracts the angular velocity (ω_e) of the grid that is used in active power control, which regulates the voltage of the

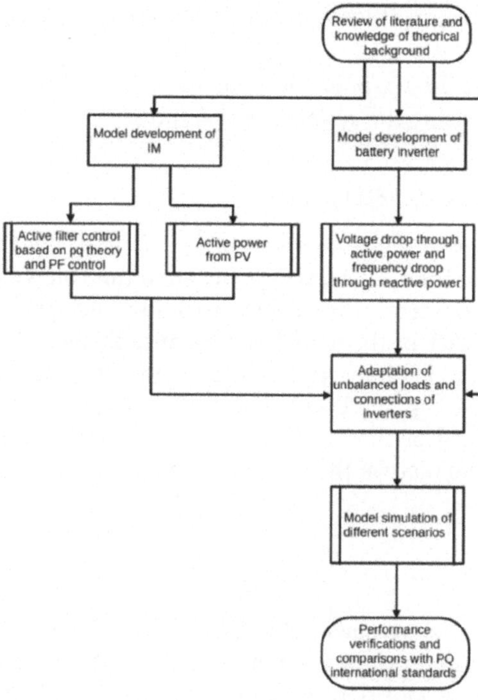

Figure 9.2 Flowchart of the proposed methodology.

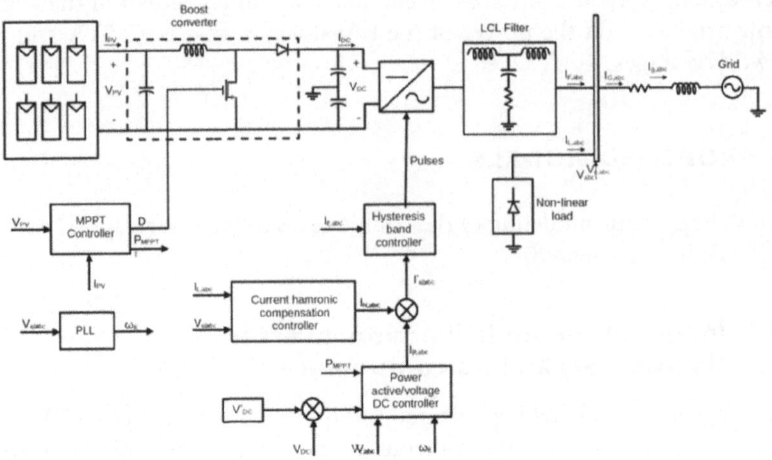

Figure 9.3 Proposed multifunctional inverter model.

dc-link (V_{DC}); this power control generates the reference current $I_{P,abc}$. On the other hand, compensating currents ($I_{h,abc}$) are calculated with the use of the *pq* theory, and the input variables are the voltage ($V_{c,abc}$) in the PCC and the current ($I_{L,abc}$) of the non-linear load. Finally, a hysteresis band controller is used to generate the current references ($I^{*}_{c,abc} = I_{h,abc} + I_{P,abc}$) and send them to the IGBTs gates in the six-pulse three-phase inverter bridge. The models of each block comprising the system are then presented.

9.3.1.1 DC-DC converter with P&O MPPT

As seen in the previous chapter, there are different sets of dc-dc converters. Because the output voltage of the PV array is lower than the voltage of the dc-link of the MI, a boost converter was used in this study. Such choice was made also due to the simplicity of the model and because this is one of the most widely used converters in commercial and research applications. The implemented MPPT method was the P&O. The MPPT output is the duty cycle that is sent to a PWM generator and then, to the Mosfet gate of the boost converter. Thus, the voltage on the PV array is adjusted to obtain the MPP.

9.3.1.2 CCVSI with LCL filter

The frequency inverter in the developed MI is CCVSI type. Reference currents are obtained applying the *pq* theory, where it is possible to generate harmonic currents in phase opposition to those requested by the non-linear load, as well as to compensate the reactive power using instantaneous

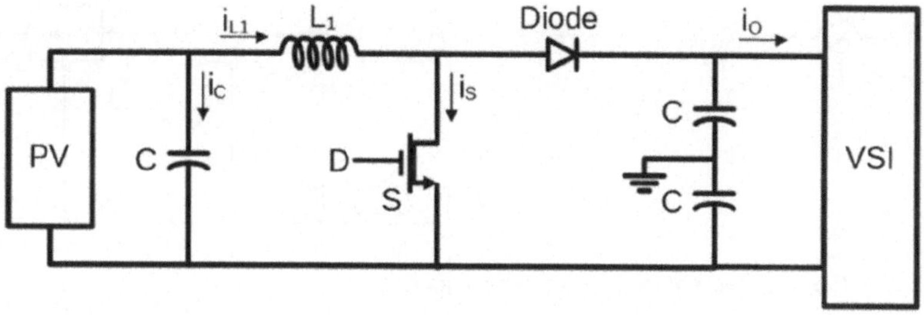

Figure 9.4 Boost converter model connected to a VSI.

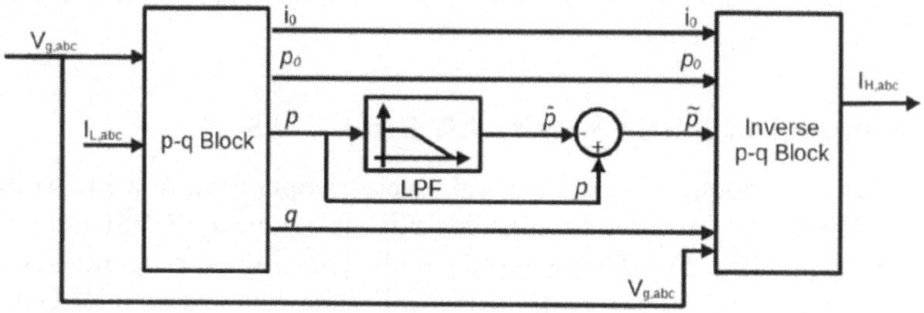

Figure 9.5 Current harmonic compensation controller block.

reactive power information of the load and compensate the phase imbalance from the zero sequence currents.

Figure 9.5 shows the current harmonic compensation controller block, which contains the implementation of the active filter. The pq block has the transformation equation of the system abc coordinates to the $\alpha\beta 0$ and of the instantaneous powers p, q, and 0. Subsequently, these powers pass by a filter, remaining only the oscillating component, which, shall be sent to the inverse block pq. In this last block, the inverse transformation shall be made, thus generating the reference currents I_r^* again in the abc coordinates.

Active power control and injection is performed based on the equations seen in section 1.1.2. The power reference for the inverter will be $P = P_c + P_{dc}$, where P_{dc} is the power measured before the capacitor, which in the case of this work will be the maximum power of the PV array and P_c will be the active power reference obtained from voltage regulation with the use of a PI controller.

In the control proposed by [5], where the v^2 magnitude that is the dc-squared link voltage, is controlled by a feed-forward scheme that aims to improve the system response. This practice is common because the W energy stored in the capacitor is proportional to the square of the voltage $W(t) = (1/2)Cv^2(t)$.

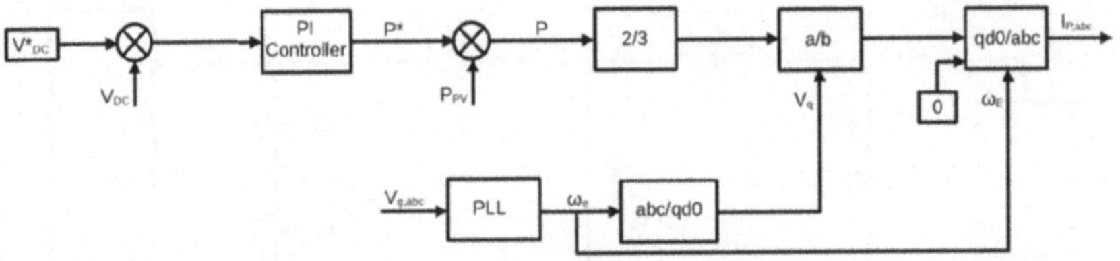

Figure 9.6 Control system of active power injection.

In this work, by mathematical simplification, V_{DC} voltage was used directly, instead of using the same squared one and the control implemented can be seen in Figure 9.6.

9.3.2 Model of a VCVSI with LC filter

Nowadays it is common to see several inverters operating in a low-voltage feeder. Therefore, in order to consider this situation, a VCVSI frequency inverter controlled by voltage is modeled. This inverter operates when the microgrid is connected to the BPS and when it is islanded. When the microgrid is islanded, this inverter provides voltage and frequency references. For that reason, this specific inverter has no other functionalities than delivering active power to the grid, with frequency and voltage equal to the BPS.

In this system, the inverter obtains the references of voltage and current of the grid through a PLL, when it is connected to the BPS. If there is a need to operate islanded, it will generate its references by itself. When VCVSI operates in the islanded microgrid, the references are obtained by the voltage and frequency drop control technique [4, 6]. In [5]; a control system for voltage-controlled inverters is presented through the SRF. The equivalent circuit representing the AC side of the frequency inverter is given by Figure 9.7. The voltage equations in *abc* coordinates are given by 9.16. Equation (9.17) is obtained through the Park transform, considering the system running in steady state mode, and in the synchronous referential frames, and also considering that the resistances are negligible. In that conditions, the control system can be represented by (9.18).

$$\begin{bmatrix} v_{ga} - v_{fa} \\ v_{gb} - v_{fb} \\ v_{bc} - v_{fc} \end{bmatrix} = \begin{bmatrix} R_l & 0 & 0 \\ 0 & R_l & 0 \\ 0 & 0 & R_l \end{bmatrix} \begin{bmatrix} i_a \\ i_b \\ i_c \end{bmatrix} + \begin{bmatrix} L_l & 0 & 0 \\ 0 & L_l & 0 \\ 0 & 0 & L_l \end{bmatrix} \frac{d}{dt} \begin{bmatrix} i_a \\ i_b \\ i_c \end{bmatrix} \tag{9.16}$$

$$\begin{bmatrix} v_{gq} - v_{fq} \\ v_{gd} - v_{fd} \end{bmatrix} = \begin{bmatrix} R_l & L_l\omega_e \\ -L_l\omega_e & R_l \end{bmatrix} \begin{bmatrix} i_q \\ i_d \end{bmatrix} + \begin{bmatrix} L_l & 0 \\ 0 & L_l \end{bmatrix} \frac{d}{dt} \begin{bmatrix} i_q \\ i_d \end{bmatrix} \tag{9.17}$$

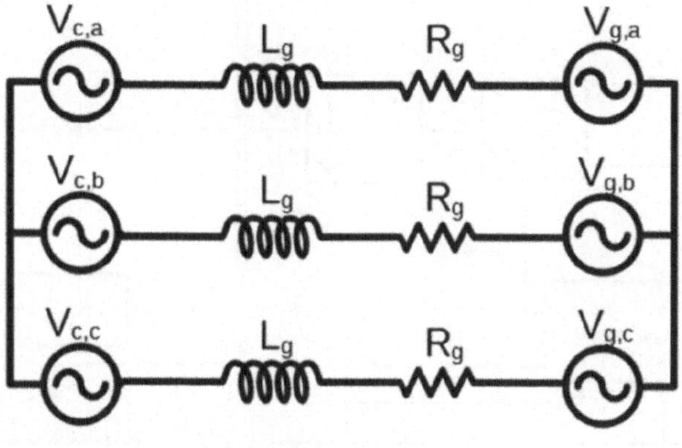

Figure 9.7 Equivalent circuit of the frequency inverter by AC side.

$$\begin{bmatrix} v_{fq} \\ v_{fd} \end{bmatrix} = \begin{bmatrix} \omega L_l i_d + v_{gd} \\ \omega L_l i_q + v_{gq} \end{bmatrix} \tag{9.18}$$

The reference currents are obtained from the real and reactive power using (9.14), (9.15), and (9.18), more details of the mathematical deduction, and of the models of VSI can be seen in [5, 7].

9.3.2.1 *Microgrid connected to the BPS: power control*

The VCVSI model can be seen in Figure 9.8, while its control system is presented in Figure 9.9. Equation (9.18) system was used for its implementation. Again, a PLL was used to obtain the angular frequency of the ω_e and perform Park's transformation and inversion. In this model, the reference real (P) and reactive (Q) power values are adjusted, and the values of the $V_{c,abc}$ and $I_{c,abc}$ are measured for system feedback. With this information, the values of V^*_{abc} are calculated which will be sent to the inverter through a sinusoidal PWM generator.

9.3.2.2 *Islanded microgrid: voltage and frequency drop through power control*

The power flow approach normally adopted for HV and MV grids cannot be applied to these grids, since the ratio of reactance and resistance X/R is very low and the microgrid has special characteristics, such as radial distribution, very short length, and a variety in conductor types and sections.

Considering Figure 9.10, where R_g and X_g is the resistance and the reactance of the grid, the power flow ($S = P + jQ$) in the line from the point A to the point B is given by (9.19) and (9.20).

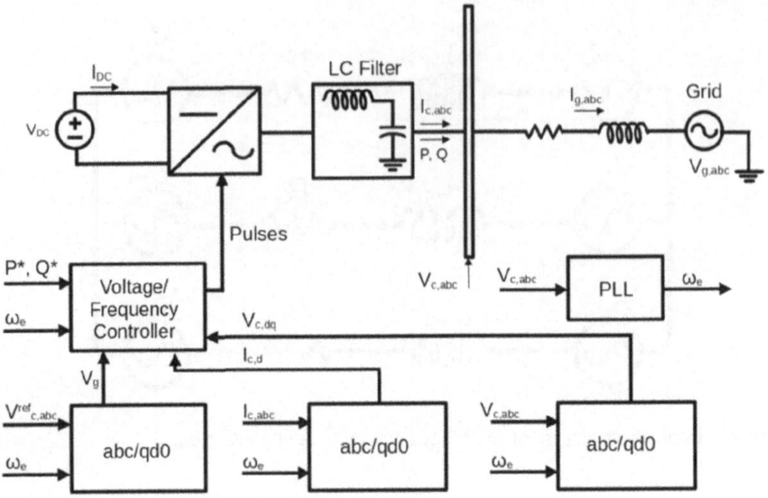

Figure 9.8 Model of the voltage-controlled inverter.

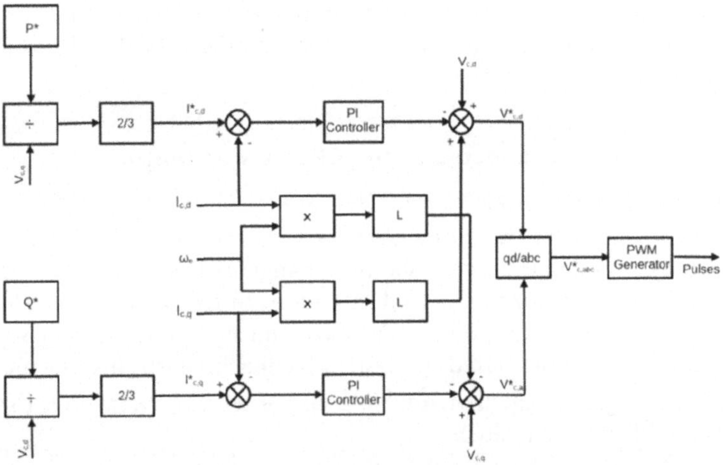

Figure 9.9 Control system of the voltage-controlled inverter.

$$P = \frac{V_c}{R_g + X_g}\left[R_g\left(V_c - V_g\cos\delta\right) + X_g V_g \sin\delta\right] \tag{9.19}$$

$$Q = \frac{V_c}{R_g + X_g}\left[-R_g V_g \sin\delta + X_g\left(V_c - V_g\right)\cos\delta\right] \tag{9.20}$$

Typically, in LV distribution grids $R_g >> X_g$. In this case, neglecting X_g and if δ is small, then $\sin(\delta) = \delta$ and $\cos(\delta) = 1$. Hence (9.19) and (9.20) can be rewritten as (9.21) and (9.22).

$$P = \frac{V_c^2}{R_g} - \frac{V_c V_g}{R_g} \tag{9.21}$$

$$Q = \frac{V_c V_g}{R_g} \delta \tag{9.22}$$

In this case, the difference $V_c - V_g$ is exceedingly small, and it is possible to control the real power P with the voltage V_c and the reactive power Q with the angle δ and with the frequency f [6]. Thus, it is possible to implement a voltage droop control through real power and a frequency droop control through reactive power.

9.4 CASE STUDY

As seen in previous chapter, many studies have dedicated to the development of power electronics devices and control techniques for the operation of PVs. However, the use of such systems in LV grids with large penetration of power electronic loads and sources has been poorly explored so far. In this section, an LV distribution grid model, where non-linear loads and two frequency inverters that will operate in parallel and without communication between them, will be used as case study.

Aiming at exemplifying the application of PVs and demonstrating their performance, computer simulations were carried out with assistance of the MATLAB/Simulink software. In such simulations, all the components of inverters and of the LV grid to which they are connected were modeled. The LV grid considered in this study was that of CIGRÉ European Commercial

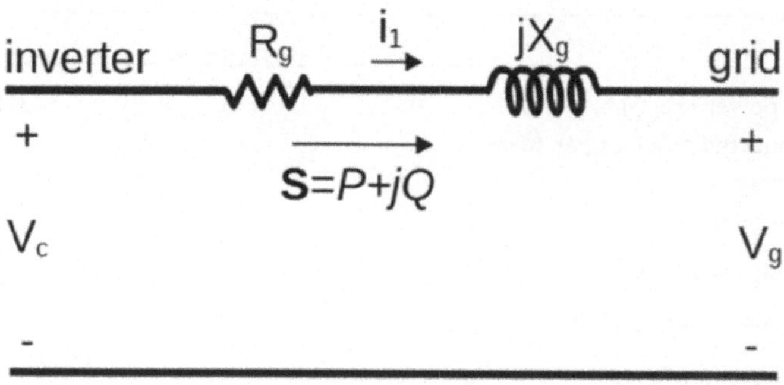

Figure 9.10 Power flow from inverter to grid.

Overhead [8] with four wires. The methodology adopted in the simulations may be seen in the Figure 9.2.

9.5 PV ARRAY USED IN THE CASE STUDY SIMULATIONS

The configured model of MATLAB/Simulink, which uses the circuit of a diode and two resistances, and has preset parameters from several PV panel manufacturers, was used in this work. The chosen PV panel was SunPower SPR-315E-WHT-D, for which the model is preset in MATLAB. The parameters of the PV arrays used may be seen in the PV table.

9.6 DC-DC CONVERTER DESIGNED

The input inductor was sized for the continuous conduction mode, where V_s and D is the voltage and duty cycle of the MPP of the array, ΔI_L is the maximum current ripple that was adjusted by 20%, and f_S is the switching frequency of the adjusted PWM at 10 kHz, and its value is 82 mH.

9.7 CIGRÉ LV TEST GRID

From a current scenario of energy transition, in which DERs start to play a key role, the need to create parameters has arisen, for the validation of grid models and smart microgrids. In the CIGRÉ Task Force (TF) report C6.04.02 [8] modeling and creation of benchmarks were proposed for different types of grids, such as for high voltage (HV) transmission, as well as for distribution MV grids and distribution LV grids with frequency of

Table 9.1 Parameters of PV module

Parameter	*Value*
Maximum power (P_{max})	315 W_p
Series-connected modules per string	6
Parallel strings	8
Cells per module	96
Open circuit voltage (V_{oc})	64.6 V
Voltage at maximum power point (V_{mp})	54.7 V
Current at maximum power point (I_{mp})	5.76 A
Short-circuit current (I_{sc})	6.14 A

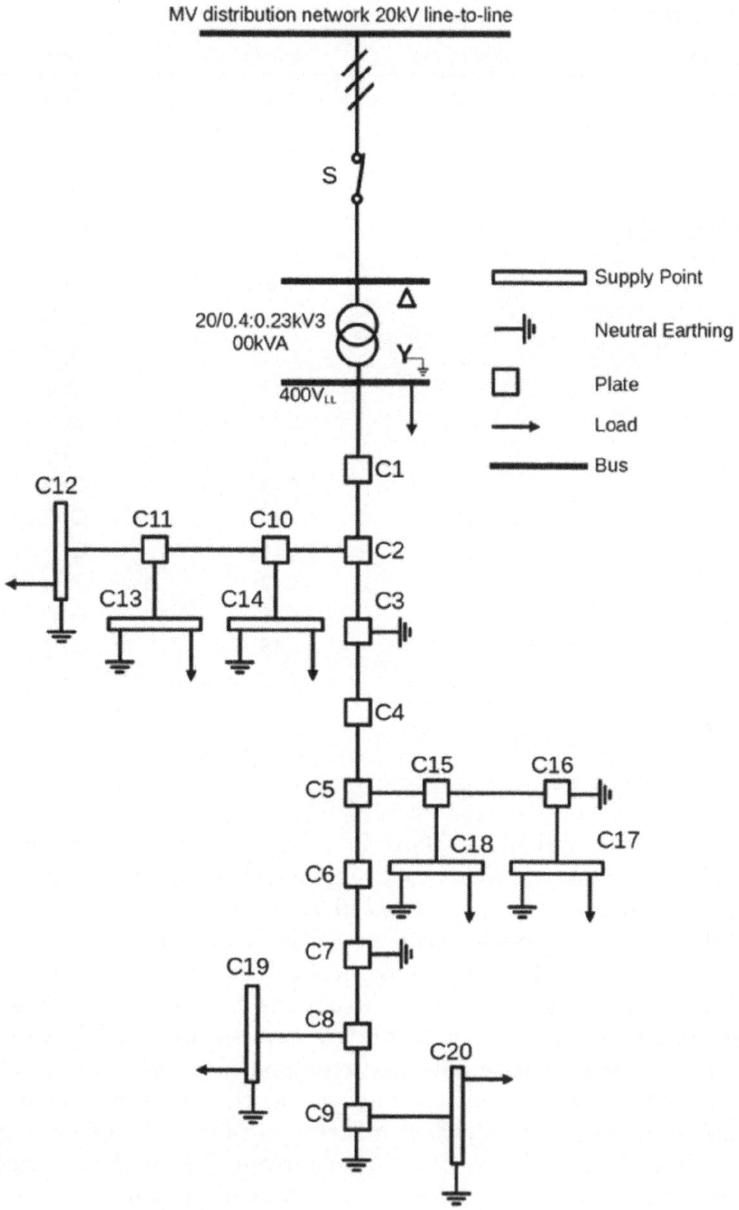

Figure 9.11 Commercial European CIGRÉ overhead network.

Table 9.2 Connections and parameters of the European commercial LV grid

Line segment	*Node from*	*Node to*	*Conduct. Id*
1	C1	C2	OH1
2	C2	C3	OH1
3	C3	C4	OH1
4	C4	C5	OH1
5	C5	C6	OH1
6	C6	C7	OH1
7	C7	C8	OH1
8	C8	C9	OH1
9	C3	C10	OH2
10	C10	C11	OH2
11	C11	C12	OH3
12	C11	C13	OH3
13	C10	C14	OH3
14	C5	C15	OH2
15	C15	C16	OH2
16	C15	C17	OH3
17	C16	C18	OH3
18	C8	C19	OH3
19	C9	C20	OH3

Source: [8]

50 Hz (European) and 60 Hz (USA). In the report, there are three different LV grids: residential, industrial, and commercial. The grid used in the case study presented here will be the 400/230 V 50 Hz European overhead commercial one. The grid topology can be seen in Figure 9.11.

All circuits of the LV commercial network are three-phase and 30 m long. The electrical parameters of the circuits are provided in Table 9.2. Commonly, power system simulators are developed to work with three-phase three-wire systems and have few component models with the ability to simulate four-wire systems. To solve this issue, the neutral conductor is considered solidly grounded, which makes it possible to use the Kron reduction to simplify a four-wire system by a three-wire equivalent circuit [8]. However, this model omits essential information regarding currents that circulate through neutral in imbalanced and non-linear loads. Therefore, the most complete model was proposed to verify these neutral currents, since imbalance of loads and harmonic distortions are common in LV grids. The complete data of the European overhead LV grid, including neutral, is shown in Table 9.3. More information on the grid as well as the list of loads can be seen at [8].

Table 9.3 Connections and parameters of the European commercial LV grid

Conductor ID / *Installation*		*A*	*B*	*C*	*N*
OH1 / 3-ph	A	0.540 + j0.777	0.049 + j0.505	0.049 + j0.462	0.049 + j0.436
	B	0.049 + j0.505	0.540 + j0.777	0.049 + j0.505	0.049 + j0.462
	C	0.049 + j0.462	0.049 + j0.505	0.540 + j0.777	0.049 + j0.505
	N	0.049 + j0.436	0.049 + j0.462	0.049 + j0.505	0.540 + j0.777
OH2 / 3-ph	A	1.369 + j0.812	0.049 + j0.505	0.049 + j0.462	0.049 + j0.436
	B	0.049 + j0.505	1.369 + j0.812	0.049 + j0.505	0.049 + j0.462
	C	0.049 + j0.462	0.049 + j0.505	1.369 + j0.812	0.049 + j0.505
	N	0.049 + j0.436	0.049 + j0.462	0.049 + j0.505	1.369 + j0.812
OH3 / 3-ph	A	2.065 + j0.825	0.049 + j0.505	0.049 + j0.462	0.049 + j0.436
	B	0.049 + j0.505	2.065 + j0.825	0.049 + j0.505	0.049 + j0.462
	C	0.049 + j0.462	0.049 + j0.505	2.065 + j0.825	0.049 + j0.505
	N	0.049 + j0.436	0.049 + j0.462	0.049 + j0.505	2.065 + j0.825

Source: [8]

On this grid, a MI and non-linear loads are connected. This LV grid is simulated both connected to the MV system and islanded. During the islanded, a VCVSI connects a Battery Energy Storage System (BESS) modeled as a DC source to the grid. The inverter control can be severely affected by PQ disturbances, such as harmonic distortion of currents and voltages. The simulation of the case study is performed in time domain to obtain the harmonic distortion observed in steady state operation. To achieve that, the models were developed in MATLAB/SIMULINK with the Simscape/Electrical/Specialized Power Systems package.

9.8 MICROGRID MODELING

Power system simulators offer very few models to represent four-wire LV systems. In the model proposed here, the MATLAB/Simulink block called mutual inductance is used. With this block it is possible to indicate a

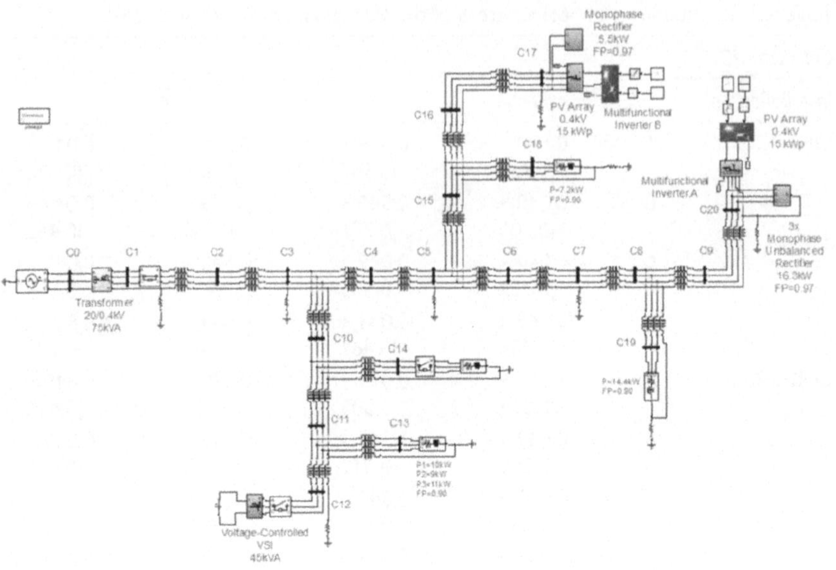

Figure 9.12 LV CIGRÉ network modeled in MATLAB/Simulink and adapted for the case study.

general matrix of mutual impedance of multiple windings, with balanced or unbalanced mutual coupling. In this way, it is also possible to represent a four-wire LV distribution network. Therefore, it is possible to include the neutral in the model and use the most complete impedance matrix presented in Table 9.3. The CIGRÉ LV grid shown in Figure 9.11 is modeled in MATLAB/Simulink as shown in the Figure 9.12. The LV grid is connected at C1 to the 20/0.4 kV transformer and then at C0 to the BPS.

In the original CIGRÉ grid, there is a generic load also connected at the C1 node. This load represents other circuits that are connected to the same transformer. To better represent the PQ disturbances and indices, this load was neglected, and the rated power of the transformer was reduced to 75 kVA.

A VCVSI as show in the Figure 9.8 was connected at C12. This inverter shall remain in standby mode when the feeder is connected to BPS and as main power source when the microgrid is islanded. Three single-phase rectifiers, each with four diode bridge and RL load and MI from a PV system, were connected at node C20. Figure 9.12 shows the complete proposed model.

For this case study, two scenarios were created, namely Scenarios 1 and 2. In Scenario 1, it is admitted that the LV grid is working in a conventional manner, i.e., connected to the MV distribution system of the BPS. In Scenario 2, an islanded operation was simulated, and in this case, in both the scenarios, at node C20 a three-phase non-linear load is connected at t

= 0.1 s and an MI (called MI A) is connected at t = 0.3 s. At node C17, at t = 0.5, another non-linear load, this time single phase, is turned on; and at t = 1 s another MI (MI B) comes running. During grid-connected operation, the VCVSI remains connected in parallel with the main network. When the microgrid is islanded the VCVSI shall deliver the voltage and frequency reference required to operate the other inverters.

9.9 RESULTS

As mentioned in the previous section, two scenarios were simulated. The first, Scenario 1, with the LV grid connected to the MV grid of the utility, and the second, Scenario 2, with the main grid disconnected. The results for both working conditions are presented below. The TDD was calculated with nominal demand equals the apparent nominal power of the transformer (75kVA). The imbalance $I(\%)$ was calculated according [9] i.e. $I(\%) = V_{neg} / V_{pos}$, where V_{neg} is the negative sequence of voltage and V_{pos} is the positive. The imbalance of currents is calculated of similar form. All the results are presented for phase A.

9.9.1 Scenario 1: Microgrid connected to MV grid

The operation of Scenario 1 considers that the LV grid is working connected to the grid of the utility. In addition, the VCVSI operates in parallel with the grid, delivering 30.4 kW with PF = 1. The currents and voltages of the grid are represented in Figure 9.13a and 9.13b, as well as the load currents (Figure 9.13c), and the currents injected by multifunctional inverter A at the node C20 (Figure 9.13d). Three non-linear and imbalanced single phase loads are connected to node C20 in t = 0.1 s, which causes a distortion in the voltage (THDv = 3.97%) (9.13a) and the current (TDD = 43.4%) (Figure 9.13b) waveform at this point. The MI A is connected at t = 0.3 s, after a short transient the currents of the grid become sinusoidal and balanced, with the THDv = 1.2% and TDD = 5.3%.

Similar behavior happens at the node C17, where the load, a single phase rectifier, is connected in t = 0.5s, which causes great imbalance between phases. Figure 9.14a shows the voltages in the point of common coupling (PCC) with harmonic distortion (THD = 2.4%) before the operation of MI A. In t = 1s, the multifunctional inverter B is connected, which reduces the THD to 0.7% at this point and the voltage imbalance, which was 0.85%, decreased to 0.16%. In Figure 9.14b, it is possible to see that the current, after the connection of MI B, becomes virtually sinusoidal and balanced. TDD at this point is reduced from 100% to 0.1%. Figures 9.14c and 9.14d show the load currents and compensating currents delivered by MI B.

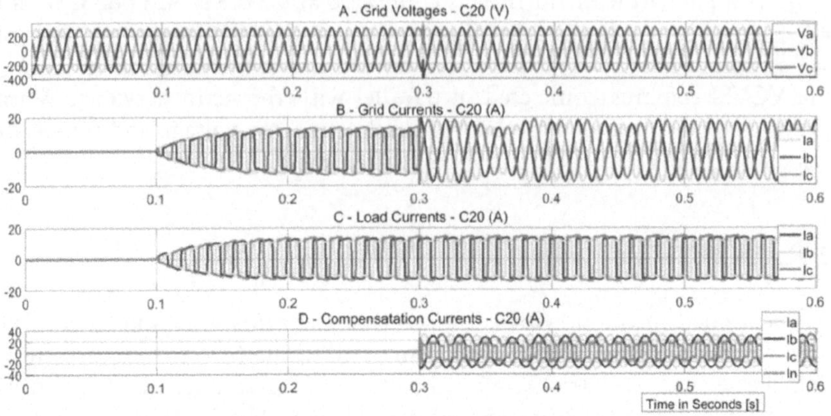

Figure 9.13 Voltages and currents at node C20 – MI A – Scenario 1.

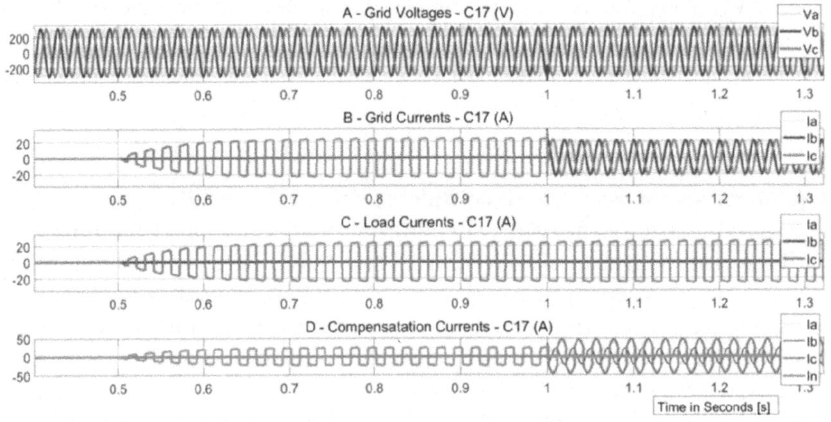

Figure 9.14 Voltages and currents at the node C17 – MI B – Scenario 1.

In the LV side of transformer (node C1), in Figure 9.15, it is possible to see the waveforms of voltages (9.15a) and currents (9.15d) and rms of voltages (9.15b) and currents (9.15d), at the moment when the non-linear load is connected in C1 (t = 0.1 s) and also when MI A starts to operate (t = 0.3 s). At this point, with the load in steady state, between t = 0.1 s and t = 0.3 s, THDv and TDD at this point was 2.0% and 4.5%, becoming 0.45% and 0.40%, respectively, after compensation, and the voltage imbalance was 0.1% and turned into 0.08%.

The apparent, real, and reactive power delivered from transformer, as well as FP can be seen in Figures 9.16a, 9.16b, 9.16c and 9.16d. It is possible to see a reduction of the powers delivered from the grid during the MIs operation. At t = 1.8 s the VCVSI was disconnected, which the grid close

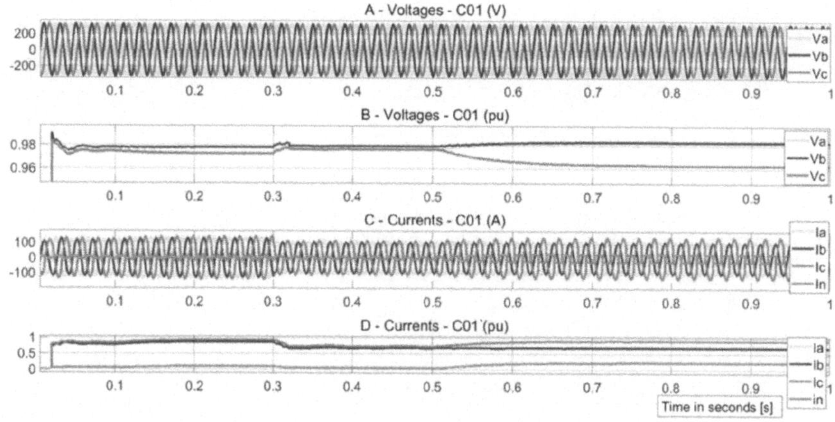

Figure 9.15 Voltages and currents at node C1 in Transformer (LV side) – Scenario 1.

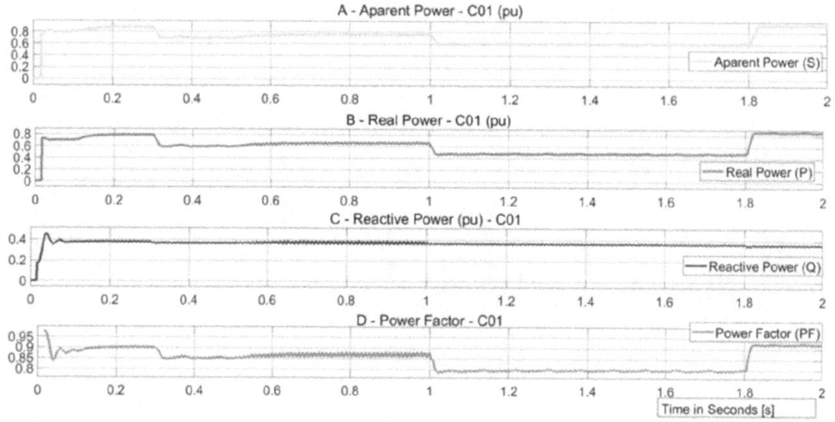

Figure 9.16 Apparent, real power, reactive power (in pu) and power factor at node C1 in Transformer – Scenario 1.

to an overload. If the other inverters were previously turned off, the transformer will operate above its rated capacity.

Now, at node C12, in Figure 9.17 it is possible to see the instantaneous voltages (Figure 9.17a), rms voltages (Figure 9.17b), instantaneous currents (Figure 9.17c), and rms currents (Figure 9.17d) that are delivered by the VCVSI in parallel with the grid. Note that the waveforms in the VCVSI are affected by the harmonic distortion produced by the load.

Figure 9.18 summarizes the results of THDv and TDD for the Scenario 1.

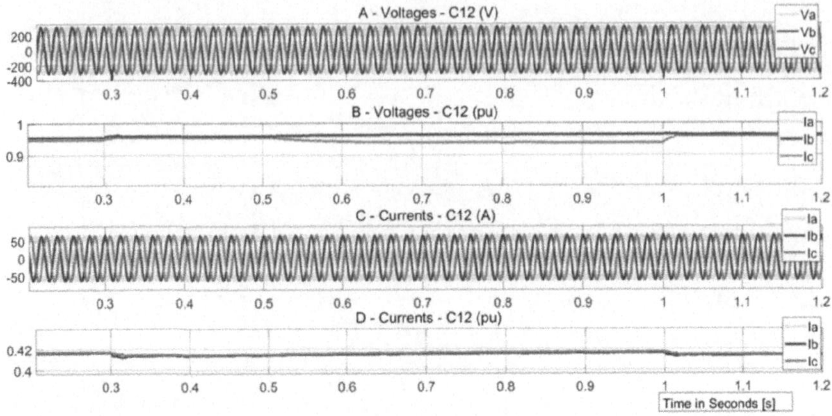

Figure 9.17 Voltages and currents at node C12 – VCVSI – Scenario 1.

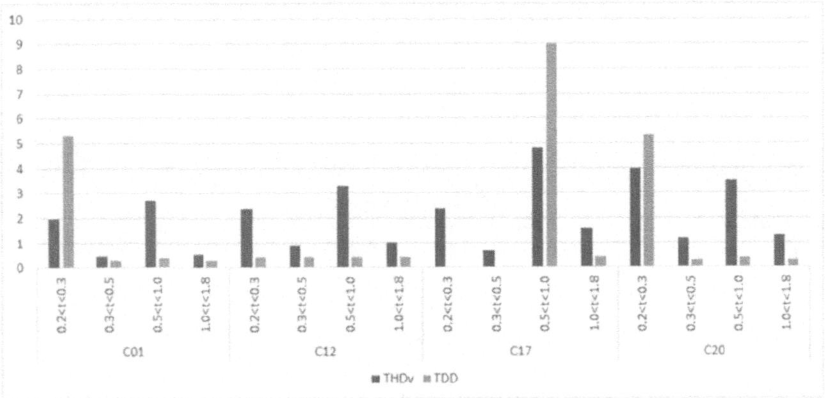

Figure 9.18 THDv and TDD values calculated in Scenario 1.

9.9.2 Scenario 2: operation connected to MV network

In the microgrid operation, VCVSI assumes the function of generating the voltage and frequency reference for synchronism of the remaining inverters. In Figure 9.19, it is possible to see the waveform of voltage (Figure 9.19a) and current (Figure 9.19b) in the C20 point PCC, as well as the waveform of current in the load (Figure 9.19c) and the compensation currents of MI A (Figure 9.19d). Now, both MI A and the three single phase loads remain connected all the time. However, in C12, there is a non-linear single phase load that generates a strong imbalance in voltage. In t = 1 s the MI B starts to operate, which improves the imbalance and the voltage levels in the

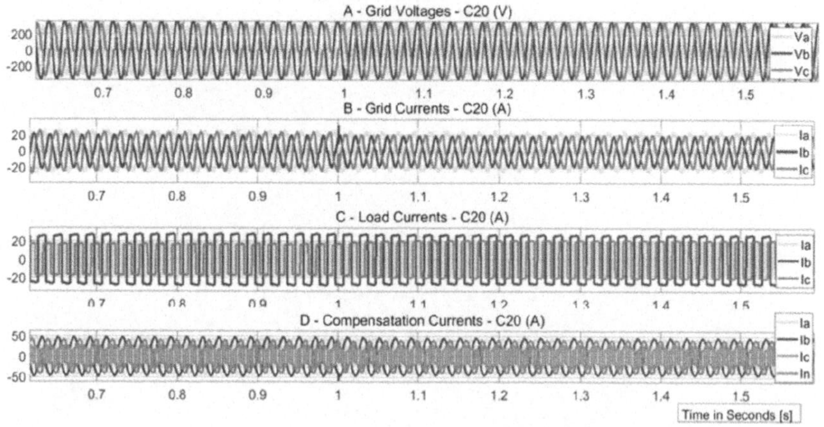

Figure 9.19 Voltages and currents at the node C20 – MI A – Scenario 2.

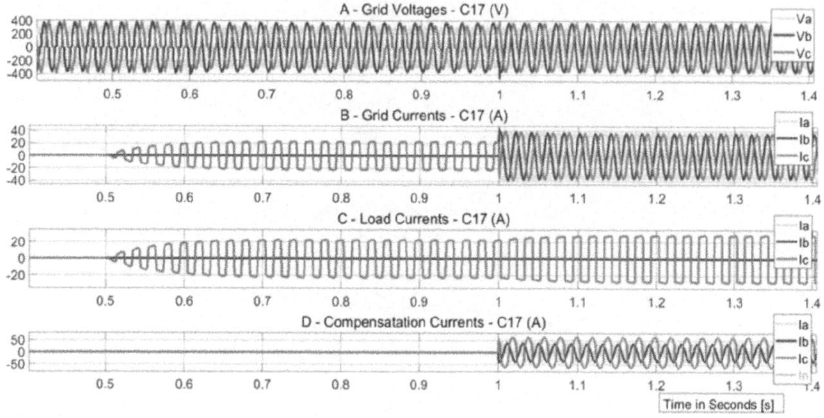

Figure 9.20 Voltages and currents at the node C17 – MI B – Scenario 2.

entire grid. This fact also has effects on the client PQ connected in C20, as it may be seen in the graphic.

Figure 9.20 shows the instantaneous voltage in the PCC in C17 (MI B) (Figure 9.20a), the current absorbed from the grid at this point (Figure 9.20b), the current of the non-linear single phase load (Figure 9.20c) and the compensating currents of MI B Figure 9.20d. As it may be seen, voltage experienced important improvement both in terms of amplitude and in phase balance (before the correction it was 2.5% and after the correction it was 1.5%). It is also possible to notice that the current of the grid, which was concentrated in a single phase and strongly distorted, assumed a sinusoidal characteristic and entered into a three-phase balanced situation.

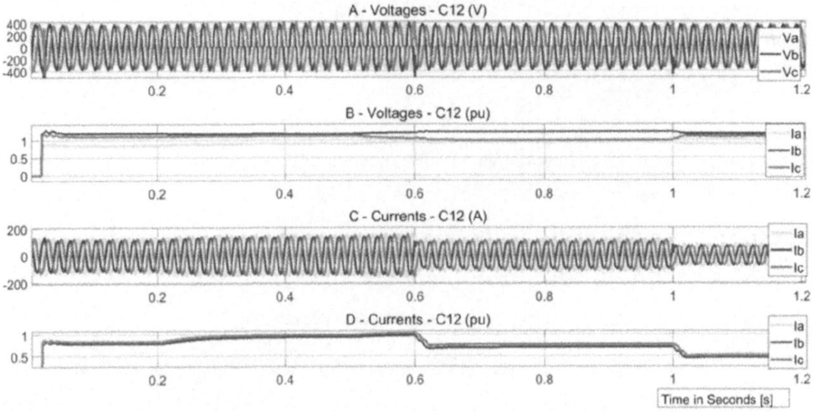

Figure 9.21 Voltages and currents at the node C12 – VCVSI – Scenario 2.

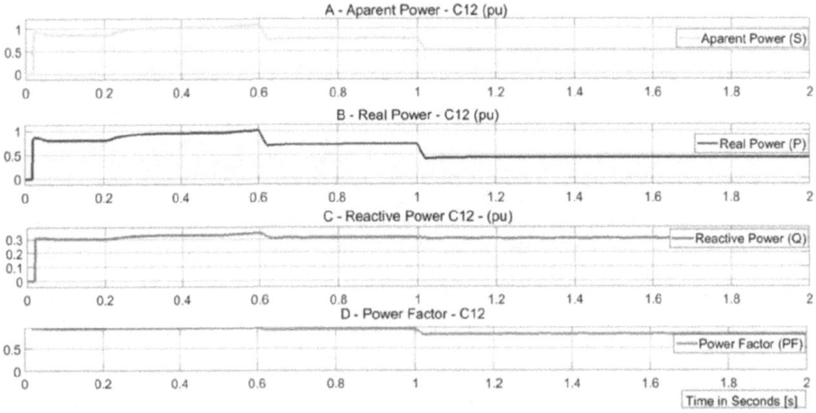

Figure 9.22 Apparent, real power, reactive power (in pu) and power factor at node C12 – VCVSI – Scenario 2.

Figure 9.21 shows the behavior of instantaneous voltages (Figure 9.21a) and its rms values (Figure 9.21b) in the VCVSI, where it is possible to notice an important improvement both in voltage levels and in phase balance with the operation of the MIs. Figures 9.21c and 9.21d show, respectively, the instantaneous currents and rms currents that leave the inverter. Figure 9.22 shows the apparent, real, and reactive powers and the power factor delivered by the VCVSI. This graph shows the importance of the MIs that contribute to the generation of electricity for the system, thus reducing the power demand from the VCVSI. Figure 9.23 summarizes the results of THDv and TDD for Scenario 2.

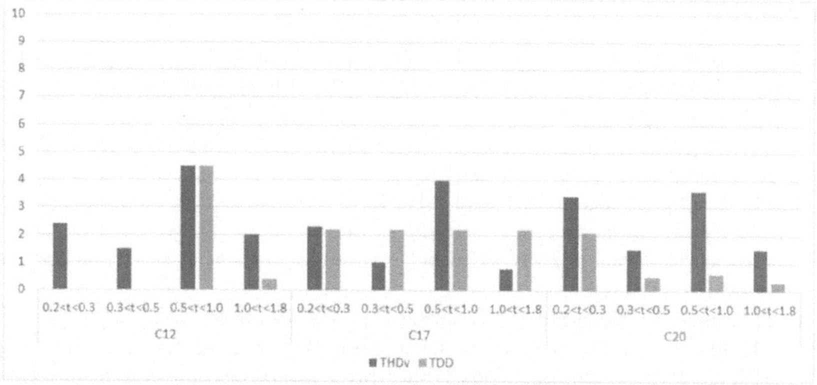

Figure 9.23 THDv and TDD values calculated in Scenario 2.

The results show that the network can continue operating normally even in the presence of a heavily unbalanced and distorted load. Comparing the distortions between Scenarios 1 and 2, it is noted that the values were practically similar, although there are more harmonic distortions in Scenario 2. A greater imbalance was also noted in the voltages generated in Scenario 2 due to the nature of the load and voltage control strategy of the VCVSI.

9.10 CHAPTER REMARKS

With technological advances, new ways of operating the power grids have become prominent both in terms of availability and reliability. Among them, the benefit of operating in intentionally and temporarily isolated microgrids is notorious. However, this type of operation brings some challenges, for example, when this network operates with loads that generate high harmonic distortions and current or voltage unbalance.

This chapter has presented an electric model of a four-wire LV feeder where several MI were connected. Furthermore, simulation of unbalanced loads were considered as well as two MIs and a VCVSI type inverter.

Two scenarios were simulated. One when the microgrid operates connected to the MV grid and other when the MV grid is unavailable and the VCVSI assumes the voltage and frequency reference for proper functioning of the MIs. Significant improvements were observed in the PQ. MIs reduced the overloading of the distribution transformer, not only with the regular load of the grid in the fundamental frequency, but also regarding its harmonic content. When operating as islanded microgrid, the VCVSI operated normally within acceptable PQ values even in the presence of heavily distorted and imbalanced loads.

REFERENCES

1. S. Y. Mousazadeh, M. Savaghebi, A. Beirami, A. Jalilian, J. M. Guerrero, and C. Li, "Control of a multi-functional inverter for grid integration of pv and battery energy storage system," *2015 IEEE 10th International Symposium on Diagnostics for Electrical Machines, Power Electronics and Drives (SDEMPED)*, pp. 474–480, 2015.
2. H. Akagi, E. Watanabe, and M. Aredes, *Instantaneous Power Theory and Applications to Power Conditioning*, ser. IEEE Press Series on Power Engineering. Wiley, 2007. [Online]. Available: https://books.google.com.br/books?id=yO0EUhCUQaAC
3. P. Krause, O. Wasynczuk, S. Sudhoff, and S. Pekarek, *Analysis of Electric Machinery and Drive Systems*, ser. IEEE Press Series on Power Engineering. Wiley, 2013. [Online]. Available: https://books.google.com.br/books?id=ocr9wSmSZEsC
4. A. H. Silveira, F. B. Libano, R. C. Leborgne, and M. J. S. Ramos, "A novel application of multifunctional inverters to enhance power quality of smart microgrids: an analysis on a low voltage and four-wire grid," *Energy Systems*, March 2021. [Online]. Available: https://doi.org/10.1007/s12667-021-00434-y
5. Egea-Alvarez, A, Junyent-Ferré, A and Gomis-Bellmunt, O 2012, Active and Reactive Power Control of Grid Connected Distributed Generation Systems. in Modeling and Control of Sustainable Power Systems: Towards Smarter and Greener Electric Grids. vol. 96, Green Energy and Technology, vol. 96, Springer, Berlin, pp. 47–81. https://doi.org/10.1007/978-3-642-22904-6_3
6. Laaksonen, H. and Kauhaniemi, K., 2009. Voltage and frequency control of low voltage microgrid with converter based DG units. *Int. J. Integr. Energy Syst.(IJIES)*, 1(1), 47–60.
7. C. Bajracharya, "Control of vsc-hvdc for wind power," *Norwegian University of Science and Technology*, 2008. [Online]. Available: https://ntnuopen.ntnu.no/ntnu-xmlui/handle/11250/256495
8. C. I. D. G. R. É. CIGRE, *Benchmark Systems for Network Integration of Renewable and Distributed Energy Resources: Task Force C6.04*, ser. [Brochures thématiques]. CIGRÉ, 2014. [Online]. Available: https://books.google.com.br/books?id=v3PcoQEACAAJ
9. IEEE-Std1159, "Ieee recommended practice for monitoring electric power quality," *IEEE Std 1159-2019 (Revision of IEEE Std 1159-2009)*, pp. 1–98, 2019.

Section IV

Data science and cyber security in smart grids

Chapter 10

A bi-level programming approach for solar power integration in distribution feeder

Mukul Dixit, Anuradha, and Nitin Chand

10.1 INTRODUCTION

As it is well known, the higher R/X ratio causes the distribution system to lose more money than the transmission system. The main aim of all electrical power companies is to reduce such power losses. Optimal DG placement at the ideal location with the highest rating helps to fully satisfy the load demand, increase the voltage level, and also reduce the network power loss. The power loss of the entire network increases and the network's voltage level is perturbed if the rating and position of the DG are improperly chosen. There are two types of power losses such as (i) real power loss and (ii) reactive power loss.

$$P_{loss} = \sum_{j=1}^{br} I_j^2 R_j \tag{10.1}$$

$$Q_{loss} = \sum_{j=1}^{br} I_j^2 X_j \tag{10.2}$$

Where, R_j and X_j are the branch's resistance and reactance, respectively, and j stands for the branch number. I_j is the branch current connecting the two buses. In addition, numerous writers have used a variety of intelligence strategies to solve the optimal DG placement problem in published literature studies.

10.2 LITERATURE REVIEW

Many writers have written numerous studies on the DG placement problem in the distribution system during the last few years. Such DGs may be renewable based and non-renewable based. Using an analytical approach [1], optimal DG placement in balanced distribution system for power loss reduction has been applied to 15-bus and 33-bus distribution networks. To enhance

DOI: 10.1201/9781003623168-14

the voltage profile and to lower down the power losses of the system, DG integration at various points in the distribution system has been done using the MINLP technique [2]. A novel voltage stability index-based approach to DG integration is the most effective for reducing power losses in distribution system with load increment. When compared to previous sensitivity-based strategies, this methodology produces more numerical results [3]. For the accurate placement and correct rating of DG in the distribution network, as well as to lower power system losses, an improved non-dominated sorting GA-II with a multi-objective fitness function has been shown [4]. To cut down power loss and increase the cost-benefit ratio while utilizing GA, the ideal DG rating and position for a meshed power system were established [5]. In order to improve the voltage profile and reduce power system losses of the system, a new power stability index has been developed. This method of analysis has been used on distribution systems with 12 and 69 buses [6]. In order to decrease network voltage variation, increase voltage level, and reduce power losses, a novice multi-object quasi-opposition TLBO has been presented for the best placement and rating of DG in distribution systems [7]. In this article, author uses the sensitivity-based approach to find the best site for solar power and the GABC optimization algorithm to find the best rating. PV is installed in the distribution system at both single and locations with a unity power factor. The ultimate goal is to reduce power loss and improve the network's voltage profile. Compare the system power loss both with and without solar power integration.

The aforementioned work and research papers which are available in the literature utilized various meta-heuristic as well as artificial intelligence techniques for the integration of solar power at optimal locations under varying load levels. But in this work, a novel approach is used for solar power integration under uncertain conditions of solar and load demand. Such collective impact of this work makes it different from other authors' work. Thus, it can be said that, this is a research gap that found from the published literature. The following are the topics that the authors covered:

- A novel GABC-based two-stage method has been developed for the identification of optimal buses with optimal number of PV module in distribution system.
- Formulation of 24-hour uncertain load demand for each bus to analyze the effect of solar and load demand uncertainties on distribution systems.

10.3 PROBLEM FORMULATION

The deployment of solar power in RDN to reduce the overall actual power losses and enhance the voltage profile of the system while subject to specific operating constraints are the study's key goals. It has been assumed that the system is balanced.

The following equation defines the objective function for the best solar installation as the total real power loss (10.3).

$$Minimize \quad P_{Tloss} = \sum_{n=1}^{br} P_{loss,n} \tag{10.3}$$

Below mentioned equation shows that the solar power capacity is smaller than the system's overall load:

$$\sqrt{P_{PV}^2 + Q_{PV}^2} < \sqrt{P^2 + Q^2} \tag{10.4}$$

Equation (10.5) specifies the maximum voltage allowed at each bus within the permitted range:

$$V_{\min} \le V_i \le V_{\max} \tag{10.5}$$

Where V_{min} and V_{max} are the system voltage's minimum and maximum values, respectively, and V_i is the bus of ith bus voltage magnitude.

Consider a distribution line between the two nodes p and q, which is depicted in Figure 10.1 and has a load (P_L+jQ_L) as well as resistance and reactance (R_{pq} and X_{pq}, respectively). The equations shown below (10.7–10.9) [8] can be used to compute the active and reactive power flow between the two nodes as well as the receiving end voltage.

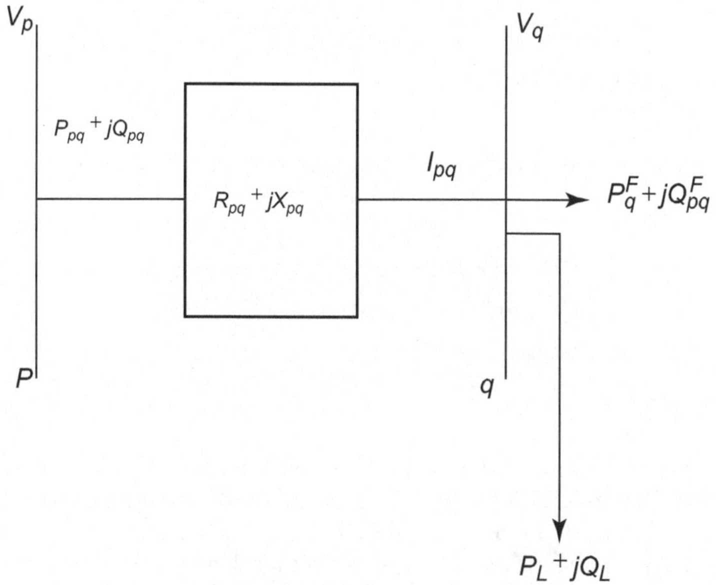

Figure 10.1 Equivalent diagram of two-bus system.

$$P_{pq} = P_q^F + P_L + R_{pq}\frac{\left(P_{pq}^2 + Q_{pq}^2\right)}{V_p^2} \tag{10.6}$$

$$Q_{pq} = Q_q^F + Q_L + X_{pq}\frac{\left(P_{pq}^2 + Q_{pq}^2\right)}{V_p^2} \tag{10.7}$$

$$V_q^2 = V_p^2 - 2\left(P_{pq}R_{pq} + Q_{pq}X_{pq}\right) + \left(R_{pq}^2 + X_{pq}^2\right)\frac{\left(P_{pq}^2 + Q_{pq}^2\right)}{V_p^2} \tag{10.8}$$

Where P_{pq} and Q_{pq}, respectively, represent the total active and reactive powers flowing through the branches.

10.4 SOLAR GENERATION MODEL

The PV module's output power is primarily dependent on the intensity of the solar radiation. Knowing the behavior of solar radiation at a certain spot for each hour would reveal that it typically followed a binomial distribution. The two straight unimodal relations are connected to this model [9, 10]. Both unimodal makes use of the beta probability density function (PDF), which is explained in more detail below:

$$f_b(s) = \begin{cases} \dfrac{\Gamma(\alpha+\beta)}{\Gamma(\alpha)+\Gamma(\beta)} \times s^{(\alpha-1)}(1-s)^{(\beta-1)} & \mathit{for}\ 0 \le s \le 1, \alpha \ge 0, \beta \ge 0 \\ 0 & \mathit{otherwise} \end{cases} \tag{10.9}$$

Here, the letter *s* stands for solar radiation (kW/m^2). The beta distribution function's parameters are calculated using (10.10) and (10.11) accordingly.

$$\beta = (1-\mu_s) \times \left(\frac{\mu_s \times (1+\mu_s)}{\sigma_s^2} - 1\right) \tag{10.10}$$

$$\alpha = \frac{\mu_s \times \beta}{1-\mu_s} \tag{10.11}$$

The entire day is divided into 24-hour time periods for the purpose of obtaining the probability values. Depending on the amount of solar radiation, each hour has its own likelihood. The hourly *s* and *s* of the day are assessed using previous data. It is assumed that there are 20 stages of operation every hour, each with a step of 0.05 kW/m^2. The probability values for each hour of the day with 20 states are calculated using the values of *s* and

s (10.9). Accordingly, utilizing (10.12) [9], the output power of PV module (PV_{out}) is calculated for that specific hour.

$$PV_{out}(s) = N \times F_F \times V_y \times I_y \tag{10.12}$$

$$V_y = V_{oc} - V_k \times T_{cy} \tag{10.13}$$

$$I_y = s\left[I_{sc} + I_k\left(T_{cy} - 25\right)\right] \tag{10.14}$$

$$F_F = \frac{V_{MPT} \times I_{MPT}}{V_{oc} \times I_{sc}} \tag{10.14 (a)}$$

$$T_{cy} = T_A + s\left(\frac{N_{OT} - 20}{0.8}\right) \tag{10.15}$$

Using (10.16), the total expected output power (EOP) of the PV module can be assessed for any given time period t.

$$EOP_{PV}(t) = \int_0^1 PV_{out}(s) \times f_b(s)ds \tag{10.16}$$

The EOP of PV module at various levels of si is calculated by (10.9)–(10.16). The output power produced hourly wise by PV modules is shown in Figure 10.2.

10.5 VOLTAGE DEVIATION

The most extreme necessary sign to assess the power attribute is a deviation in bus voltage magnitude. If there are significant disruptions in the voltage profile, the system appears to failing in that instance. Each bus's overall 24-hour voltage variation is calculated using (10.16).

$$V_{deviation} = \sum_{t=1}^{24}\sum_{i=1}^{NB}\frac{\left|V_{rated} - V_i\right|}{V_{rated}} \tag{10.17}$$

10.6 OPTIMIZATION TECHNIQUE: GABC ALGORITHM

GABC optimization algorithm is a variant/modified version of ABC technique; it is motivated through the social behavior of honey bees to find the food source position. This algorithm is formed by Zhu and Kwong in 2010 [11]. It is associated with three types of artificial bees named as employed bees, onlooker bees, and scout bees. The location of various food sources indicates a likely result of the considered optimization task. The nectar quantity of the food source clarifies the eminence or capability of the result.

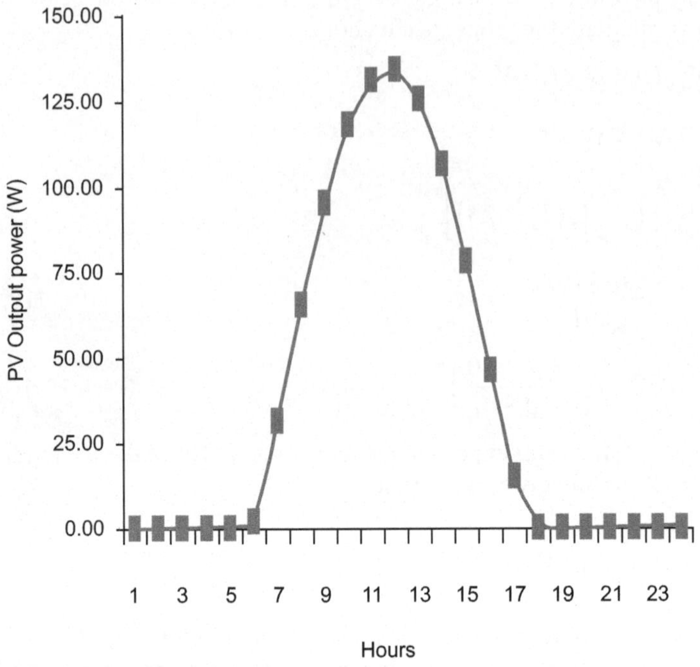

Figure 10.2 Output power of PV module for each hour.

In addition, the steps for the implementation of the GABC technique are mentioned below:

Step 1: Initialize the control variables of GABC algorithm such as population size, limit, and *MCN* etc. Normally, the optimization task is designed in terms of the objective function.

Step 2: Generate initial population: GABC algorithm initiates with population of food sources *Ppop = (p1, p2,...,pK). Ppop* is denoted as an array of dimension $K \times d$, and the rows indicate the food source solutions and columns illustrate the variables of such solutions. It can be represented using (10.18).

$$P_{pop} = \begin{bmatrix} p_1^1 & p_2^1 & \cdots & p_d^1 \\ p_1^2 & p_2^2 & \cdots & p_d^2 \\ \cdot & \cdot & & \cdot \\ \cdot & \cdot & \cdots & \cdot \\ \cdot & \cdot & & \cdot \\ p_1^K & p_2^K & \cdots & p_d^K \end{bmatrix} \tag{10.18}$$

Here, is the ith variable in the jth solution. These solutions are depending upon the random generation of variables. It can be defined using (10.19).

$$p_i^j = p_{i,\min} + rand(0,1) \times \left[p_{i,\max} - p_{i,\min} \right] \tag{10.19}$$

$$j \in 1,2,....,K$$

Step 3: Employed bee phase: In this stage, each food source is allocated for an employed bee. The total numbers of employed bees present in the hive are equal to the food sources. The target of each and every employed bee is to search out better food source location in the available search space. The search equation for determining the food source location can be expressed by (10.20):

$$p_i^{'j} = p_i^j + rand(0,1) \times \left[p_i^j - p_i^m \right] \tag{10.20}$$

Where m and $j \in (1,2,....,K)$; p^k is chosen from the population (P_{pop}) in random manner, $j \neq m$; $i \in (1,2,....,d)$. $rand(0,1)$ is the random number between 0 and 1. $p_i^{'j}$ is the new food source nearer to p_i^j.

Step 4: Onlooker bee phase: In this section, the employed bees return to their hive after searching out the food source and share the information with the onlooker bees related to the food source location through waggle dance. As per the information received, the onlooker bees go there and search out for better quality food source that is having a good amount of nectar. The search equation of the onlooker bee to determine the food source location will remain the same as employed bee, as given in (10.20). The tendency of onlooker bee is to attain better quality food source in the search space. Therefore, the probability values can be determined for the selection of better food source using equation (10.21).

$$Probability_j = \frac{fitness_j}{\sum_{i=1}^{K} fitness_i} \tag{10.21}$$

Where, $fitness_j$ indicates the objective function value of food source j, and the probability value $Probability_j$ depends upon the objective function $fitness_j$ value.

Step 5: Scout bee phase: Whenever the employed and onlooker bees could not improve the quality of food source location, then the food source is abandoned. After that, the scout bees explore the new food source location in the search space randomly using (10.19).

Step 6: In this step, they remember the best result found so far and discard the worst ones.

Step 7: Stopping criteria: If the stopping condition is satisfied i.e., *iter* = *Max iter*, then the program will terminate.

10.7 SIMULATION RESULT AND DISCUSSION

The effect of solar power on a 28-bus distribution system has been examined in this section. In order to eliminate the full power loss, the proposed integrated technique has also been implemented on an 11 kV 33-bus radial distribution feeder. Thirty-three buses, 32 branches, 1 major feeder, and 3 laterals make up this system. It has a base of 100 MVA and runs at 11 kV. Figure 10.3 displays the single-line schematic for this test system. Figure 10.4 illustrates the variable load demand for each bus over a 24-hour period. The MATLAB software (R2009) version, which is compatible with

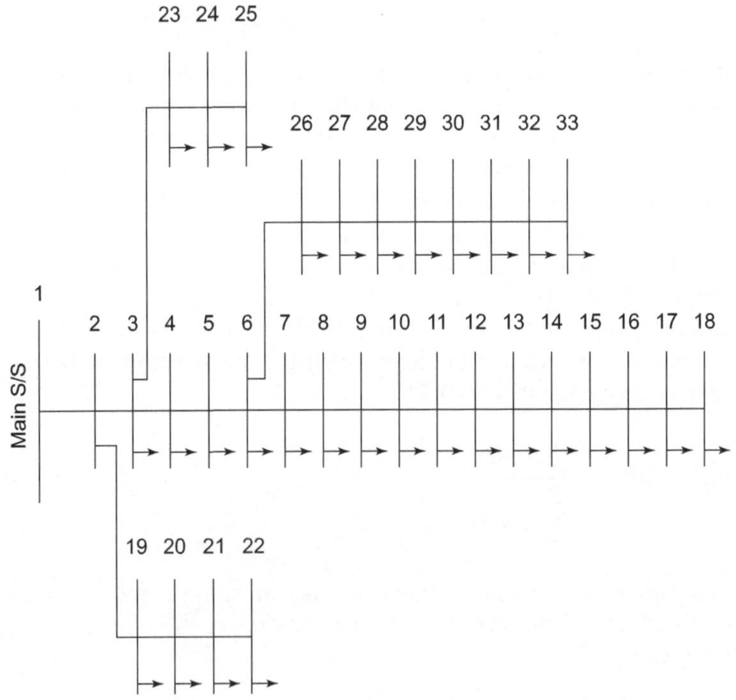

Figure 10.3 33-bus radial distribution system.

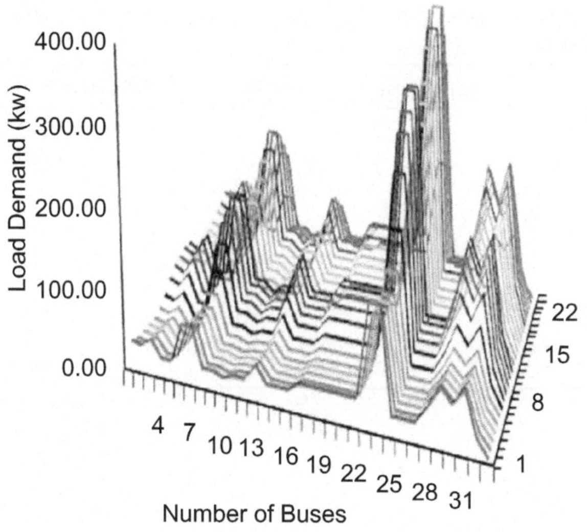

Figure 10.4 Uncertain load demand of each bus.

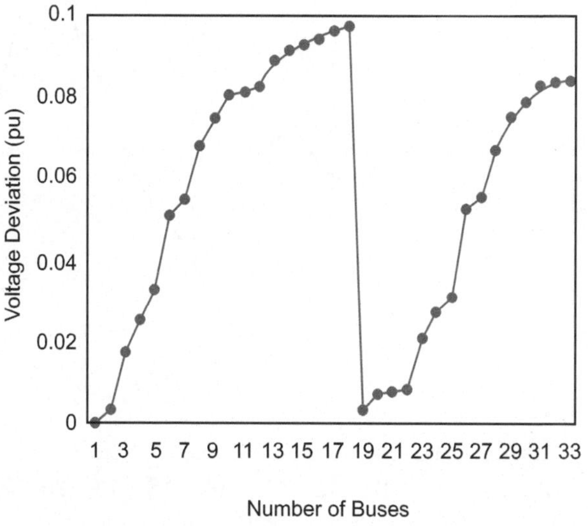

Figure 10.5 Voltage deviation of each bus.

Intel®CoreTMi5 CPUs, is used for the entirety of the simulation work. Sensitive buses of the distribution system are determined via voltage deviation, and those buses having a large voltage deviation can be considered as sensitive buses for PV installation. Figure 10.5 indicates the sensitive level of each bus of the network.

Table 10.1 Simulation of 33-bus radial distribution system after solar power installation.

Items	*Before solar power installation*	*After solar power installation*
Real power loss (KW)	2114.73	1840.61
Reactive power loss (KVAR)	1433.59	1259.99
Optimal location of PV	–	16 17 18
No. of PV module	–	5130 617 8482

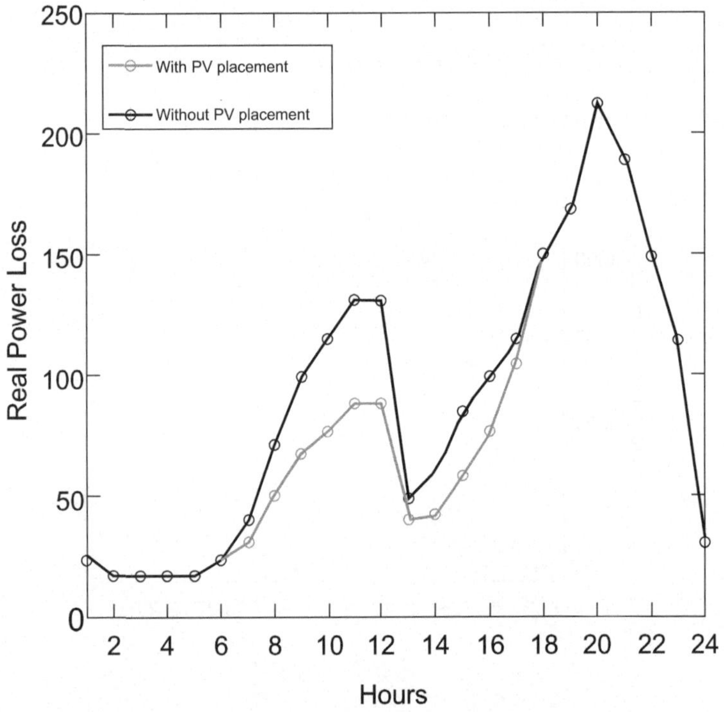

Figure 10.6 Real power loss of each hour for 33-bus distribution system.

Table 10.1 indicates the simulation results before and after inclusion of solar power in distribution network. The active and reactive power loss of the network becomes 1840.61 kW and 1259.99, respectively, after solar power integration. The optimal buses for PV installation are 16, 17, and 18, and the respective number of PV modules required on these locations is 5,130, 617, and 8,482. Hour-wise active power and reactive power that has

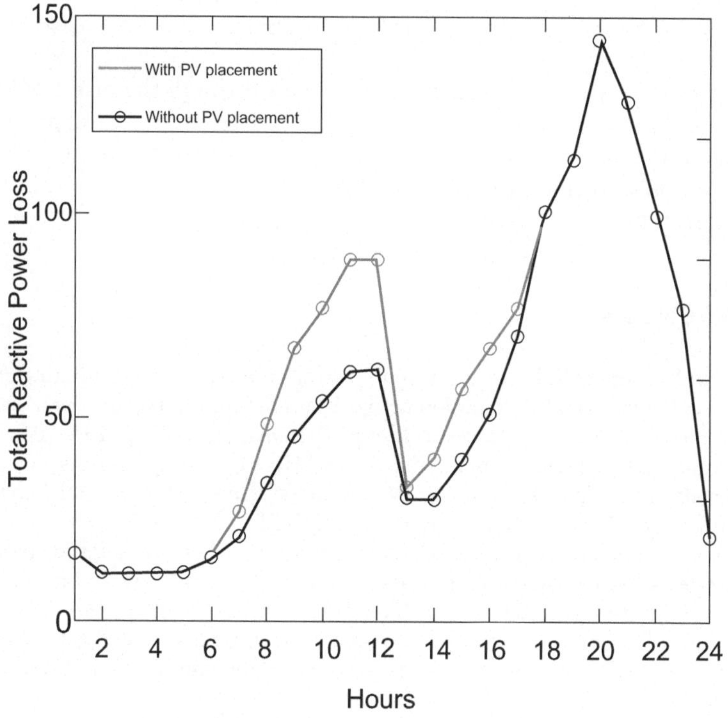

Figure 10.7 Reactive power loss of each hour for 33-bus distribution system.

been occurred before and after solar power integration are shown in Figure 10.6 and Figure 10.7, respectively.

10.8 CONCLUSION

In this work, a complex model of distribution system has been presented with the inclusion of solar power under uncertain load demand of 24 hours. GABC-based two-stage method is proposed and demonstrated successfully on a standard 33 bus test network for determining the optimal buses and number of PV module. Moreover, following are the essential facts that are obtained from this study as discussed below:

- Beta PDF has been applied effectively for assessing the probability as well as EOP of solar for every hour. Then, consequently decide the finest number of PV module in distribution system.
- From the simulation outcomes, it is analyzed that after solar power integration, the total power loss is reduced significantly to a great value.

- From Table 10.1, performance comparison can be made between before and after solar power integration in the distribution network.
- From the attained simulation outcomes through the proposed technique, it can be said that the proposed technique is capable and suitable to solve typical optimization problems. Furthermore, it can be said that GABC algorithm finds the finest solution in lesser iterations.

REFERENCES

1. Viral, R. and D.K. Khatod, An analytical approach for sizing and siting of DGs in balanced radial distribution networks for loss minimization. *International Journal of Electrical Power & Energy Systems*, 2015. **67**: p. 191–201.
2. Kaur, S., G. Kumbhar, and J. Sharma, A MINLP technique for optimal placement of multiple DG units in distribution systems. *International Journal of Electrical Power & Energy Systems*, 2014. **63**: p. 609–617.
3. Murty, V.V. and A. Kumar, Optimal placement of DG in radial distribution systems based on new voltage stability index under load growth. *International Journal of Electrical Power & Energy Systems*, 2015. **69**: p. 246–256.
4. Sheng, W., et al., Optimal placement and sizing of distributed generation via an improved nondominated sorting genetic algorithm II. *IEEE Transactions on Power Delivery*, 2014. **30**(2): p. 569–578.
5. Akorede, M.F., et al., Effective method for optimal allocation of distributed generation units in meshed electric power systems. *IET Generation, Transmission & Distribution*, 2011. **5**(2): p. 276–287.
6. Aman, M., et al., Optimal placement and sizing of a DG based on a new power stability index and line losses. *International Journal of Electrical Power & Energy Systems*, 2012. **43**(1): p. 1296–1304.
7. Sultana, S. and P.K. Roy, Multi-objective quasi-oppositional teaching learning based optimization for optimal location of distributed generator in radial distribution systems. *International Journal of Electrical Power & Energy Systems*, 2014. **63**: p. 534–545.
8. Haque, M., Efficient load flow method for distribution systems with radial or mesh configuration. *IEE Proceedings-Generation, Transmission and Distribution*, 1996. **143**(1): p. 33–38.
9. Teng, J.-H., et al., Optimal charging/discharging scheduling of battery storage systems for distribution systems interconnected with sizeable PV generation systems. *IEEE Transactions on Power Systems*, 2013. **28**(2): p. 1425–1433.
10. Hung, D.Q., N. Mithulananthan, and K.Y. Lee, Determining PV penetration for distribution systems with time-varying load models. *IEEE Transactions on Power Systems*, 2014. **29**(6): p. 3048–3057.
11. Zhu, G. and S. Kwong, Gbest-guided artificial bee colony algorithm for numerical function optimization. *Applied Mathematics and Computation*, 2010. **217**(7): p. 3166–3173.

Chapter 11

Implementing brownout through fair threshold allocation

Anshul Agarwal and Kedar Khandeparkar

11.1 INTRODUCTION

On a global scale, the residential sector has witnessed a sharp increase in energy consumption due to the rapid growth of population and various technologies. However, the majority of nations lack the necessary infrastructure to meet these rising demands, and the gap between power consumption demand and supply in these nations is widening over time.

Globally, the residential sector contributes the most to a country's total electricity consumption [1–3]. Consequently, the majority of research focuses on reducing the energy demand of residential residents and their peak power consumption.

For the electric grid to function properly, the gap between power supply and demand must be as small as possible to prevent grid failure and instability [4, 5]. Recent research has focused on demand side management (DSM) programs that attempt to regulate the power demands of households to address these issues [6, 7].

When power utility companies are unable to provide the necessary amount of electricity to homes, they employ a technique known as rolling blackouts. It implies that the utilities disconnect the power supply to different parts of the distribution region for distinct time periods. During blackouts, circuit breakers are activated, allowing for the disconnection of some system loads. This helps utilities manage the increased demand for electricity. Thus, residents of numerous cities in these nations regularly experience power outages.

Although blackouts are simple to implement, they have the following disadvantages: (a) residents are inconvenienced and (b) if the circuit breakers are not opened in time, a large, undesirable blackout may occur.

To address these blackout-related issues, a more intelligent solution known as *brownout* can be implemented. In contrast to blackouts, it permits a drop in voltage, which reduces the load on the grid and allows residents to use some of their appliances. However, the implementation of this brownout scheme necessitates investment and upgrade of the electric network, resulting in an increase in capital expenditures. It may also cause

DOI: 10.1201/9781003623168-15

equipment failure [8, 9]. Therefore, a more pragmatic approach is required for the effective implementation of brownout.

This chapter presents a novel strategy for the efficient implementation of brownout. It consists of allocating load thresholds to the distribution region (consisting of households) during the times when the power demands of the households cannot be fulfilled by the power utility company. It ensures that (a) households are allocated load thresholds in a fair manner and (b) minimal violation of this allocated threshold should be observed across all the households. Following are the contributions of this chapter:

- Load threshold allocation algorithms that use different heuristic criteria to ensure fairness in allocation to the households.
- A novel optimal algorithm that ensures threshold allocation has minimum percentage violation in all the time slots. It also avoids starvation of households with respect to threshold allocation since it uses multiple heuristic criteria for threshold allocation. This is the primary contribution of this chapter.
- The developed algorithms have been tested on a real world dataset consisting of twenty households.

11.2 RELATED WORK

Demand side management (DSM) is gaining popularity for regulating the demand and consumption of customers. It applies to smart grids that employ Information and Communication Technologies (ICT) to monitor the status of the grid and the behavior of customers. Albadi & El-Saadany [10] and Strbac [4] have discussed different DSM techniques. Using minimum sensors to reduce building energy consumption is proposed by [11]. Elma and Selamogullari have proposed a home energy management (HEM) algorithm based on the voltage control method [12]. However, this algorithm for controlling voltage has no effect on the power consumption of electronic loads such as printers and televisions. Zhou et al. [13] and Abouelela & Abouelela [14] discuss alternative strategies for implementing HEMs. Detecting faulty HVACs to reduce energy wastage is discussed in [15]. Costanzo et al. [16] have presented a novel framework for autonomous DSM with the layers load balancing, admission control, and DR management. It has been primarily concerned with the architecture and development of a solution for only two layers. Shafie-Khah et al. have proposed a novel decentralized DR model that is dependent on bidirectional communications. It is also possible to reduce peak power demand by requesting that customers move their appliances in a methodical manner. Several methods have been proposed [17–23] to reduce the peak demand for electricity. Mahmood et al. [24] have proposed an autonomous scheme for residents with installed power meters and energy control units. By scheduling home appliances, the scheme reduces power consumption costs and peak power demand. But (a) these techniques require active participation from consumers and providers, which can be challenging in real-world scenarios; (b) these techniques have not been successfully applied on a large scale to residential customers; (c) scheduling home appliances may cause increased inconvenience for residents because they want to schedule the appliances according to their needs.

Rastegar et al. propose an optimization formulation for minimizing the cost of consumption [25]. Using a nonlinear control method, Yang et al. [26] have proposed a price-based demand response (DR) scheme. It is a nonlinear power management system with price feedback. Alsalloum et al. [27] discuss a Stackelberg game theory model for optimizing price and energy demands via interaction between prosumer and provider. Mishra & Parida [28] propose a new pricing structure with an integrated feature to capture the benefits of both real-time pricing (RTP) and time of use (TOU).

Bin-Halabi et al. [29] have proposed an algorithm that proportionally allocates the utility-specified threshold to consumers' appliances. The techniques discussed by [30, 31] classify loads as either essential or non-essential. These systems aim to first allocate power to essential loads, and then distribute the remaining power to non-essential loads. Salimian & Aghamohammadi [32] have discussed predicting blackouts with decision trees (constructed using a combination of parameters such as frequency and voltage index). This chapter's algorithm provides a pragmatic approach to the equitable allocation of thresholds. Additionally, it ensures that the proportion of violations attributable to households in a region is minimal at all times.

11.3 METHODOLOGY

Whenever there is a shortage of power supply relative to demand in a given time slot, utilities define a maximum power limit (called a load threshold) that can be consumed without penalty during this time slot. If the region's power consumption exceeds this limit, the utility will impose a penalty. This load threshold allocated by the utility to a distribution region in time slot t is denoted as D_t in this chapter. The developed algorithms then distribute the local load threshold to individual households. It is indicated by the notation $d_{t,h}$ for household h in time slot t. Table 11.1 provides specifics about the various notations utilized in this chapter.

The utility assigns a fixed load threshold to the distribution region for a given time slot. Before the start of the time slot, it is necessary to determine the load threshold that will be assigned to each household. This ensures that the threshold does not change during the allotted time period.

Since multiple algorithms are proposed in this chapter, it is imperative to have a metric that can be used to compare different approaches for load threshold algorithm. This metric is termed as mean percentage violation (*MPV*). Let us consider an example to understand how this metric is calculated. Assume a household is allocated a load threshold of 200 Watts for the time slot t. Now if the power consumption of this house exceeds 200 Watts during time slot t, then these will be treated as violation. For a given

Table 11.1 Semantics of notations

Notation	*Semantics*
D_t	Load threshold allocated by the utility to the distribution region for time slot t
$\{H\}$	Set of households present in the distribution region
t	Time slot
$\|S\|$	Slot size
$l_{t,h}$	Power consumption of household h in time slot t
$l'_{i,h}$	Power consumption of household h at time instance x
l_h^{max}	Maximum power that may be consumed by household h
$algo$	Load threshold allocation algorithm
$d_t^{algo}[h]$	Local load threshold allocated to household h by algorithm $algo$ in time slot t
rr^{turn}	Household id that is selected for maximum load threshold allocation while implementation of round robin based threshold allocation
$vcnt_h$	Number of violations incurred by household h

time slot and house, the ratio of (a) summation of power consumption of this house at time instances for which violation has occurred and (b) total power consumption of this house in the given time slot – is calculated. Taking mean of this violation ratio across the entire neighborhood will give the value of *MPV*. This is denoted by Eq. (11.1) as follows:

$$MPV_t = \frac{\sum_{h\in\{H\}} \left\{ \frac{\sum_{j\in\{t\}} l'_{j,h}}{\sum_{i\in\{t\}} l'_{i,h}} \mid l'_{j,h} > d_t[h] \right\}}{|H|} \tag{11.1}$$

where MPV_t represents MPV for time slot t, $l'_{i,h}$ denotes power consumption of house h at time instance i and $d_t[h]$ represents load threshold allocated to house h in time slot t. Suppose there are two algorithms $\{algo_i, algo_j\}$ that allocate load thresholds to households in a neighborhood. Now consider the following scenario:

$$MPV_t^{algo_i} < MPV_t^{algo_j}$$

It denotes that the value of MPV of algorithm $algo_j$ in time slot t is higher than that of algorithm $algo_i$ in the same time slot; This implies that the threshold allocated by $algo_i$ leads to lesser violations as compared to $algo_j$ and thus, is preferred over $algo_j$.

The proposed infrastructure required for this brownout implementation is shown in Figure 11.1 and is briefly described below. When a region's

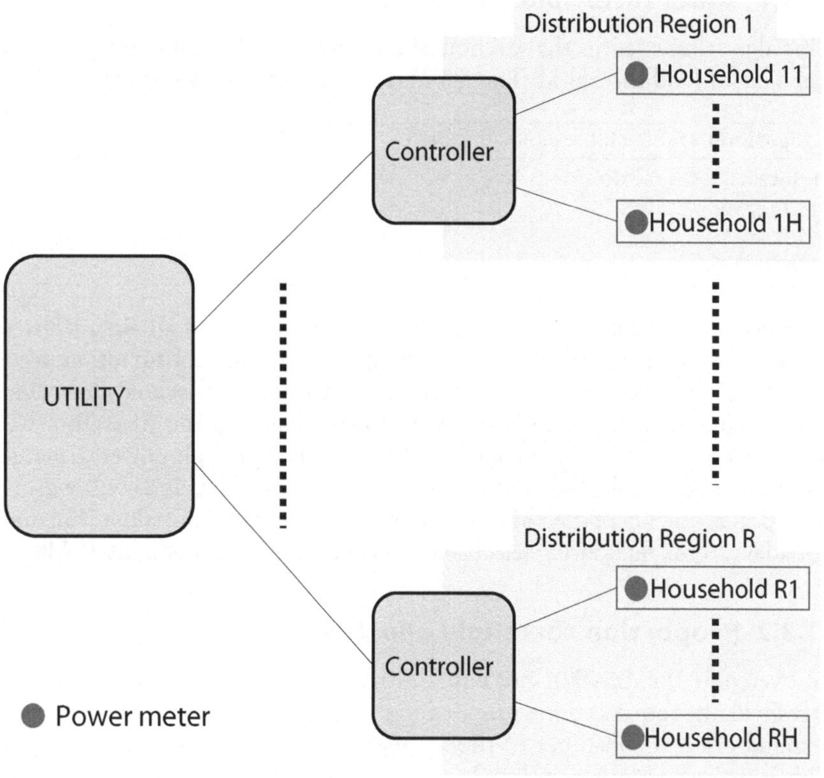

Figure 11.1 Proposed layout for implementing the novel load threshold allocation algorithm.

power demand is high enough to cause the grid to become overloaded, the utility decides to restrict the region's power consumption. In a neighborhood, a controller is present that receives data on energy consumption from the power meters installed in each home. It assigns the threshold load to each household in the region using the algorithms developed in this chapter. If a household's power consumption exceeds the load threshold set by the controller, then the controller transmits this information to the utility and the household is penalized. This assists the utility in regulating the energy demand of residential customers.

A load threshold allocation should ensure that all households in the distribution region are treated fairly. Fairness in threshold allocation is necessary because it prevents certain households from starving. The developed algorithms utilize various heuristic criteria to fairly assign thresholds to households. All algorithms discussed in this chapter receive as input the time t and threshold D_t allocated to the region by the utility. Their specifics are described below.

11.3.1 Equal threshold allocation

This algorithm ensures that all households in a distribution region are allocated equal load threshold. The details are presented in Algorithm 1.

Algorithm 1: Equal threshold allocation

foreach $h \in \{H\}$ **do**
  $d_t^{equal}[h] \leftarrow \frac{D_t}{|H|}$
return d_t^{equal}

In terms of implementation, this algorithm is the simplest of all algorithms. It functions well when households have comparable energy consumption needs during a given time period. In scenarios in which some households have high power demands and others have lower power demands, this algorithm will result in a high *MPV* of the region, as households with high power demands will have higher violations and households with lower demands will waste a (significant) portion of their allotted thresholds. Therefore, this algorithm does not adapt to the fluctuating (electricity consumption) needs of households.

11.3.2 Proportion threshold allocation

To overcome the drawback of equal distribution, in this algorithm, power threshold allocation is done proportionally. The ratio is calculated as power demand of each consumer to the aggregate demand of the neighborhood. This is presented in Algorithm 2.

Algorithm 2: Proportion threshold allocation

foreach $h \in \{H\}$ **do**
  $d_t^{prop}[h] \leftarrow \left(\frac{l_{t,h}}{\sum_{h' \in \{H\}} l_{t,h'}} \right) \times D_t$;
return d_t^{prop}

Unlike the equal allocation of load threshold, this algorithm takes into account the varying energy consumption requirements of different households. Nonetheless, it has the following flaws: (a) it does not incentivize households with lower power demands, and (b) it unfairly assigns high load thresholds to households with high power demands; the consumer can thus unfairly increase its demand to get higher power threshold allocation from this algorithm.

11.3.3 Power usage ratio based threshold allocation

A household with a lower energy consumption cannot be considered superior to one with a higher energy consumption. This is due to the fact that a home with more residents, appliances, and square footage will consume more electricity than one with fewer residents, fewer appliances, and a smaller square footage. Therefore, assigning a higher threshold to households that consume less energy is not a fair method of allocation. In this regard, the primary heuristic used by Algorithm 3 to allocate load thresholds is the ratio of a household's power consumption in the preceding time slot to the maximum power it could have consumed.

Algorithm 3: Power usage ratio based threshold allocation

foreach $h \in \{H\}$ **do**

$$val_h \leftarrow 1 - \left(\frac{l_{t-1,h}}{\sum_{i=1}^{|S|} \max(l | l_{h,j} \forall j = 1 \ldots |S|)} \right);$$

$$val^{sum} \leftarrow \sum_{h \in \{H\}} val_h$$

foreach $h \in \{H\}$ **do**

$$d_t^{power}[h] \leftarrow \left(\frac{val_h}{val^{sum}} \right) \times D_t;$$

return d_t^{power}

This facilitates equitable distribution among households. On the basis of this ratio, households are ranked, and those with lower values are assigned higher load thresholds. Therefore, this algorithm rewards households whose power consumption behavior is more prudent and frugal – a feature absent from proportional distribution. This results in fewer violations of household-specific thresholds.

11.3.4 Round robin threshold allocation

Algorithm 4 allocates load thresholds in a round robin manner, explained as follows:

- rr^{turn} is initialized with the minimum household id in the respective distribution region.
- in a given time slot t, rr^{turn} is initialized with the threshold that is minimum of available threshold and maximum power demand of this household.
- if any threshold is available to be distributed amongst other households, it is then distributed amongst them using any allocation algorithm.
- rr^{turn} is then initialized with the id of next household (using *Next* function) in the respective distribution region.

Algorithm 4: Round Robin threshold allocation

$max^l_h \leftarrow \max(l_{t',h} | \forall t' \in \{T\}, h = rr^{turn})$;
if $max^l_h \geq D_t$ **then**
  $d^{rr}_t[rr^{turn}] \leftarrow D_t$
  foreach $\bar{h} \in \{H \setminus rr^{turn}\}$ **do**
    $d^{rr}_t[\bar{h}] \leftarrow 0$
else
  $d^{rr}_t[rr^{turn}] \leftarrow max^l_h$
  $avail \leftarrow D_t - max^l_h$
  foreach $\bar{h} \in \{H \setminus rr^{turn}\}$ **do**
    $d^{rr}_t[\bar{h}] \leftarrow \left(\frac{l_{t,\bar{h}}}{\sum_{h' \in \{H \setminus rr^{turn}\}} l_{t,h'}} \right) \times avail$
$rr^{turn} \leftarrow Next(\{H\})$

This process is repeated for every time slot. Unlike other algorithms, it gives a fair chance to every household at least once in every $|H|$ time slots to fulfill maximum power demand. But the disadvantages of this approach are as follows:

- MPV of households that are allotted the remaining threshold will be very high since the remaining threshold is generally very less.
- household with id as rr^{turn} that is allowed to satisfy its maximum power demand may waste the allotted threshold since this household may not consume its maximum power at this given time slot.
- if the number of houses is very large in number, then a household will have a higher waiting time when it can satisfy its full power demand.

11.3.5 Violation-based threshold allocation

Algorithm 5 uses the violation count as a heuristic to allocate threshold – as one of the requirements for load threshold allocation algorithms is minimal violation across all households. It keeps track of the number of instances in the past in which the household's power consumption exceeded the load threshold allotted for the respective time slot. This is denoted by $vcnt_h$, where h is the household. On the basis of this number, households with a greater number of violations are assigned relatively lower load thresholds.

Algorithm 5: Violation based threshold allocation

foreach $h \in \{H\}$ **do**
    foreach $i \in \{t-1\}$ **do**
        if $l'_{i,h} \geq d_{t-1}[h]$ **then**
            $vcnt_h \leftarrow vcnt_h + 1$
foreach $h \in \{H\}$ **do**
    $ratio_h^{vio} \leftarrow \left(1 - \left(\frac{vcnt_h}{|S|}\right)\right)$
foreach $h \in \{H\}$ **do**
    $d_t^{violate}[h] \leftarrow \left(\frac{ratio_h^{vio}}{\sum_{h' \in \{H\}} ratio_{h'}^{vio}}\right) \times D_t$

This algorithm gives preference to households with a history of committing fewer violations. Thus, it encourages households to avoid exceeding assigned load limits. However, if a household has a very high violation rate during a specific time slot, the algorithm will punish this household for a very long period of time. Thus, even if the household's behavior improves, the fair allocation of threshold to this household will take a considerable amount of time.

function *get_MPV(t)*
foreach $algo \in \{algorithms\}$ **do**
    foreach $h \in \{H\}$ **do**
        $v_1^h \leftarrow 0$
        foreach $i \in \{t\}$ **do**
            if $l'_{i,h} > d_t^{algo}[h]$ **then**
                $v_1^h \leftarrow v_1^h + l'_{i,h}$
        $val_h \leftarrow \frac{v_1^h}{\sum_{i \in \{t\}} l'_{i,h}}$
        $MPV_t^{algo}[h] \leftarrow val_h \times 100.0$
return *MPV*
end function

11.3.6 Optimal threshold algorithm

Algorithm 6: Optimal Threshold Allocation

$optimal_t^{algo} \leftarrow \phi$

$\forall h \in \{H\} : optimal_t^{d}[h] \leftarrow 0$

$mean_t^{mpv} \leftarrow 0$

$MPV = get'MPV(t)$

foreach $algo \in \{algorithms\}$ **do**

  $mean_t^{algo} \leftarrow mean\left(MPV_{t,h}^{algo} \middle| h \in \{H\}\right)$

$optimal_t^{algo} \leftarrow argmin_{algo}\left(mean_t^{algo} \middle| algo \in \{algorithms\}\right)$

foreach $h \in \{H\}$ **do**

  $optimal_t^{d}[h] \leftarrow d_t^{optimal_t^{algo}}[h]$

$mean_t^{mpv} \leftarrow \min\left(mean_t^{algo} \middle| algo \in \{algorithms\}\right)$

The utility prefers a load threshold allocation algorithm that has a lower *MPV*. A particular algorithm among the algorithms discussed thus far may have a low *MPV* value in one time slot, but none of them guarantees a minimal *MPV* in all time slots. Before the start of each time slot, it simulates these algorithms and calculates their *MPV* values (using the function *get_MPV*); it then selects the allocation algorithm with the smallest *MPV* value. This results in a load threshold allocation with minimal *MPV* across all time slots. Set $\{algorithms\}$ is initialized with the list of threshold allocation algorithms. The output of the 6 algorithm is as follows:

- $optimal_t^{algo}$: algorithm *algo* with minimum *MPV* value in time slot *t* that is selected by the optimal algorithm
- $optimal_t^{d}[h]$: load threshold allocated by the optimal algorithm to household *h* in time slot *t*
- $mean_t^{mpv}$: mean of the *MPV* value of the optimal algorithm in time slot *t* across all the households

The primary benefit of this algorithm is that it guarantees a minimum amount of *MPV* in each time slot. In addition, it is more comprehensive in terms of load threshold allocation fairness because it takes into account different algorithms in different time slots; thus, a household whose threshold is not allocated fairly due to a particular algorithm will be allocated fairly by another algorithm. Therefore, it prevents households from going without load threshold allocation.

11.4 RESULTS AND EVALUATION

This section describes and compares the outcomes of applying various algorithms developed in this chapter to an actual dataset.

11.4.1 Dataset

Dataset description: Smart* Dataset [33] is the dataset used to evaluate the developed algorithms. It contains minute-by-minute electricity consumption data for over 400 anonymous households. To understand the efficacy of the developed algorithms, a 20-household distribution region is considered. In the case study examined in this chapter, a total of 24 one-hour time slots are examined. The total power consumption of a given time slot is calculated as the median of the power consumption values that exceed the 75th percentile for that time slot. This is a more accurate representation of the power consumption in that time slot than the mean or median value. As shown in Figure 11.2, the load threshold allocated by the utility to the distribution region is simulated as 70–80% (uniform random sampling) of the total aggregate power of the distribution region in the given time slot. Figure 11.3 depicts the hourly consumption of electricity by each of the 20 households on the first day of February. The results obtained after applying the developed algorithms to the households are evaluated as follows.

11.4.1.1 *Performance of algorithms in the distribution region*

MPV (Eq. (11.1)) is utilized to compare the performance of the developed algorithms. A household-wide algorithm with a lower *MPV* value is

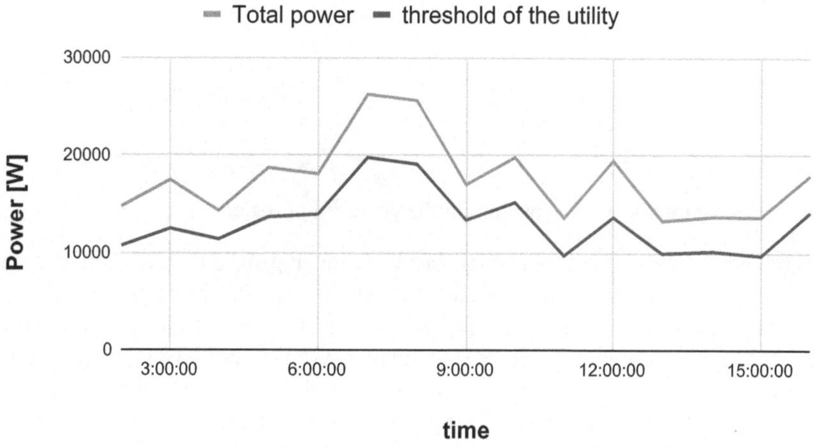

Figure 11.2 Threshold power allocated by the utility and total power of the distribution region.

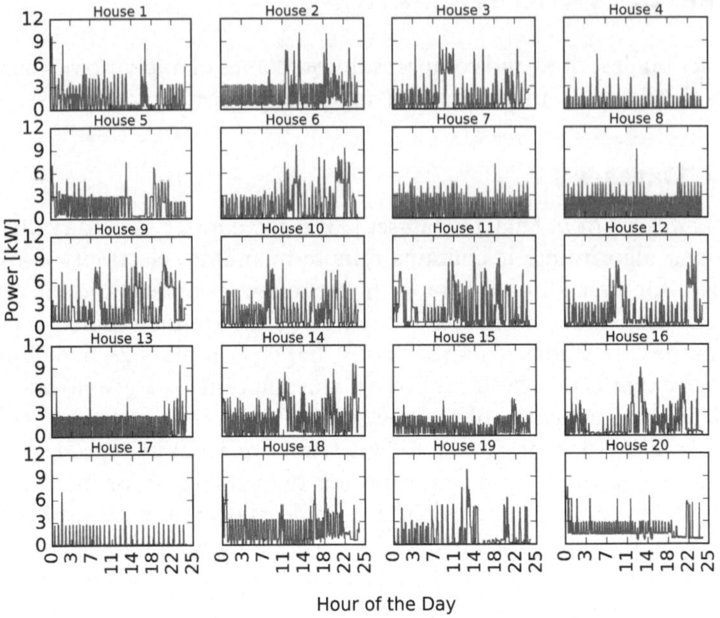

Figure 11.3 Power consumption of all the households for a given day.

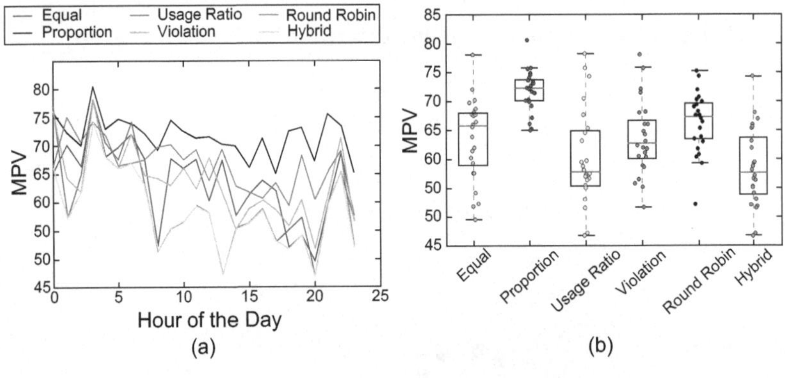

Figure 11.4 Comparison of algorithms in the distribution region.

preferred over one with a higher *MPV* value. Figure 11.4a depicts a summary of algorithms when applied to all of the region's households. It demonstrates that *MPV* of the optimal algorithm is the smallest across all time slots. Thus, the objective of having lowest *MPV* is achieved by this novel algorithm. Figure 11.4c displays a box plot depicting the *MPV* value for each algorithm. It indicates the algorithmic variation in relation to their *MPV* values. The following conclusions can be drawn from this graph:

- Proportion-based allocation has the lowest standard deviation but the highest median of all algorithms. This indicates that the *MPV* values for all time slots are high. Even an outlier exists with a *MPV* value greater than 80%.
- The second quartile ranges of the round robin and violation algorithms are nearly identical, but the median of the round robin algorithm is higher than that of the violation algorithm.
- The usage algorithm has the lowest median, which is why it is chosen the most frequently by the optimal algorithm.
- Because the optimal algorithm has the lowest *MPV* value across all time slots, its second quartile range is less than that of the usage algorithm.

In Table 11.2, the various algorithms selected by the optimal algorithm in each time slot are described in detail. The optimal algorithm allocates load thresholds based on the power usage ratio 54.17% of the time. Equal

Table 11.2 Algorithms selected by the optimal algorithm in each time slot

Hour of the Day	*Algorithm with minimal MPV value*	*MPV value*
0	Equal	65.6
1	Usage	57.24
2	Violate	61.96
3	Round robin	74.3
4	Equal	68.1
5	Violate	66.05
6	Round robin	66.98
7	Usage	63.19
8	Usage	51.52
9	Usage	55.38
10	Usage	56.72
11	Usage	59.49
12	Usage	58.17
13	Usage	47.18
14	Violate	55.14
15	Usage	56.39
16	Usage	58.79
17	Usage	53.1
18	Equal	51.86
19	Equal	54.21
20	Usage	46.77
21	Equal	59.34
22	Usage	65.36
23	Round robin	52.1

threshold allocation is chosen 20.83% of the time, while round robin and violation-based threshold allocation are chosen 12.5% of the time each. Therefore, the equal allocation method is superior to the violation and round robin allocation methods. The proportion allocation algorithm is never chosen by the optimal algorithm; consequently, allocation based on their relative power demand does not significantly decrease the *MPV* value. In comparison to other heuristics for threshold allocation, it can be concluded that allocation based on the power usage ratio of each household has the greatest effect on the *MPV* value of the region. Thus, to minimize the *MPV* value across all households, the optimal approach allocates threshold primarily based on the power usage ratio of the households, as the usage algorithm dominates threshold allocation in the majority of time slots.

11.4.1.2 Performance of algorithms at household level

Figure 11.5 depicts the *MPV* value for each household according to the algorithms. Although the optimal algorithm has the lowest *MPV* in each time slot for each distribution region, it may not be the lowest at the household

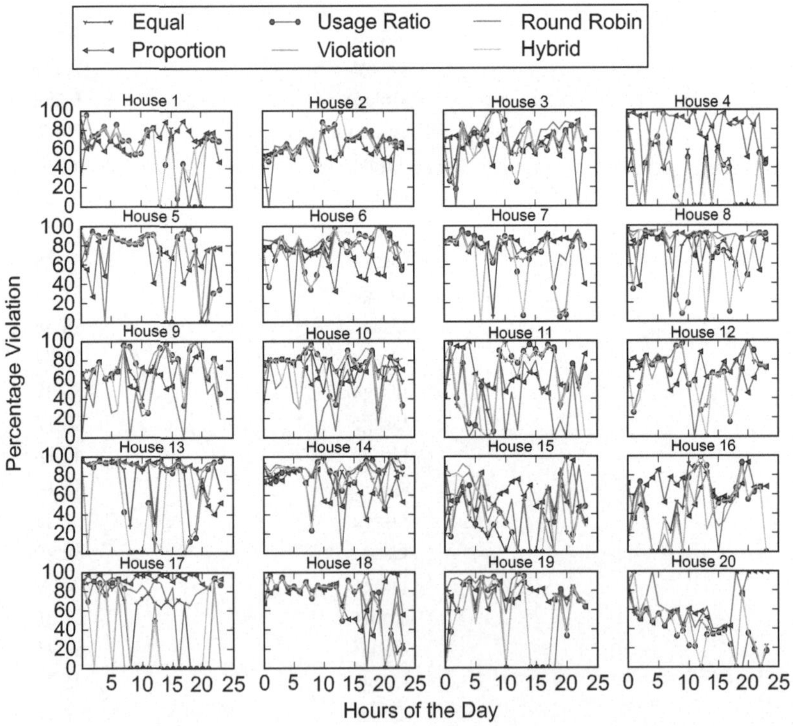

Figure 11.5 Percentage violation of each algorithm during different time slots of the day for various houses.

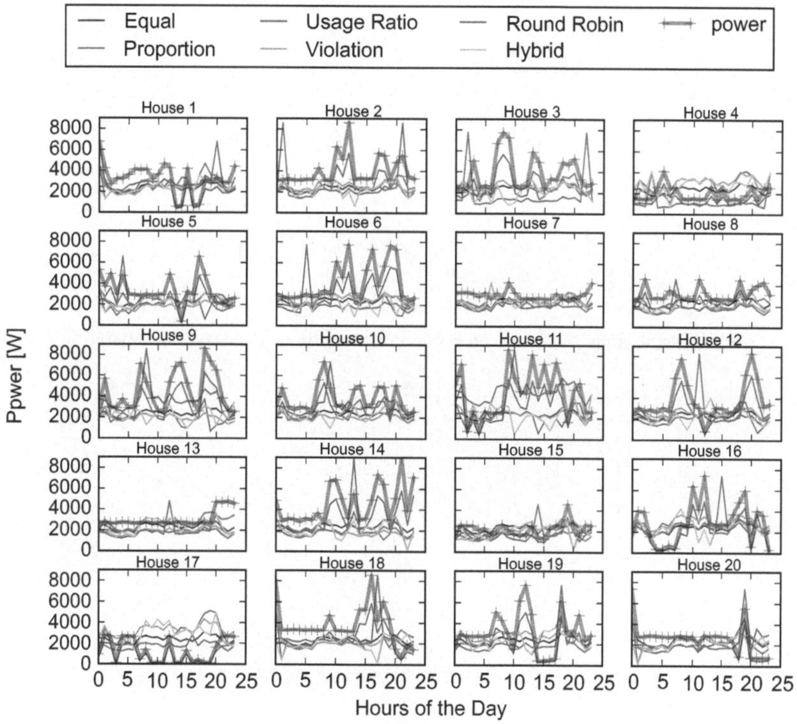

Figure 11.6 Load thresholds allocated by the algorithms to different houses (thick line denotes the power consumption of the respective house).

level. Consider, for instance, House 1. During the first eight hours, the optimal algorithm does not have the minimum *MPV* value for House 1. Figure 11.4 depicts the mean *MPV* of all households for a given time slot, which is the lowest for the optimal algorithm in each time slot. Figure 11.6 illustrates the time instances in which each household violated the load threshold, with the thick line representing the power consumption of each residence and the other lines representing the load thresholds allocated by various algorithms.

11.4.1.3 Overall behavior of the algorithms

Figure 11.7 depicts a box plot for each algorithm applied to all households. It indicates the range of *MPV* values. The following are the conclusions drawn from this boxplot:

- From Figure 11.3, it can be seen that the power consumption of House 4 is generally lower than that of other houses during all time slots. As shown in Figure 11.7, the *MPV* values for this residence's equal, usage, and violation algorithms are low. This is due to the fact that these

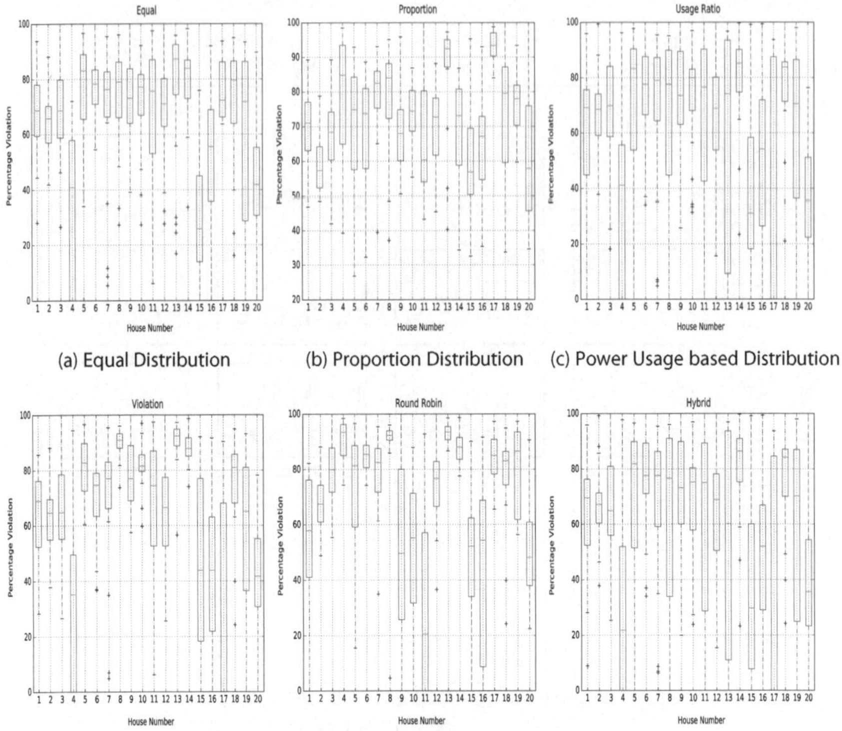

(a) Equal Distribution (b) Proportion Distribution (c) Power Usage based Distribution

Figure 11.7 Box plots denoting MPV of the algorithms.

algorithms incentivize this low consumption behavior, whereas round robin and proportional allocation do not.

- Equal algorithm reveals high *MPV* values for the majority of houses (Figure 11.7a). This is due to the fact that it does not account for the energy needs of households. Consequently, Houses 4, 15, 16, 19, and 20 with a lower power consumption profile have lower *MPV* values. The majority of the allocated threshold for these residences is lost due to their low consumption. As a result, other homes have high *MPV* values. Figure 11.7b depicts that the proportion allocation algorithm has a high violation rate for all houses. Therefore, it is never chosen by the optimal algorithm for threshold allocation to regional households.
- Figure 11.7c depicts that the *MPV* values for the majority of houses are on the low end for the algorithm based on the power usage ratio. This algorithm intelligently considers the ratio of power consumed to the maximum power consumed by the household. Consequently, it modifies the threshold allocation such that lower *MPV* values are observed in the majority of households and is chosen the most frequently by the optimal algorithm (Figure 11.7f).
- Similar to the equal allocation algorithm, the violation-based algorithm results in lower *MPV* values for Houses 4, 15, 16, 19, and 20 (Figure 11.7d). As a result of these households' lower energy

consumption, fewer violations are observed in these homes. In contrast, the number of violations is high for other households. This algorithm is only selected 12.5% of the time by the optimal threshold allocation algorithm.

- Round robin algorithm has *MPV* on the higher side (Figure 11.7e). This is primarily because in each time slot, household rr^{turn} that is selected to fulfill its maximum demand might not want to consume maximum power at that time slot; consequently, the remaining threshold after the allocation to this selected household is generally very less to satisfy the demands of other households satisfactorily. This leads to high *MPV* values for the remaining houses.

11.4.2 Evaluation of algorithms using other metrics

Until now, the algorithms have been evaluated using *MPV* as the evaluation metric. *MPV* represents an average percentage of algorithm violations across all the homes in the neighborhood for different time slots throughout the day. However, averaging the violations does not always reveal the significant violation that may have occurred at a specific residence. Consequently, the algorithms have also been evaluated based on a standard statistical metric known as the L2 norm, and the results will be represented as *L2viol*. It represents the squared difference between the percentage of violations that occurred in each home and the lowest possible violation rate for each home. It is represented by the equation 11.2.

$$L2viol_t = \sqrt[2]{\left(\sum_{\forall h \in \{H\}} \left(viol_t[h] - viol_t^{min}[h] \right) \right)} \tag{11.2}$$

where $viol_t[h]$ represents percentage violation of house h in time slot t, and $viol_t^{min}[h]$ represents minimum percentage violation of house h in time slot t. In the best-case scenario, the minimum percentage of violations across all residences and time periods of the day could be zero. The obtained results are depicted in Figure 11.8.

In comparison to other algorithms used for load threshold allocation, the value of the metric *L2viol* is lowest for the optimal algorithm across all time slots of the day, as depicted in the figure. *L2viol* can also be calculated for various algorithms on a given day by applying metric L2 norm to the *MPV* values of each algorithm for various time slots. This metric is denoted by the symbol *L2viol′*.

Table 11.3 displays the value of the *L2viol′* metric for various algorithms on a given day. From Table 11.3, the following are the observations:

- Proportion algorithm has the highest *L2viol′* metric value signifying that it leads to the highest overall violations.
- Amongst all the load threshold algorithms (except the optimal algorithm), the usage ratio algorithm has the minimum *L2viol′* value.

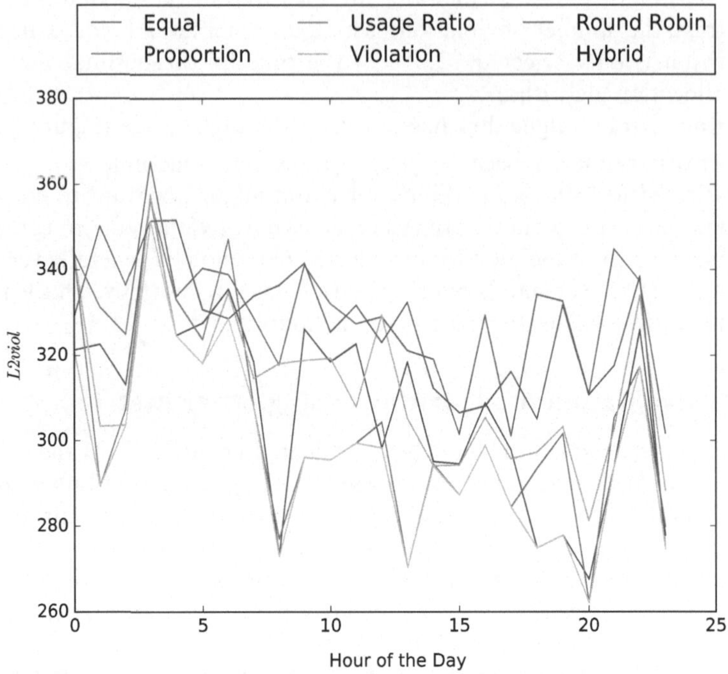

Figure 11.8 $L2viol$ value of load threshold algorithms for different time slots.

Table 11.3 L2viol' values of different load threshold algorithms for the given day

Algorithms used for load threshold allocation	***L2viol'*** *value for the given day*
Equal	313.36
Proportion	352.07
Usage Ratio	297.60
Violation	313.21
Round Robin	326.39
Optimal	288.66

- Round robin algorithm has a higher $L2viol'$ value as compared to equal and violation algorithms.
- Equal and violation algorithms have an almost similar value of $L2viol'$ metric.
- The optimal algorithm has a minimum metric value.

These violation patterns are also observed in Figure 11.4c for the MPV metric. As a result, it can be confirmed that the optimal algorithm allocates the load threshold with the least violations across all time slots, relative to other algorithms.

11.5 CONCLUSION

It is essential to ensure that a minimal percentage of the allocated threshold for households is violated during all time slots. Additionally, households should be allocated thresholds fairly. In this regard, this chapter has developed and tested (using a dataset from the real world) five algorithms that use various heuristics to allocate thresholds fairly to households. After evaluating all of the results, it is possible to conclude that the novel optimal algorithm ensures minimal percentage violations (using *MPV* and *L2viol* metrics) in each time slot and prevents households from getting starved in terms of threshold allocation. The results encourage us to adopt this approach in the practical and effective implementation of brownouts for managing the electric grid's peak power consumption demand.

REFERENCES

1. EIA - Electricity Data Eiagov 2018. https://www.eia.gov/electricity/monthly/epm_table_grapher.php?t=epmt_5_01
2. Amatya V B, Chandrashekar M and Robinson J B 1993 Residential sector energy-supply-demand analysis: A modelling approach for developing countries and fuelwood-supply sustainability in nepal. *Energy* 18(4): 341–354. http://www.sciencedirect.com/science/article/pii/036054429390069P
3. Subbiah R, Pal A, Nordberg E, Marathe A and Marathe M 2017 Energy demand model for residential sector: A first principles approach. *IEEE Transactions on Sustainable Energy* 8: 1215–1224
4. Strbac G 2008 Demand side management: Benefits and challenges. *Energy Policy* 36(12): 4419–4426. http://www.sciencedirect.com/science/article/pii/S0301421508004606
5. Lawton P 2012 Balancing the energy network. *Ingenia* 53: 20–26
6. Gellings C W 2009 *The Smart Grid: Enabling Energy Efficiency and Demand Response*. Fairmont Press, Lilburn.
7. Gellings C W 1981 Power/energy: Demand-side load management: The rising cost of peak-demand power means that utilities must encourage customers to manage power usage. *IEEE Spectrum* 18(12): 49–52
8. Blume S and Sons J W 2016 *Electric Power System Basics: For the Nonelectrical Professional*. IEEE Press series on power engineering, Wiley-Interscience. https://books.google.co.in/books?id=lxFrnQAACAAJ
9. Wyatt A 1986 *Electric Power: Challenges and Choices*. The Book Press Limited, Toronto
10. Albadi M H and El-Saadany E F 2007 Demand response in electricity markets: An overview. In: *2007 IEEE Power Engineering Society General Meeting*, Tampa, FL, pp. 1–5. doi: 10.1109/PES.2007.385728
11. A. Agarwal and K. Ramamritham 2020 Sensor minimization for energy management in smart buildings. In: *2020 IEEE First International Conference on Smart Technologies for Power, Energy and Control (STPEC)* pp. 1–6. https://doi.org/10.1109/STPEC49749.2020.9297755
12. Elma O and Selamogullari U S 2015 A new home energy management algorithm with voltage control in a smart home environment. *Energy* 91: 720–731

13. Zhou S, Wu Z, Li J and ping Zhang X 2014 Real-time energy control approach for smart home energy management system. *Electric Power Components and Systems* 42(3–4): 315–326. https://doi.org/10.1080/15325008.2013.862322
14. Abouelela M A and Abouelela M 2016 Home energy management system (hems) for fair power distribution. *Advanced Science Letters* 22: 2638–2641
15. Agarwal A. 2022 A novel approach to save energy by detecting faulty HVACs. *Journal of the Institution of Engineers Series B* 103: 305–311. https://doi.org/10.1007/s40031-021-00666-7
16. Costanzo G T, Zhu G, Anjos M F and Savard G 2012 A system architecture for autonomous demand side load management in smart buildings. *IEEE Transactions on Smart Grid* 3(4): 2157–2165
17. Shafie-Khah M, Talari S, Wang F and Catalão J P S 2020 Decentralised demand response market model based on reinforcement learning. *IET Smart Grid* 3(5): 713–721
18. Akasiadis C and Chalkiadakis G 2017 Cooperative electricity consumption shifting. *Sustainable Energy, Grids and Networks* 9: 38–58. http://www.sciencedirect.com/science/article/pii/S2352467716301850
19. Ebrahimi J, Abedini M and Rezaei M M 2020 Optimal scheduling of distributed generations in microgrids for reducing system peak load based on load shifting. *Sustainable Energy, Grids and Networks* 23: 100368. http://www.sciencedirect.com/science/article/pii/S235246772030299X
20. Stamatescu G, Stamatescu I, Arghira N and Fagarasan I 2019 Data-driven modelling of smart building ventilation subsystem. *Journal of Sensors* 2019: 1–14
21. Jain A, Smarra F, Behl M and Mangharam R 2018 Data-driven model predictive control with regression trees—an application to building energy management. *ACM Transactions on Cyber-Physical Systems* 2(1). https://doi.org/10.1145/3127023
22. Irwin D, Iyengar S, Lee S, Mishra A, Shenoy P and Xu Y 2017 Enabling distributed energy storage by incentivizing small load shifts. *ACM Transactions on Cyber-Physical Systems* 1(2). https://doi.org/10.1145/3015663
23. Kumar G K, Maniadarsh S, Thungeshwaran R, Ashwin M and Jayabarathi R 2020 Remotely controllable consumer perspective demand response using genetic algorithm. In: *2020 Fourth International Conference on I-SMAC (IoT in Social, Mobile, Analytics and Cloud) (I-SMAC)*, Palladam, India, pp. 904–908
24. Mahmood A, Ullah M, Razzaq S, Basit A, Mustafa U, Naeem M and Javaid N 2014 A new scheme for demand side management in future smart grid networks. *Procedia Computer Science* 32: 477–484. http://www.sciencedirect.com/science/article/pii/S1877050914006504, the 5th International Conference on Ambient Systems, Networks and Technologies (ANT-2014), the 4th International Conference on Sustainable Energy Information Technology (SEIT-2014)
25. Rastegar M, Fotuhi-Firuzabad M and Lehtonen M 2015 Home load management in a residential energy hub. *Electric Power Systems Research* 119: 322–328

26. Yang J, Zhang G and Ma K 2016 A nonlinear control method for price-based demand response program in smart grid. *International Journal of Electrical Power & Energy Systems* 74: 322–328. http://www.sciencedirect.com/science/article/pii/S0142061515003105
27. Alsalloum H, Merghem-Boulahia L and Rahim R 2020 Hierarchical system model for the energy management in the smart grid: A game theoretic approach. *Sustainable Energy, Grids and Networks* 21: 100329. http://www.sciencedirect.com/science/article/pii/S2352467719304837
28. Mishra M K and Parida S K 2020 A game theoretic approach for demand-side management using real-time variable peak pricing considering distributed energy resources. *IEEE Systems Journal* 16: 1–11
29. Bin-Halabi A, Nouh A and Abouelela M 2018 Interactive energy management system to avoid rolling blackouts. In: *2018 5th International Conference on Electric Power and Energy Conversion Systems (EPECS)*, Kitakyushu, Japan, pp. 1–7
30. Kelkar S, Kothari N and Ramamritham K 2015 Brownout energy distribution scheme for mitigating rolling blackouts. In: *Proceedings of the 2015 ACM Sixth International Conference on Future Energy Systems*, e-Energy '15, Association for Computing Machinery, New York, NY, pp. 193–194. https://doi.org/10.1145/2768510.2770937
31. Raj B D, Kumar S, Padhi S, Sarkar A, Mondal A and Ramamritham K 2018 Brownout based blackout avoidance strategies in smart grids. In: *Proceedings of the Ninth International Conference on Future Energy Systems*, e-Energy '18, Association for Computing Machinery, New York, NY, pp. 456–458. https://doi.org/10.1145/3208903.3212059
32. Salimian M R and Aghamohammadi M R 2018 A three stages decision tree-based intelligent blackout predictor for power systems using brittleness indices. *IEEE Transactions on Smart Grid* 9(5): 5123–5131
33. Barker S, Mishra A, Irwin D, Cecchet E and Shenoy P 2012 Smart*: An open data set and tools for enablingresearch in sustainable homes. In: *Proceedings of the ACM SustKDD, August 12, 2012*, Association for Computing Machinery, New York, NY

Chapter 12

Using a novel game theory auctioning framework for trading energy in smart grids

Anshul Agarwal

12.1 INTRODUCTION

According to a report by the International Energy Agency, non-renewable energy sources account for 80 to 85% of all the electricity produced in India [1]. Furthermore, peak consumption hours are when power demands are at their highest [2]. In order to meet peak demand, utilities must operate their backup power plants, which are not environmentally friendly and raise the cost of electricity per unit. Many developing nations lack the infrastructure necessary to meet these expanding demands. They consequently experience a serious power shortage. In this regard, the use of renewable energy sources as a substitute for traditional energy sources is encouraged. This is accomplished by combining both renewable and traditional energy sources and implementing a two-way or bi-directional communication architecture in smart grids [3]. Prosumers (those who produce and consume energy [4])) and utilities both gain from the smart grid in different ways. Prosumers, for instance, can earn from trading the extra energy they produce locally [5, 6]. Various policies and programs are also put into place by the government to encourage and promote the usage of renewable energy sources [7, 8]. The infrastructure for two-way communication offered by the smart grid makes this possible. The research and studies cited in [9] offer a summary of the many technologies and cutting-edge communication needed to establish a smart grid system. Because smart grids are complicated systems, reliable solutions and methods are required to make the energy trading process easier to execute [10].

The smart grids can implement peer-to-peer energy trading. A detailed review of the recent advances in the application of game-theoretic methods to local energy trading scenarios is presented in [11]. An extensive description of a complete game theory-based energy trading framework is presented. Game-theoretic approach is proposed by Liu et al. [12] to optimize the storage capacity and energy consumption considering the uncertainty of distributed generation. Sun et al. [13] have proposed a distance-oriented method for maximizing coalition utilities to guide the order of transactions in a coalition, and the corresponding coalition utility, thus, would be

 DOI: 10.1201/9781003623168-16

fairly allocated by the Shapley value. Li et al. [14] discuss a tri-layer noncooperative energy trading approach among multiple grid-tied multi-energy microgrids (MEMGs) in the restructured integrated energy market. In the literature, several difficulties related to power limit distribution and PMU allocation in smart grids have been discussed [15–17]. A three-level gameplay-based intelligent structure to evaluate individual and collaborative strategies of electricity manufacturers, considering network and physical constraints, is proposed by Moafi et al. [18]. Two iterative algorithms are proposed in [19] for the implementation of energy trading such that an equilibrium state exists in each of the games. A hierarchical system architecture model is proposed by Zhang et al. [20] to identify and categorize the key elements and technologies involved in P2P energy trading. Prosumers can communicate on a P2P market platform using either a centralized approach [21] or a decentralized approach [22]. To reduce power losses, authors of [23] have suggested installing distributed generation (DG) and capacitors in the distribution system at the best possible location and size. Authors of [24] have suggested a novel strategy to reduce the amount of energy used by the computer center. But the discussed approaches do not integrate P2P energy with auction and forecasting models – which help in higher savings and increased user satisfaction.

To make P2P techniques more pragmatic for consumers to adapt to, there is still room for development. In this regard, the primary contribution of this chapter is the development of a P2P energy trading framework that incorporates an auction-based game theory method. This boosts cost savings for prosumers and consumers compared to trading with the primary energy provider or utility directly. Demand, storage capacity, and transmission line costs are other factors that are taken into account while developing the novel framework. To further analyze the performance of P2P energy trading framework using auctions, it is integrated with forecasting models that predict energy price, production, and consumption – this serves as the second major contribution of this chapter. To test the effectiveness of the developed framework, a simulation test bed that consists of trading energy among a set of hundred buildings and an energy supplier or utility has been discussed.

12.2 METHODOLOGY

Our novel framework allows individuals who both produce and consume energy (called "prosumers") and consumers to trade energy with each other and with their primary energy provider using various game theoretic auction models. For evaluation of the framework, a simulation test bed featuring a grid of buildings connected to a main energy supplier has been created. The simulation examines four different peer-to-peer (P2P) auction models.

In addition, the framework has also been integrated with forecasting models that predict energy prices, production, and consumption of buildings. It will be later discussed how the integration of forecasting models leads to a change in the performance of the trading approach. The developed framework consists of several modules, which are discussed below.

12.2.1 Smart grids

The various components of the smart grid include energy storage units, smart meters, demand supply management, electric vehicles, smart appliances, renewable energy resources, self-consumption, and transmission automation, as shown in Figure 12.1. Smart Grid uses sensors throughout the transmission and distribution to collect the data and provide bi-directional communication between users. The users in smart grid are consumers, producers, and prosumers. The prosumers are the ones who can produce and consume the energy. These prosumers when integrated into smart grid with the use of distributed energy resources help in balancing the demand supply management as it will increase the possibilities of distributed generation; for example, consuming the energy generated from a resource located nearby (like the solar panel of the home) rather than the

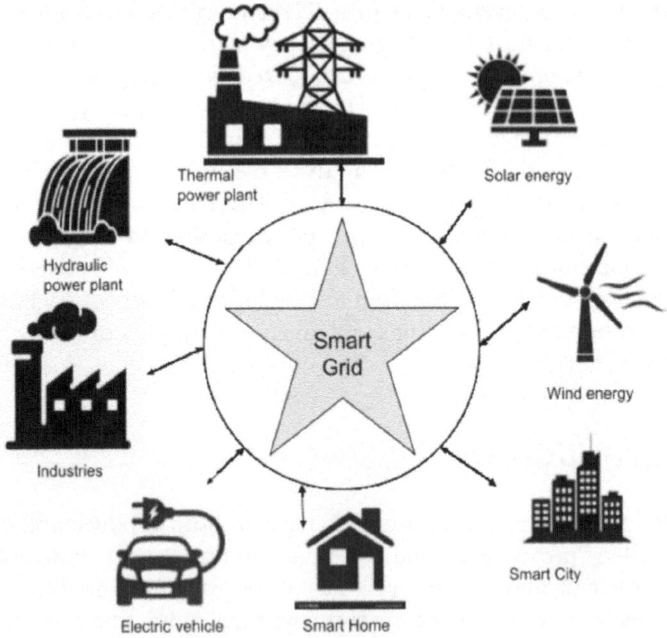

Figure 12.1 Peer-to-Peer energy trading model.

one generated at the main utility might be more efficient and economical as the distance from generation to consumption is reduced. The concept of two-way communication helps in improving electric power interactions for both utilities and end users (consumers and prosumers). This also provides more visibility and control over the consumption of energy. In this regard, a peer-to-peer (P2P) energy trading model has been discussed in this work; it is represented in Figure 12.2.

For the current work, a set of hundred buildings are simulated in a grid of 10 by 10. Buildings in the designed grid are positioned along its width and height at specific locations (x, y). There is a primary energy supplier for the grid at location (–1, –1). Consumers and prosumers calculate their (Manhattan) distance from the utility when transacting with it. There is an assumption that sending power over a greater distance will be more expensive since it will be less efficient due to higher losses; this leads to increases in line payment costs. The following formula is used to determine the distance $d(i,j)$, which represents the separation between the seller at position (x_i, y_i) and buyer at the position (x_j, y_j):

$$d(i,j) = |x_i - x_j| + |y_i - y_j| \tag{12.1}$$

Having calculated the distance between buyer and seller, the next step is to calculate building's excess energy, which is based on its production and consumption. After self-consumption, if the building still has excess energy, it stores it in an energy storage system. The grid receives the excess energy that cannot be stored by the energy system of buildings. The buildings exchange energy with one another using the P2P energy trading framework.

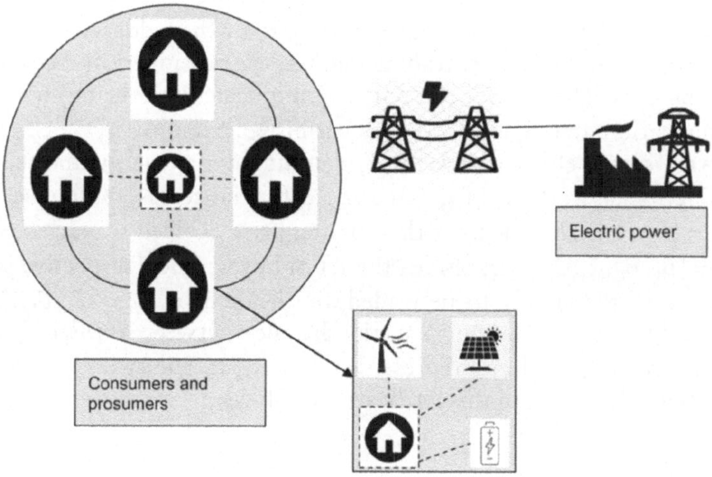

Figure 12.2 Peer-to-Peer energy trading model.

Depending on how much energy each building can store (E_a), the buildings are classified as either buyers or sellers. Buildings are classified as buyers if $E_a < 0$, otherwise they are considered as sellers. The building typically purchases energy directly from the grid once per day if P2P trading is unable to meet its demand.

12.2.2 Game theory approach: auction mechanisms

A game is defined as a competitive activity where the players or agents compete to maximize their own profit according to a set of rules. Thus, it consists of a set of players, the corresponding actions or strategies available for each player and the utility or payoff function that determines the profit of each player when a particular action or strategy is chosen. In this regard, auctions can also be considered as games since the bidders participating in the auctions are considered as players or agents of the game; the bid(s) placed by the bidders represent their actions in this auction game; and the profit or utility is generally defined as the difference between the bid and the corresponding valuation that the bidder has associated with the item being auctioned. Thus, in this work, auction is used as the primary game theory tool for providing a pragmatic solution to the problem of energy trading.

For presenting a complete analysis of the energy trading market, there are four different auction models that are discussed under the category of open bid and closed bid auctions. Properties and rules are different for each auction type, and thus, the strategies of buyers and sellers also change. This results in different price allocations to the energy being auctioned.

12.2.2.1 Open bid auctions

In these types of auctions, the bids put forward by the bidders are known to all the bidders. In the energy trading market, the prosumers and consumers are the bidders. The bidders can only communicate with the buyers and no communication among the bidders is permitted. Set $S = \{S_1, S_2, \ldots S_n\}$ represents the set of sellers; each seller S_i contains stored energy that the seller is willing to sell as $E_a = \{E_{a1}, E_{a2}, \ldots, E_{an}\}$ for the corresponding seller S_i. Set $P_s = \{P_1, P_2, \ldots, P_n\}$ denotes the starting price decided by each seller S_i. Set $B = \{B_1, B_2, \ldots, B_m\}$ represents the set of buyers; their respective private valuations of the energy to be traded are denoted by $P_b = \{P_1, P_2, \ldots, P_m\}$, respectively. Using Equation 12.1, the distance between a seller S_i and a buyer B_j is calculated and represented as $d(i,j)$. The two types of open bid auctions discussed in this work are as follows:

12.2.2.1.1 English auction model

English Auction is an ascending dynamic auction. It follows an open bid auction mechanism. This auction follows an iterative algorithm. A starting bid is set and anybody is able to bid on the auction for that price or greater. After the first bid, bidders have to raise the price in an amount greater than a predetermined minimum increment. After each bid, the bidders drop out until only one bidder remains. The last one to remain is the winner and pays an amount equal to its bid placed at the last round. If nobody bids after the start of the auction, then the price is lowered by a lowering increment that is fixed till it reaches the fixed price set by the seller; still if no one bids, then the auction just ends.

The minimum increment (MI_i) for each bid is calculated based on the initial price and an increment percentage (IP) is set to 2%; this implies a two percentage increase in the initial price. It is calculated as follows:

$$MI_i = P_i' \times IP \tag{12.2}$$

where P_i' is the initial price of seller $S_i \in S$. In addition, a lowering increment (LI) is also initialized to decrease the price if there are no bids after the start of the auction.

The set B is iterated, and a Buyer say B_j starts the auction and its minimum cost value MC_j is set to its private value. Each seller $S_i \in S$ is at a certain distance $d(i,j)$ from the buyer B_j and so the cost $C(i,j)$ is calculated by the buyer B_j to each of these sellers based on distance and price of that seller that is open to the buyers as:

$$C(i,j) = P_i' + \left(d(i,j) \times d_p\right) \tag{12.3}$$

where d_p is the line payment cost.

The buyer B_j can offer bid Bid to a particular seller Si only if: $C(i,j) < MC_j$. The bid Bid is calculated as follows:

$$Bid = P_i + MI_i \tag{12.4}$$

The bid is placed by the buyer B_j if $Bid > MB_i$, where MB_i is the minimum bid. The process is iterative, with each buyer adhering to the same set of guidelines, and the bids are updated and raised in the order described above. If the new buyer submits a higher bid, the previous buyer withdraws, leaving the seller and the buyer who submitted the highest bid. After the auction is over, the seller gives the item (that was auctioned) to the highest bidder. After establishing the starting bid, if no bids are received, the seller reduces its value as follows:

$$P_i' = P_i' - LI \tag{12.5}$$

12.2.2.1.2 Dutch auction model

Dutch auction uses the concept of a descending price auction. It comes under the category of open bid auction type. Up until a willing buyer makes a bid, the seller sets a starting bid and reduces the price by decrementing the starting bid's value. The auction is won by the first bidder, who also gets the specified amount of energy for the bid price. The maximum starting price or bid that will start the auction is set by each seller in set S. The current starting bid SB_i is calculated as follows:

$$SB_i = P_i' + \left(D \times d_p\right) \tag{12.6}$$

$$P_i = SB_i \tag{12.7}$$

where D is the maximum distance possible between a buyer and a seller. The bidding ends when it is equal to a reserved price RP_i that is set by the seller. Initially, $RP_i = P_i'$. The value of the lower increment (LI_i) is set as 0.05; this implies that the original price decreases by five percentage. The buyer B_j checks which seller has not sold the energy yet and calculates the cost $C(i,j)$ based on distance and price from seller S_i (as calculated in Equation 12.3).

Each buyer in the buyer set B calculates its minimum cost, which is its private value that they are ready to buy the energy amount for. This value is based on its place or position in the grid. The set is iterated and a Buyer say B_j starts the auction and its minimum cost value MC_j is set to its private value P_j. The buyer offers the bid to a particular seller S_i only if $C(i,j) < MC_j$.

Seller S_i sells the energy to the first buyer to place a bid after satisfying the above condition. If the energy is not sold after every iteration, the seller compares the lowered price to its reserve price. If this lowered price is less than the reserve price, the auction ends; otherwise, the lowered price is now the seller's current price, and the bidding goes on. The buyer compares its price to the updated value once more and adjusts the bid as necessary.

12.2.2.2 Sealed-bid auctions

In this auction type, bids are submitted simultaneously but in a secretive manner. This implies that no bidder is aware of the bids submitted by other bidders. The seller will gather all bids before awarding the auction to the highest bidder, who will then pay the price of its bid. The bids are submitted in only one iteration of this mechanism. The sellers do not start the auction with a starting price or bid because it is a sealed bid. The buyers will submit their bids to the seller while adhering to certain guidelines.

12.2.2.2.1 First Price auction model

The term *blind auction* is used to describe the First Price auction. The sealed-bid auction method is used in this auction. The buyer's bid is slightly less than its true private value because the winning bidder will be required to pay his or her price or placed bid. The buyer determines the expected cost of each seller based on the estimated price. The buyer uses this price as the basis for deciding whether to submit a bid to the seller. Using private value information P_j and taking the seller's distance $d(i,j)$ into account, the buyer submits a bid $Bid(i,j)$ to the relevant seller S_i, which is calculated as follows:

$$Bid(i,j) = P_j + \left(d(i,j) \times d_p\right) \tag{12.8}$$

This procedure is followed by all the buyers, and they place their bid to the respective sellers accordingly. The seller then compares the highest bid value that it has received with its initial price; if the value of the bid is greater than the initial price, then the buyer placing this highest bid wins and pays its respective bidding price.

12.2.2.2.2 Vickrey auction model

The Vickrey auction, which uses the sealed-bid mechanism, is based on the concept of second price auction. The winner (buyer) pays an amount equal to the second highest bid in the entire auction. The procedure for gathering bids and starting the auction is the same as for a First Price auction, with the exception that when a bidder places a bid, it bids its true private value P_j. In the Vickrey auction model, the second highest bid will now be paid by the auction winner (that placed the highest bid).

After the conclusion of each of the aforementioned auctions, the total amount of energy and the balance of buyers and sellers are updated based on the amount of energy sold.

12.3 RESULTS AND DISCUSSION

The effectiveness of the developed framework is tested on the simulation test bed (details as discussed in Section 12.2.1). The buildings in smart grids trade energy using both mechanisms during the model's 100-day simulations. Results are compared for primarily two cases: (a) direct trading with the utility or main energy supplier in the smart grids and (b) P2P trading. Smart grids have been simulated for 100 days. An auction is chosen from the English, Dutch, First Price, and Vickrey auctions for each simulation. Support vector regression (SVR) mode is chosen as the forecasting model in this study. The results are analyzed for three types of models that consist of forecasting models and auctions. The details of these models are discussed as follows:

1. *M0 model*: The auctioning is conducted among the buildings or users without a forecasting model
2. *M1 model*: A forecasting model for price prediction is used while conducting auctions. Each bidder, B_j determines the minimum cost MC_j that is its true value on which they are ready to bid and purchase the amount of energy being auctioned. For evaluating the developed models, Support Vector Regression (SVR) is used as the forecasting model.
3. *M2 model*: When auctioning energy, the building uses an energy production and an energy consumption prediction model. In this model, the buildings predict the energy production and consumption values of the grid for that period, after the total energy TE is calculated. If it is anticipated that more energy will be produced, then that additional energy will be added to the amount of energy stored for trading; if it is anticipated that more energy will be consumed, then this additional energy will be subtracted from the stored energy value. Buildings or prosumers will use this total calculated energy amount to engage in P2P energy trading.

Buildings or prosumers are the sellers of energy when it is sold to the utility, and the utility acts as the buyer. The following information is gathered for all prosumers or buildings participating as sellers in the energy trading activity with the utility based on the amount of energy sold.

AGS: Amount of money earned in selling energy to the grid
LGS: Line payment cost of selling energy to the grid
EGS: Energy sold to the grid

Average grid sell cost GCS per unit of energy for all the buildings is calculated as:

$$GCS = (AGS + LGS) / EGS \tag{12.9}$$

Buildings or prosumers are the buyers when energy is purchased from a utility, and the utility acts as the seller. The following information is gathered for all prosumers or buildings participating as buyers in the trading of energy to the utility based on the quantity of energy purchased from the grid.

AGB: Amount of money spent buying energy from the grid
LGB: Line payment cost buying energy from the grid
EGB: Energy bought from the grid

Average grid buy cost GCB per unit energy for all the buildings is calculated as:

$$GCB = (AGB + LGB) / EGB \tag{12.10}$$

Thus, the average grid cost GC is computed as follows:

$$GC = mean(GCS, GCB) \tag{12.11}$$

When energy is traded using a peer-to-peer (P2P) model, sellers with excess energy sell it to buyers who require it to meet their needs. For all the prosumers or buildings, the following information is gathered based on the volume of energy traded in P2P.

DA: Amount of money traded in auction
EA: Energy traded in auction
LA: Line payments cost in auction while selling the energy

For all buildings, the average auction buy cost AB per unit of energy is calculated as follows:

$$AB = (DA + LA) / EA \tag{12.12}$$

For all buildings, the average auction sell cost AS is determined as follows:

$$AS = DA / EA \tag{12.13}$$

In P2P energy trading, the average auction cost AC is calculated as follows:

$$AC = mean(AS, AB) \tag{12.14}$$

Using the average values calculated above, the following metrics are used to evaluate the performance of the developed framework when P2P energy trading is adopted as opposed to direct trading with the utility.

12.3.1 Average cost % advantage

The first metric demonstrates the advantage of trading energy through auctions over trading with the grid in terms of the average cost of energy trading. The calculation for this metric is as follows:

Auction Buying % advantage:

$$\alpha B = (GCB / AB) \tag{12.15}$$

Auction Selling % advantage:

$$\alpha S = (GCS / AS) \tag{12.16}$$

Auction Trading % advantage:

$$\alpha A = (GC / AC) \tag{12.17}$$

where GC represents average grid cost (Equation 12.11) and AC denotes average auction cost (Equation 12.14).

Figure 12.3 represents the percentage advantage collected while buying energy using P2P trading model as compared to buying from utility in smart grids directly. Highest advantage while buying the energy is observed for the English auction; it is followed by the Vickrey, First Price, and Dutch auctions. It can be observed that integration with the forecasting model to predict the prices (M1 model) gives higher advantage to all the different auction types. On the other hand, use of forecasting for energy production

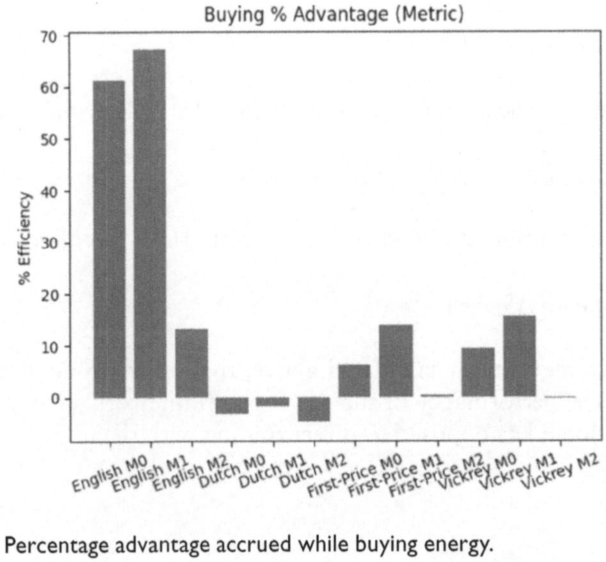

Figure 12.3 Percentage advantage accrued while buying energy.

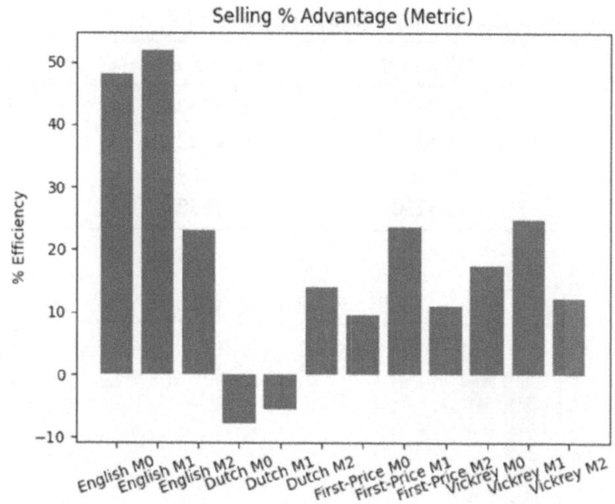

Figure 12.4 Percentage advantage accrued while selling energy.

and consumption has an adverse effect and leads to reduction in advantage metric for all auction types.

Figure 12.4 represents the percentage advantage accrued while selling energy to the grid. It can be observed that the English auction produces the maximum percentage of advantage as compared to other types of auctions; it is followed by the Vickery, the First Price, and lastly the Dutch auction – that has the minimum value of the advantage metric. However, when forecasting models for energy production and consumption are used (M2 model), the value of the advantage metric for the Dutch auction performs better than that of the First Price and the Vickery auction. It can also be observed that using the M1 model (that is using the forecasting model to predict the prices) helps the prosumers while selling the energy; and using the forecasting model to predict consumption and production, however, does not tend to make any significant difference in the overall metric value.

In terms of the average cost of the energy traded in smart grids, Table 12.1 and the graph in Figure 12.5 illustrate the benefit of trading using an auction mechanism. In the absence of a forecasting model, the average cost efficiency of the English auction is 54.79%, but it increases to 59.55% when a price forecasting model is implemented. Nonetheless, the integration with the energy generation and consumption model results in a reduction in efficiency. The English auction is followed by the Vickrey auction, which again demonstrated an increase in average cost efficiency when integrated with the M1 model. The efficiency of the First Price auction is 18.72%, which is comparable to the efficiency of the Vickrey auction for the M1 model. The Dutch auction performs poorly with the M0 and M1 models, but efficiency improves when the M2 model is implemented. This metric takes into

Table 12.1 Values of auction trading percentage advantage

Auction Type	*M0 Model*	*M1 Model*	*M2 Model*
English	54.7954	59.5540	18.1101
Dutch	−5.5392	−3.6322	4.5686
First Price	7.8485	18.7264	5.7266
Vickrey	13.3230	20.1658	5.9960

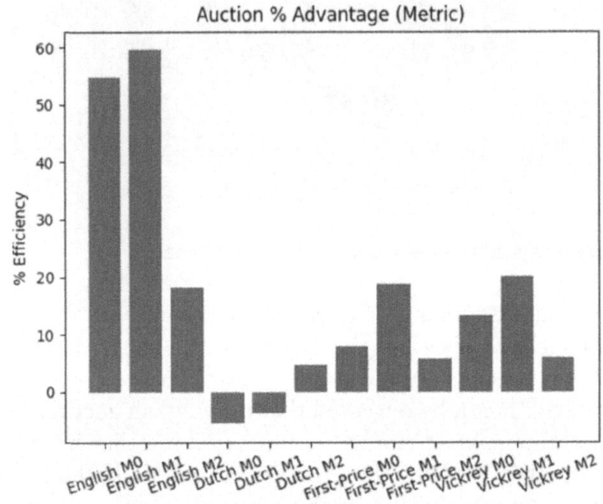

Figure 12.5 Auction trading percentage advantage.

account the average grid cost of energy and the average cost of energy when trading using an auction model to determine the efficiency in terms of cost savings per kiloWatt-Hour.

12.3.2 Savings accrued while trading energy

Auction Trading $ savings: This metric calculates the overall sum of money saved by auctioning versus the case where all trading is conducted directly with the primary energy provider or utility in smart grids. The metrics are calculated as follows: Auction Buying $ savings:

$$\gamma B = (GCB - AB)^{*} DA \tag{12.18}$$

where DA represents the amount of money traded in auction.

Auction Selling $ savings:

$$\gamma S = (GCS - AS)^{*} DA \tag{12.19}$$

Auction overall \$ savings:

$$\gamma A = (GC - AC)^* DA \qquad (12.20)$$

In order to perform the uniform comparison of all the auction scenarios, this section presents savings obtained under different auctions in terms of cost savings in percentage values.

Auction Buying % savings: This metric represents the percentage of money saved while buying energy using an auction as compared to buying energy directly from the grid. It is represented by Equation (12.21).

$$\beta B = (\gamma B / (AGB + DA + LGB + LA)) \qquad (12.21)$$

Figure 12.6 shows the value of this metric for different auctions with M0, M1, and M2 models. It helps us to analyze how much money can be saved while buying energy through auction as compared to trading with the smart grid directly. It can be observed that for all the auctions integrating with the M2 model, the metric value increases only for the English auction and not for other types of auction. In fact, for other auctions, the M1 model gives the highest savings while buying energy. It may be noted that, overall, English auctions are producing the highest savings, followed by the First Price, and the Vickery auction; similar to the advantage metric, even here the Dutch auction performs the worst. In addition, the use of forecasting models for predicting energy production and consumption leads to lowest savings of about –0.39% (negative savings imply that the amount was paid rather than saved).

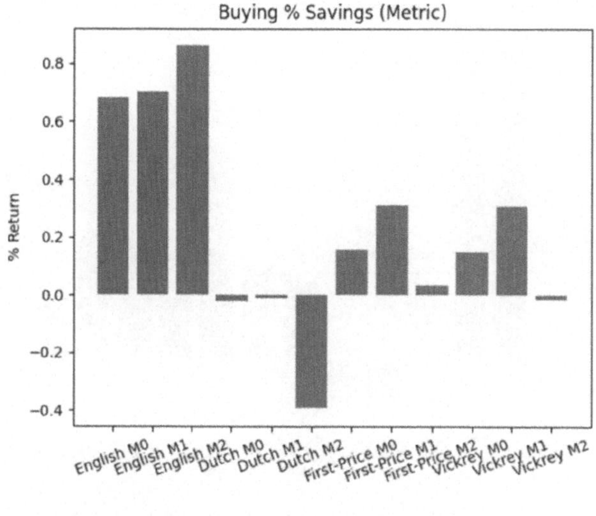

Figure 12.6 Percentage of savings obtained while buying energy.

Auction Selling % savings: This metric represents the percentage of money saved while selling energy to the grid. It is calculated as follows:

$$\beta S = \left(\gamma S / \left(AGS + DA + LGS\right)\right) \tag{12.22}$$

Figure 12.7 represents the values of percentage savings or returns obtained under different auction scenarios. The graph shows that selling the energy using the M1 model leads to the highest return for all the auctions, except for the Dutch auction – where the M2 model is the only case where actual savings are obtained. For Dutch auctions with the M0 and M1 models, negative returns are accrued. The M2 model proves to be more efficient in helping the prosumers in Dutch auctions to make money. Thus, the M1 model is preferred for maximum savings in all the auction scenarios except Dutch, where the M2 model is preferred. Overall, the English auction produces the highest percentage returns, followed by the First Price and Vickrey auctions.

Auction Trading % savings: This is an important metric which shows the total percentage of amount that is saved by various buildings through trading with auctions instead of trading with the main energy supplier. It is calculated as follows:

$$\beta A = \left(\gamma A / \left(AGS + AGB + DA + LGS + LGB + LA\right)\right) \tag{12.23}$$

From Table 12.2 and Figure 12.8, it can be inferred that the English auction is clearly the most popular auction type, followed by First Price and

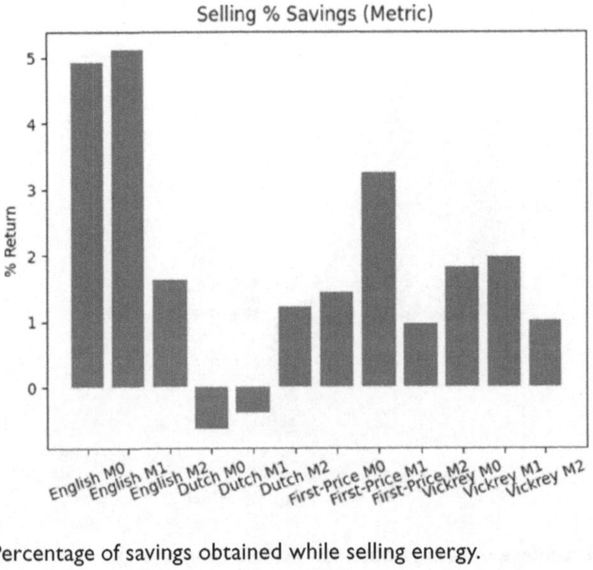

Figure 12.7 Percentage of savings obtained while selling energy.

Table 12.2 Values of auction trading percentage savings

Auction Type	*M0 Model*	*M1 Model*	*M2 Model*
English	1.2037	1.2333	1.5869
Dutch	−0.0748	−0.0402	0.4981
First Price	0.3968	0.8270	0.6392
Vickrey	0.4115	0.7171	0.6446

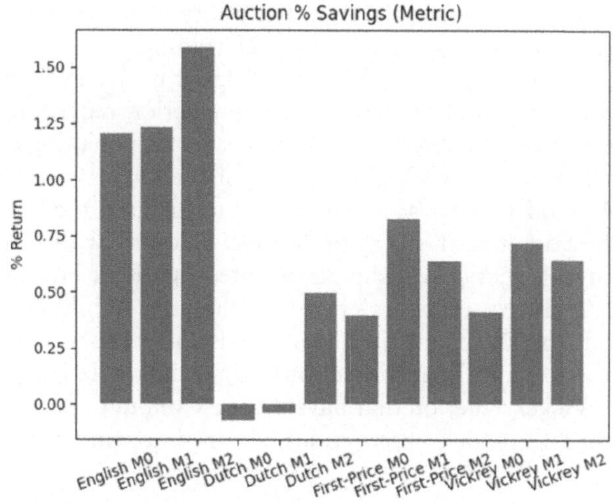

Figure 12.8 Auction trading percentage savings.

Vickrey; the Dutch auction is clearly by far the least effective. The "Auction % Savings" metric depicts the total amount of money that different buildings saved by trading with auctions rather than with their primary energy provider in smart grids. Integration of the First Price and Vickrey auctions with the M1 model works best; integration with the M2 model increases the cost savings or returns when compared to the M0 model (not using any forecasting models). We can draw the conclusion that in the case of sealed-bid auctions, the First Price auction when integrated with the M1 model has 0.82% savings, while the M2 model has 0.63% savings, both of which are higher than the efficiency when traded without using a forecasting model. When combined with the M1 model, the Vickrey auction also demonstrates improvement in cost savings percentage. Trading without a forecasting model results in cost savings of 0.41%, which increases to 0.71% with model M1 and 0.64% with model M2. When model M2 is integrated into Dutch auctions, the cost savings rise. Another interesting observation is that Dutch auctions do not give good results (except in the case of integration with the M2 model). Overall, it can be concluded that the English auction with the M2 model is the most effective auctioning method.

12.3.3 Comparison based on amount of energy traded and number of auctions

Table 12.3 details the total amount of energy traded using different auction models. Following are the observations drawn from the table:

The amount of energy bought from the utility is higher than the energy sold to the grid and energy traded using P2P trading.

In all the auction types, the M2 model leads to a higher exchange of energy. This is attributed to the fact that prediction regarding consumption and production of energy leads to better decision making due to more information and thus shows higher trading of energy.

For the M1 model, as far as energy is sold to the grid, the Dutch auction sells the highest amount of energy to the grid, followed by the Vickery, First Price, and English auctions. It should be noted that First Price and English auctions sell relatively very less energy to the grid. While buying energy from the grid, the Dutch auction again leads the scenario. However, the Dutch, First Price, and English auctions buy comparable energy from the grid. It is the Vickery auction that buys the least amount of energy from the utility. For the M2 model, observations obtained are similar to that of the M1 model.

Table 12.3 Total amount of energy traded in different auction models

	Energy Traded (kWh)		
Models	*Selling to the Grid*	*Buying from the Grid*	*P2P Trading*
English M0	1978.7898	533678.2465	101823.3919
English M1	2012.3957	532704.7854	102524.2562
English M2	2116766.5296	2629834.6501	2259492.9867
Dutch M0	29599.4815	562597.1605	22935.3835
Dutch M1	29803.1590	562930.6127	19255.9939
Dutch M2	1377134.6091	1900831.0437	1412119.5574
First Price M0	4524.4044	529003.8345	107686.1051
First Price M1	7690.7660	522702.3189	105124.0233
First Price M2	2370929.4234	2880699.5688	2624593.9617
Vickrey M0	17475.3308	354713.3638	47362.5157
Vickrey M1	17953.0995	152636.8064	28901.1180
Vickrey M2	2263883.9133	2749322.3472	2381997.6855

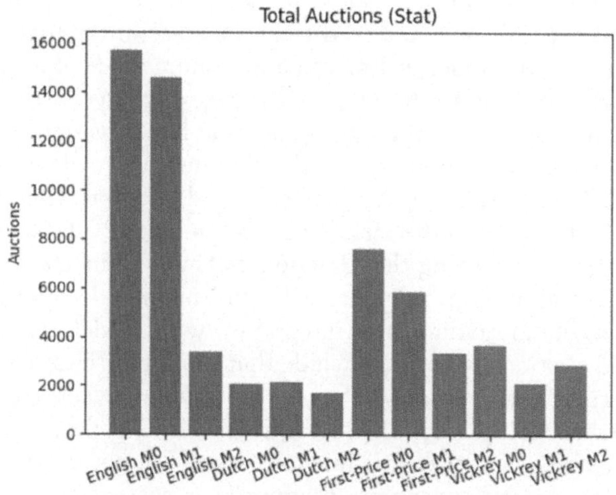

Figure 12.9 Total number of auctions conducted under each category.

For P2P trading, comparison results are different as compared to trading with utility. For M1 model, highest amount of energy is traded with First Price auction, followed by the English, Vickery, and Dutch auctions. In addition, the energy traded using the First Price and English auctions are comparable and very high as compared to the Vickery and Dutch auctions – where energy trading is lower.

To understand the effect of forecasting models on different types of auctions, Figure 12.9 represents the total number of auctions conducted under each category.

It can be inferred from the figure that English auctions conduct the highest number of auctions among all the auctions.

The M0 model conducts the highest number of auctions within an auction type, with an exception of the Dutch auction type, where the M1 model has more numbers of auctions than the M0 model.

The least number of auctions is held by Dutch auctions.

First Price auctions have the second highest number of auctions, followed by the Vickery and Dutch auctions.

12.4 SUMMARY AND CONCLUSION

After the comparative analysis, it can be observed that the Dutch auction performs worst among all the auctions with respect to advantage and savings metrics. The maximum amount of energy traded in the proposed approach is when M2 model is used by the prosumers in P2P trading. This concludes that integrating a forecasting model in the prediction of energy production and consumption helps in increasing the amount of energy traded. However, this does not lead to an increase in the total number of auctions conducted. In fact, it leads to a minimum value of advantage and savings metric value for all the auction types – except the Dutch auction. In terms of advantage metric, the English auction has consistently performed well with all the three models, but the maximum benefit is with model M1 59.55%. This is followed by the sealed-bid auctions, like the Vickrey auction with 18.73% and the First Price auction with 20.17% using model M1. However, by analyzing the "Auction % Savings" metric to determine the amount of money saved/made, it is evident that the English auction produces maximum savings after integration with model M2. In the case of sealed-bid auctions, we can conclude that the Blind (First Price) auction when integrated with model M1 has 0.83% savings, whereas the Vickrey auction with model M1 has 0.72% savings. The Dutch auction leads to minimum savings; but when it is integrated with the forecasting model for energy production and consumption, it produces savings of around 49.8%. There can be many other factors influencing the way these forecasting models affect the auction model performance. But overall, we can conclude that integration of these (M0 and M1) models in the energy trading market definitely helps in improving efficiency and cost savings.

Conclusion: This chapter has developed a novel auction-based framework for P2P energy trading in the smart grid. Different performance metrics have been discussed to compare the benefits of trading energy in the P2P model using the four different auction models versus the direct trading mechanism with the utility. As the P2P model has been shown to be more effective than direct trading with the utility, these results can be used to encourage prosumers to participate in energy trading with their peers. Overall, the results obtained strongly suggest the idea of integrating forecasting models into the energy trading framework through the use of an auction-based game theoretic approach.

REFERENCES

1. International Energy Agency India 2020 Energy Policy Review, 2020. https://webstore.iea.org/download/direct/2933?fileName=India_2020-Policy_Energy_Review.pdf.
2. Gaur K, Kumar H, Agarwal R P K, Baba K and Soonee S K 2016 Analysing the electricity demand pattern. *2016 National Power Systems Conference (NPSC)*, pages 1–6. IEEE, 2016.
3. Electricity Generation from Renewable Energy. https://education.nationalgeographic.org/resource/electricity-generation-renewable-energy-sources.
4. Cai Y, Huang T, Bompard E, Cao Y and Li Y 2017 Self-sustainable community of electricity prosumers in the emerging distribution system. *IEEE Transactions on Smart Grid*, vol. 8, no. 5, pp. 2207–2216, 2017.

5. Chen S, Shroff N B and Sinha P 2013 Energy trading in the smart grid: From end-user's perspective. *2013 Asilomar Conference on Signals, Systems and Computers*, pp. 327–331, IEEE, 2013.
6. What is electricity trading? 2020. https://www.drax.com/power-generation/what-is-electricity-trading/.
7. Renewable Energy: Distributed Generation Policies and Programs. https://www.energy.gov/eere/slsc/renewable-energy-distributed-generation-policies-and-programs.
8. Doris E, Busche S and Hockett S 2009 Net metering policy development and distributed solar generation in Minnesota: Overview of trends in nationwide policy development and implications of increasing the eligible system size cap *National Renewable Energy Laboratory*. https://doi.org/10.2172/969897.
9. Yan Y, Qian Y, Sharif H and Tipper D 2013 A survey on smart grid communication infrastructures: Motivations, requirements and challenges. *IEEE Communications Surveys & Tutorials*, vol. 15, no. 1, pp. 5–20, 2013.
10. Mittal A and Tomar A 2019 *Smart Grid: Vision for India*. Electrical India Magazine, Chary Publications, Navi Mumbai. https://www.electricalindia.in/smart-grid-vision-for-india/.
11. Pilz M and Al-Fagih L 2019 Recent advances in local energy trading in the smart grid based on game-theoretic approaches. *IEEE Transactions on Smart Grid*, vol. 10, no. 2, pp. 1363–1371, March 2019, doi: 10.1109/TSG.2017.2764275.
12. Liu X, Ji Z, Sun W and He Q 2023 Robust game-theoretic optimization for energy management in community-based energy system. *Electric Power Systems Research*, vol. 214, Part B, p. 108939, 2023, ISSN 0378-7796. https://doi.org/10.1016/j.jpgr.2022.108939.
13. Sun R, Wang L, Song W, Li G, Li Q 2023 A coalitional game theoretic energy transaction algorithm for networked microgrids. *International Journal of Electrical Power & Energy Systems*, vol. 144, p. 108494, 2023, ISSN 0142-0615. https://doi.org/10.1016/j.ijepes.2022.108494.
14. Li Z, Wu L, Xu Y, Wang L and Yang N 2023 Distributed tri-layer risk-averse stochastic game approach for energy trading among multi-energy microgrids. *Applied Energy*, vol. 331, p. 120282, 2023, ISSN 0306-2619. https://doi.org/10.1016/j.apenergy.2022.120282.
15. Agarwal A and Khandeparkar K 2021 Distributing power limits: Mitigating blackout through brownout. *Sustainable Energy, Grids and Networks*, vol. 26, p. 100451, 2021.
16. Agarwal A 2022 A novel approach to save energy by detecting faulty HVACs. *Journal of The Institution of Engineers (India): Series B*, vol. 103, no. 2, pp. 305–311, April 2022.
17. Agarwal A and Ramamritham K 2020 Tackling issues related to pmu deployment in the grid using a novel algorithm. In *2020 IEEE First International Conference on Smart Technologies for Power, Energy and Control (STPEC)*, pages 1–6, IEEE, 2020.
18. Moafi M, Ardeshiri R R, Mudiyanselage M W, Marzband M, Abusorrah A, Rawa M, Guerrero J M 2023 Optimal coalition formation and maximum profit allocation for distributed energy resources in smart grids based on cooperative game theory, *International Journal of Electrical Power &*

Energy Systems, vol. 144, p. 108492, 2023, ISSN 0142-0615. https://doi.org/10.1016/j.ijepes.2022.108492. (https://www.sciencedirect.com/science/article/pii/S0142061522004975)

19. Paudel A, Chaudhari K, Long C and Gooi H B 2019 Peer-to-peer energy trading in a prosumer-based community microgrid: A game-theoretic model. *IEEE Transactions on Industrial Electronics*, vol. 66, no. 8, pp. 6087–6097, August 2019. https://doi.org/10.1109/TIE.2018.2874578.
20. Zhang C, Wu J, Zhou Y, Cheng M, Long C 2018 Peer-to-peer energy trading in a microgrid. *Applied Energy*, vol. 220, pp. 1–12, 2018, ISSN 0306-2619. https://doi.org/10.1016/j.apenergy.2018.03.010.
21. Jadhav A M, Patne N R and Guerrero J M 2019 A novel approach to neighborhood fair energy trading in a distribution network of multiple microgrid clusters. *IEEE Transactions on Industrial Electronics*, vol. 66, no. 2, pp. 1520–1531, 2019.
22. Morstyn T, Teytelboym A and Mcculloch M D 2019 Bilateral contract networks for peer-to-peer energy trading. *IEEE Transactions on Smart Grid*, vol. 10, no. 2, pp. 2026–2035, 2019.
23. Manikanta G, Mani A, Singh H P and Chaturvedi D K 2019 Simultaneous placement and sizing of dg and capacitor to minimize the power losses in radial distribution network. In Kanad Ray, Tarun K. Sharma, Sanyog Rawat, R. K. Saini, and Anirban Bandyopadhyay, editors, *Soft Computing: Theories and Applications*, pp. 605–618, Singapore: Springer Singapore, 2019.
24. Verma K K and Saxena V 2019 Energy consumption of university data centre in step networks under distributed environment using Floyd–Warshall algorithm. In Kanad Ray, Tarun K. Sharma, Sanyog Rawat, R. K. Saini, and Anirban Bandyopadhyay, editors, *Soft Computing: Theories and Applications*, pp. 99–111, Singapore: Springer Singapore, 2019.

Chapter 13

Power quality analysis of rooftop grid-connected PV in composite climate

A case study

Saurabh Kumar Rajput, Mukta Singh, and Dharmendra Kumar Dheer

13.1 INTRODUCTION

Nowadays, the grid-connected photo voltaic system (GCPVS) are widely used all around the world and since 1997 its market has grown continuously [1]. In developed and developing countries, new renewable energy resources represent a 2.4% rapid growth among all the renewable energy resources [2]. India receives 5000 trillion kWh of solar energy in a year and 2300–3200 hours annually. Also, in India most part receives 4–7 kWh of solar radiation per square meter per day and have 250–300 sunny days per year. With this much solar photovoltaic and solar thermal energy, India can generate 35 MW per square kilometer [3]. The year 2020 was record breaking for the solar PV market, with new installations of an estimated value of 139 GW, summing up the global total at an estimated 760 GW with both on grid and off-grid capacity. Demand for solar PV is increasing and expanding since it is becoming the most suitable option for electricity generation not only in households but also in industries. Solar PV power along with the wind power helped the renewable power sector to rise in the second half of 2020 and assisted in prevailing over the COVID-19 pandemic. Solar PV accomplished the largest increase ever seen in capacity in a year [4]. In the year 2030, there will be an increase of 1.5% annually in the worldwide energy demand and the developing countries of Asia will lead this growth with a 40% increase [5]. At the global level, sustainable development is prevailing with solar technology along with other renewable energy technologies. To existing energy systems, these technologies can give small increasing capacity additions with short lead times [6]. In a grid-connected solar PV system, the inverter is an important component since it converts the generated DC power into AC power in order to match the grid frequency and voltage. Also an inverter is significant for the safety and reliable integration of the grid [7, 8]. At the Point of Common Coupling (PCC) when the inverter is connected to the grid, the power quality shouldn't deteriorate [9]. As a result, the impacts of an inverter with poor power quality should be taken seriously [10]. Appropriate power quality can be produced by using current controlled VSI (Voltage Source Inverter) [11]. Regulations

DOI: 10.1201/9781003623168-17

applied by grid codes [12] or international standard requirements like IEC 61850-7-420 standards [13] and IEEE 1547 standards [14] say that Total Harmonic Distortion should be less than 5% [10] of fundamental current for current and less than 2% for voltage according to standard IEC 61727 [15] and power factor greater than 0.9 [10]. According to the W. Mack Grady and Robert J. Gilleskie, if the total harmonic distortion increases then power factor decreases. This represents a relation of inverse proportionality [16]. Several studies have been conducted in the past on the power quality aspects of PV inverter, which show test results of power factor and current THD's mathematical models. These researches show that based on the capacity and the demand profile, the power factor can change [17, 18]. The PV also affects the energy quality parameters. To increase the power quality using compensation technique in micro grid, various devices are proposed and suppression of harmonics along with reactive power compensation is done by deploying the droop control technique [19–21].

13.2 SYSTEM UNDER STUDY

Solar PV systems directly convert sun light into electrical energy. The solar cells/modules are used in PV plants that convert sunlight into a direct current (DC) form of electrical energy, which is converted to an alternating current (AC) supply via a solar inverter. Voltage and current management at load, maximum power point tracking, and power control for grid connection are all possible with solar inverters. Hence, the PV-based electricity production majorly needs PV modules and PV inverter and transformer components (as shown in Figures 13.1 and 13.2).

The presented case study is performed on a 100 kWp grid integrated solar rooftop photovoltaic system (grid-connected solar rooftop PV system). The data is collected in the months of peak summer (May) and peak winter (December). As shown in the Figures 13.1 and 13.2, the system consists of PV array that converts solar energy into electrical energy. The PV array consists of 338 modules with *Pmax* of 320 W per module. All the modules

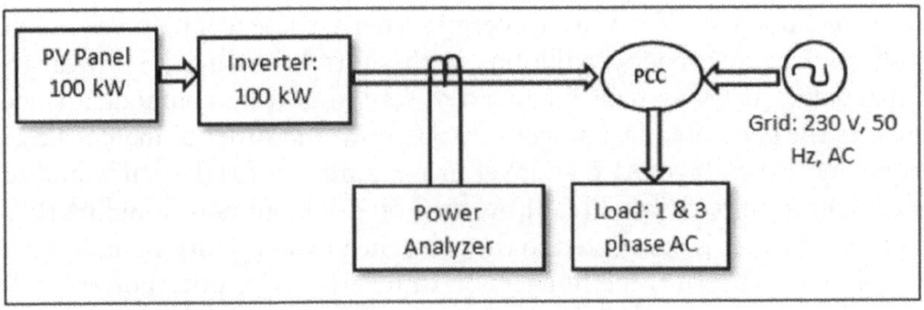

Figure 13.1 Schematic diagram of PV grid-connected system.

Figure 13.2 Experimental study and data monitoring from the study system.

are opaque type polycrystalline. For maximum sunlight/radiation from the sun, the array is set at an inclination of 28.7°C in the south facing direction. The PV array is connected to an inverter with a 100 kW rating, which converts DC power into AC to feed the AC load. Solar PV modules have an open circuit voltage of 46 V, a rated voltage of 37.7 V, a short circuit current of 9.03 A with a rated current of 8.50 A. The Fill Factor of modules is 77.04%, with an efficiency of 16.67%. The dimension of the module is 1955 × 982 × 36) mm^3. The solar PV system is integrated with a 63 kVA, 3 phase delta (high-voltage side) – 3 phase-star (low-voltage side) distribution transformer (TPDT) with oil natural air cooling (ONAN) having HV side voltage as 11 kV and LV side voltage as 433 V. For recording the data regarding the power factor and harmonic distortion in current, a power analyzer is employed in parallel to the main connection (as shown in Figure 13.2 (d)). The PCC receives power from both the PV system and the grid and supply to the AC load.

13.3 POWER QUALITY PARAMETERS OF PV INVERTER AND THEIR MATHEMATICAL MODELING

A PV system often works alongside the grid in order to provide better service. Checking the quality of the power before feeding it to the grid is important to save the system from disruptions. Any complication with the power quality is usually the disturbances in voltage, current, or frequency which produce impairment in the operations of the equipment. The PV system shouldn't get damaged due to any defects in the grid. In the same way, the failure in the PV system shouldn't harm the grid and affect the consumers [17]. For the performance study of a GCPVS system, the important power quality parameters are power factor and THDi.

13.3.1 Power factor

The power factor is the ratio of working power measured in kilowatts (kW) to apparent power measured in kilovolt-amperes (kVA), as given in (1.1). Harmonic distortions are generated in current and voltage waveforms in non-linear or switching devices. This happens when there is a reactive load since there is a phase difference between current and voltage waveform.

$$PF = \frac{P}{S} = \frac{\frac{1}{T}\int_0^t V.I\,dt}{Vrms.Irms} \tag{13.1}$$

Where V and I are the voltage and current at time t, $VRMS$ and $IRMS$ are the root mean square (RMS) of voltage and current and T is the period of the waveform which correlate to the integration time used to calculate the power [17].

The rooftop PV plant provides active power to the electrical power system, but it can't provide the reactive power that the connected electrical load needs [22]. Figure 13.3 illustrates the power triangle before and after the integration of a PV plant with an existing electrical power system. If the active and reactive power requirements of the load are P (kW) and Q (kVAR), then S (kVA) will be the apparent power (according to the power

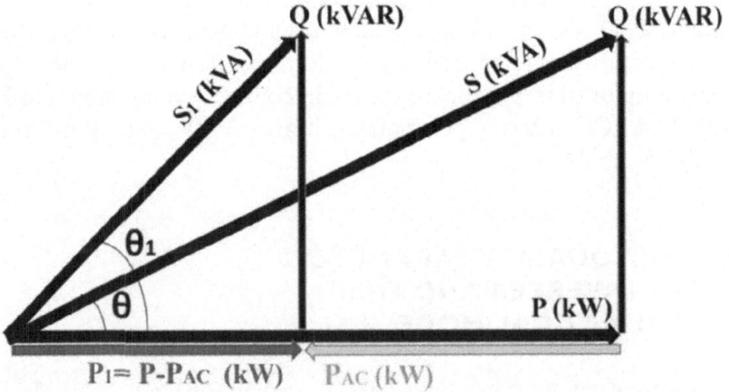

Figure 13.3 power triangle before and after the integration of a PV plant with an existing electrical power system.

triangle). After PV integration, if the real power support from the PV plant is PAC (kW), then the power factor angle will increase to 1 according to the new power triangle. As a result, the power factor will decrease from cos θ to cos θ 1 [22].

13.3.2 Total harmonic distortion

The total harmonic distortion is the ratio of RMS value of harmonic components in the current to the RMS value of the fundamental component of the current. The same is true for the total harmonic distortion in voltage, as given in (13.2) and (13.3).

$$THD_v = \frac{\sum_2^{\infty} Vn^2}{V_1} \tag{13.2}$$

$$THD_i = \frac{\sum_2^{\infty} In^2}{I_1} \tag{13.3}$$

Where THD_i is the total harmonic distortion in current, I_n is the current component of the nth harmonic, I_1 is the fundamental component of current, THD_v is the total harmonic distortion in voltage, V_n is the component of the voltage of the nth harmonic and V_1 is the fundamental component of the voltage [17].

A solar inverter is an equipment that changes electricity from direct current (DC) to alternating current (AC) [23, 24]. During this process, the inverters change the shape of the voltage wave and add THDi to the power system. THDi was found to be dependent on relative power [17, 22], which is the ratio of AC power output to rated power for an inverter, after testing ten different solar inverters [17].

13.3.3 PV inverter modeling using MATLAB

In the presented work the mathematical modeling is done based on the results obtained by testing and storing the data through a power analyzer and then by curve plotting and fitting in MATLAB. The mathematical model is the function of relative power, which is the ratio of PAC and $PRATED$, as given in (13.4) and (13.5).

$$THD_i = A\left(\frac{PAC}{PRATED}\right)^{-B} \tag{13.4}$$

Where A and B are fitting parameters. PAC is the inverter's output power and $PRATED$ is the inverter's rated power. Three approximation methods validated with three functions was presented by Hernandez and Jay for representing the inverter with grid support functions in simulation to create an inverter model [25]. According to the research study by A. Rampinelli, the mathematical model used for calculating the power factor and total harmonic distortion in current as a function of the inverter's relative power is given in (1.5) [17].

$$PF = \frac{C_o C_1 + \left[C_2 \left\{ \frac{PAC}{PRATED} \right\}^{C_3} \right]}{\left[C_1 + \left\{ \frac{PAC}{PRATED} \right\}^{C_3} \right]} \tag{13.5}$$

Where C_0, C_1, C_2, C_3 are coefficients of the model [17].

13.4 DATA ANALYSIS AND RESULTS

During the performance analysis in the peak summer season of May, the data is recorded from 6:30 am till 6 pm.

As shown in Figure 13.4 the highest power (70 kW) is produced in the afternoon time. This is because the solar intensity at this time is very high and it decreases in the morning and evening. The average power generation in May is the highest in the past one year. Figure 13.5 shows the variation

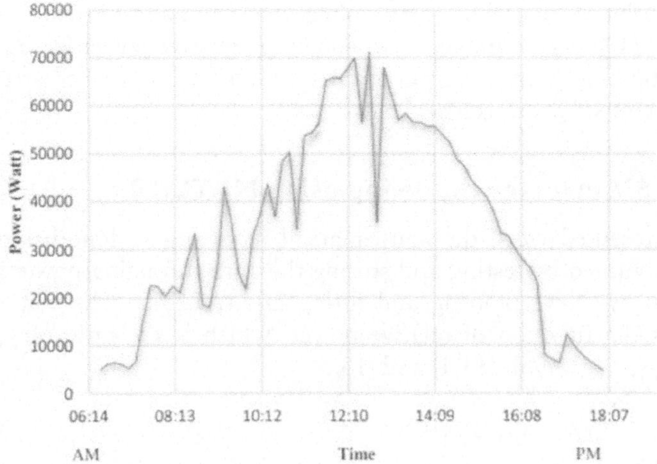

Figure 13.4 Variation of power with time (May).

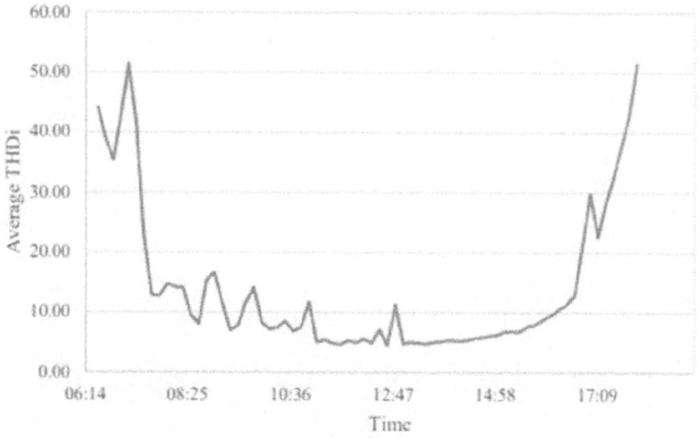

Figure 13.5 Variation of average THDi with time (May).

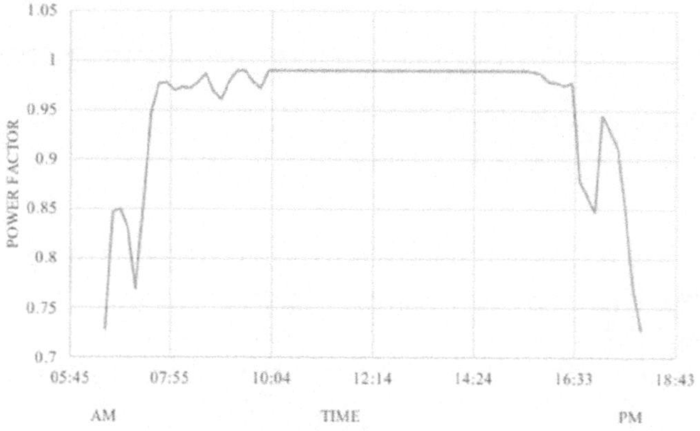

Figure 13.6 Variation of power factor with time (May).

of average THDi with the time of the day. The harmonic distortion in current is lowest in the afternoon time and higher in the morning and evening. The Figure 13.6 shows the variation of the power factor with the time of the day. Figure 13.7 depicts that the power factor value is highest in the afternoon and lesser at other times. In Figure 13.7 a comparison of average THDi with power generation (Watt) is shown for a typical summer day of May. It shows that when power produced is highest, the THDi is lowest and vice versa. In the noon time (From 11:30 am to 02:30 pm), the power output of the plant is very high and corresponding values of average THDi are low in this duration. Hence, the generation of harmful THDi is inversely proportional to the power output of the plant. Figure 13.8 shows the variation of power and power factor with the time of the day in May. The power

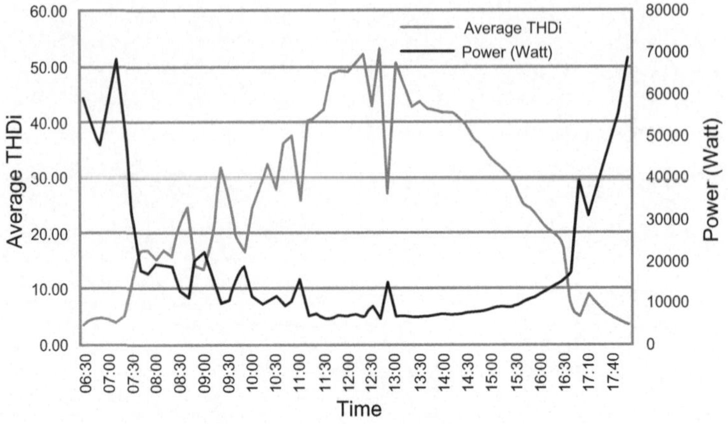

Figure 13.7 Variation of average THDi and power with time (May).

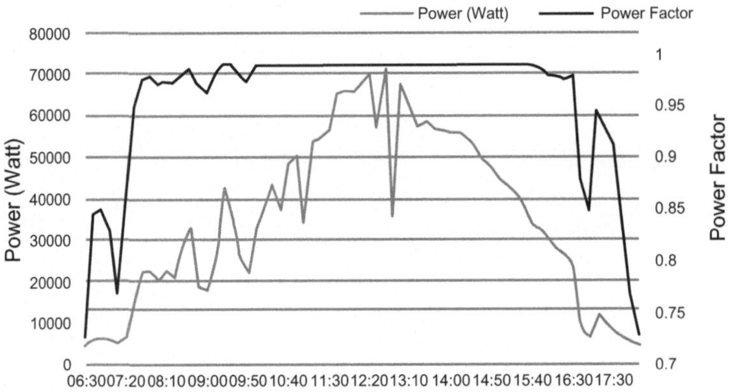

Figure 13.8 Variation of power and power factor with time (May).

factor value is highest when the power is highest; that is, the power factor is directly proportional to the power. With the increase in the power output of the plant, the power factor automatically improves. In Figure 13.9, a comparison of average THDi and power factor is made for a typical summer day of May. The results shows that when the power factor is more, THDi is less and vice-versa.

The key observations of the collected data and its analysis using a mathematical model are as follows:

(a) The average THDi (%) reduces/ improves with the increase in the power output (W) of the PV plant.

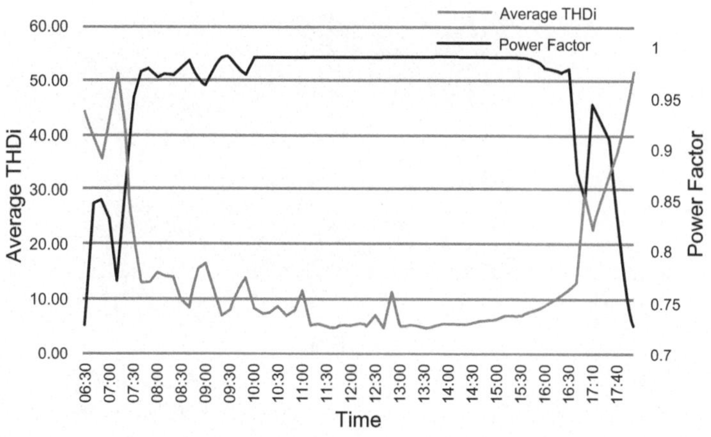

Figure 13.9 Variation of average THDi and power factor with time (May).

(b) The power factor increases/improves with the increase in the power output (W) of the PV plant.
(c) The average THDi (%) and power factor have an inverse relation with each other.

During the performance analysis in the peak winter season of December, the data is recorded from 9 am till 4:30 pm. The highest power is produced in the afternoon time and it decreases in the morning and evening. Similar to the summer season, the harmonic distortion in current is lowest in the afternoon time and higher in the morning and evening in the winter also. Also the power factor values are highest in the afternoon and lower at other times.

A comparison of PV plant performance in the summer and winter seasons is also made in the study. While comparing the data obtained in May and December, the data collected during both the seasons are at different timings because of the fact that sun rises early in summer and also sets late, whereas in winter the sun rises late and also sets early. Due to this, PV plants will be able to work for longer hours in May than in December. Through the graphs plotted it can be clearly seen that although the power quality values follow the same pattern in both the months but they differ in the time period that they provide that result in. As shown in Figure 13.10, the power factor in May (with values above 0.9 lagging) is for a much longer time than in December. Likewise, it can also be seen in Figure 13.11 that the values of THDi below the recommended value are obtained for a much shorter time in December than in May.

The key observations related to the relative power of the inverter are as follows

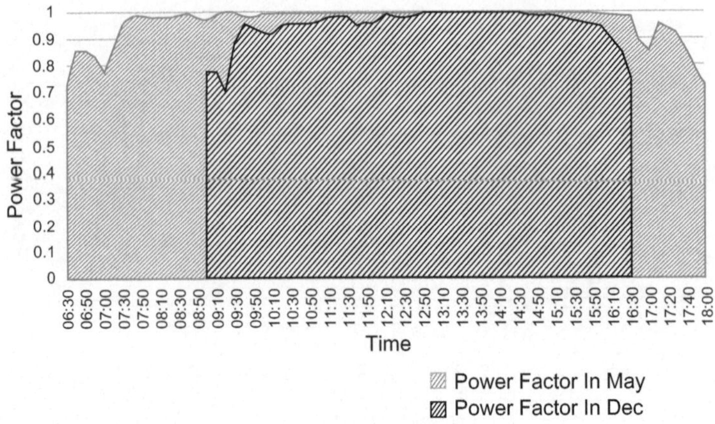

Figure 13.10 Comparison of power factor values obtained in the months of May and December.

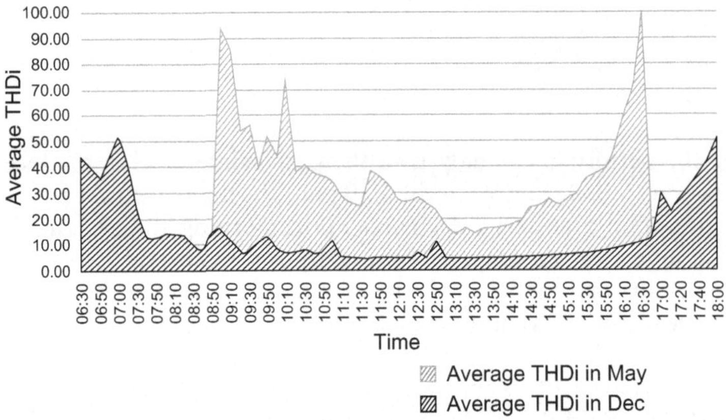

Figure 13.11 Comparison of average THDi values obtained in the months of May and December.

(a) The relative power and power factor are directly proportional to each other.
(b) The harmonic distortion in the current (THDi) and relative power are inversely proportional to each other.

The minimum, average and maximum power values (for power factor above 0.9 obtained from the testing in May and December) show that the maximum power at which power factor is above 0.9 is 71% of rated power and the minimum is 8% of rated power for summer whereas for winter maximum power is 41% of rated power and minimum is 8% of rated power.

Power for harmonic distortion in current values shows that THDi should be less than the recommended value, but above that value harmonic distortion can impact the power quality severely. At 20% of rated power THDi is within the limits but below 20% it increases beyond the limit in May, whereas in December THDi goes beyond the advised value at and below the power at 30% of rated power.

In May the time for which total harmonic distortion in current is under limit is approximately 7 hours (9:10 am–16:40 pm), whereas in December it is reduced to approximately 2 hours (12:50 pm–14:20 pm) only. This concludes that total harmonic distortion in current is found for longer times in winter than in summer. The power factor '0.99 lagging' is achieved at a minimum power of 33% of rated power in summer and in winter the minimum power is 32% of rated power. In May this high value of power factor is maintained for approximately 5.5 hours, and in December only, for 1.5 hours.

13.5 CONCLUSION

In the work examined here, tests were done on a 100 kW system and data was collected in two different seasons to cover the range of changes in the power quality parameters (THDi and Power Factor). Through the tests, the inverse proportionality relation of power factor with THDi is obtained. It is also observed that the THDi and power factor values vary according to climate change. The following are the key conclusions of the study:

(a) The relative power and power factor are directly proportional to each other.
(b) The harmonic distortion in the current (THDi) and relative power are inversely proportional to each other.
(c) Both factors (power factor and THDi) are poor during the early morning and evening times, whereas they are improved during the day time, when the sun light intensity becomes high to support the PV produced electrical power.
(d) Comparison is also presented for the summer and winter months and it is deduced that power quality is impacted more in winter months like December than in summer months like May. This is because of poor relative power values in the winter months rather than the summer months.

Based on the available generation data and climatic conditions, the presented mathematical model and its experimental validation demonstrate that the model is useful for identifying the expected degradation in current harmonics and power factor.

REFERENCES

1. IEA - International Energy Agency. Trends in photovoltaic applications. Survey report of selected IEA countries between 1992 and 2011; pp. 1 – 26, 2012.
2. Rajput, S.K. and Dheer, D.K. (2023). Impact of solar intensity on photovoltaic-generated current harmonics and transformer life: A mathematical model with experimental validation. *ASME Journal of Solar Energy Engineering*, vol. 145, no. 2, p. 021006.
3. Kumar, A., Kumar, K., Kaushik, N., Sharma, S. and Mishra, S. (2010). Renewable energy in India: Current status and future potentials. *Elsevier Renewable and Sustainable Energy Reviews*, vol. 14, pp. 2434–2442.
4. Roy, S.K., Rajput, S.K., Prakash, R. and Dheer, D.K. (2022). Performance and energy metrics analysis of stand alone rooftop PV plant. International Conference on Emerging Frontiers in Electrical and Electronic Technologies (ICEFEET), pp. 1–6, doi: 10.1109/ICEFEET51821.2022.9848352.
5. Rana, N., Kumar, M., Goel, R.K. and Jain, A. (2019). The development and prospect in the solar energy. *International Journal of Mechanical and Production Engineering Research and Development (IJMPERD)*, vol. 8, pp. 45–49.
6. Dincer, I. and Rosen, M.A. (1999). Energy, environment and sustainable development. *Applied Energy*, vol. 64, no. 1–4, pp. 427–440.
7. Hojabri, M., Ahmad, A.Z., Toudeshki, A. and Soheilirad, M. (2012). An overview on current control techniques for grid connected. *International Proceedings of Computer Science and Information Technology*, vol. 56, pp. 119–126.
8. Sujod, A.Q. and Al-Shetwi, M.Z. (2018) Voltage sag detection in grid-connected photovoltaic power plant for low voltage ride-through control. *Recent Advances in Electrical & Electronic Engineering*, vol. 12, no. 4, pp. 384–392.
9. Al-Shetwi, A.Q., Hannan, M., Jern, K.P., Alkahtani, A.A. and PG Abas, A. (2020). Power quality assessment of grid-connected PV system in compliance with the recent integration requirements. *Electronics*, vol. 9, no. 2, pp. 366–388.
10. Al-Shetwi, A.Q., Sujod, M.Z., Hannan, M.A., Abdullahd, M.A., Al-Ogaili and Jern, K.P. (2020) Impact of inverter controller-based grid connected PV system in the power quality. *International Journal of Electrical and Electronic Engineering & Telecommunications*, vol. 9, no. 6, pp. 462–469.
11. Bonaldo, J.P., Schiavon, G., Paredes, H.K.M. and Pomilio, J.A. (2017). Multifunctional operation of current controlled VSI based on the harmonic content of PCC voltage. Proceedings IEEE 8th International Symposium on Power Electronics for Distributed Generation Systems, pp. 1–6.

12. Al-Shetwi, A.Q. and Sujod, M.Z. (2018). Modeling and control of grid-connected photovoltaic power plant with fault ride-through capability. *Journal of Solar Energy Engineering*, vol. 140, no. 2, p. 021001.
13. Cleveland, F. (2018). IEC 61850-7-420 communications standard for distributed energy resources (DER). Proceedings IEEE Power and Energy Society General Meeting-Conversion and Delivery of Electrical Energy in the 21st Century, pp. 1–4, 2018.
14. Basso, T., Hambrick, J. and DeBlasio, D. (2012). Update and review of IEEE P2030 smart grid interoperability and IEEE 1547 interconnection standards. IEEE PES Innovative Smart Grid Technologies (ISGT) Europe Conference.
15. Semea, S., Lukač, N., Štumberger, B. and Hadžiselimović, M. (2017). Power quality experimental analysis of grid-connected photovoltaic systems. *Energy*, vol. 139, pp. 261–1266.
16. Grady, W.M. and Gilleskie, R.J. (1993). Harmonics and how they relate to power factor. Proceedings of the EPRI Power Quality Issues & Opportunities Conference, pp. 1–8.
17. Rampinelli, G.A., Gasparin, F.P., Buhler, A.J., Krenzinger, A. and Romero, A.C. (2015). Assessment and mathematical modeling of energy quality parameters of grid connected photovoltaic inverters. *Elsevier Renewable and Sustainable Energy Reviews*, vol. 52, pp. 133–141.
18. Oliveira, L.G.M., Boaventura, W.D.C., Amaral, G., Mello, J.V.G.C., Mendes, V.F., Macedo, W.N. and Torres, P.F. (2016). Assessment of harmonic distortion in small grid-connected photovoltaic systems. 17th International Conference on Harmonics and Quality of Power (ICHQP), pp. 810–816.
19. Volokhin, V.V., Diahovchenko, I.M. and Derevyanko, B.V. (2017). Electric energy accounting and power quality in electric networks with photovoltaic power stations. IEEE International Young Scientists Forum on Applied Physics and Engineering (YSF), pp. 36–39.
20. Mohanty, S. R., Sen, S., Kishor, N., Ray, P. K., and Singh, V. P. (2011) Harmonic compensation in distributed generation based micro-grid using droop control Technique. 5th International Power Engineering and Optimization Conference, pp. 233–237.
21. Kourtesi, S.T., Ekonomou,L., Nakulas, A., Fotis, G.P. and Zoulias, E. (2007). Modeling and simulation of a single phase photovoltaic inverter and investigation of switching strategies for harmonic minimization. 6th WSEAS International Conference on Applications of Electrical Engineering, Istanbul, Turkey.
22. Rajput, S. K. and Dheer, D. K. (2023). Forecasting power-factor reductions in rooftop PV-integrated industrial power systems: Mathematical modelling and experimental validation. *Sustainable Energy, Grids and Networks*, vol. 33, p. 100974, ISSN 2352-4677. doi: 10.1016/j.segan.2022.100974.
23. Liang, X. and Andalib -Bin- Karim, C. (2018). Harmonics and mitigation techniques through advanced control in grid-connected renewable energy sources: A review. *IEEE Transactions on Industry Applications*, vol. 54, no. 4, pp. 3100–3111. doi: 10.1109/TIA.2018.2823680.

24. Elkholy, A, and Abou El-Ela, A. (2019). Optimal parameters estimation and modelling of photovoltaic modules using analytical method. *Heliyon*, vol. 5, no. 7. doi: 10.1016/j.heliyon.2019.e02137.
25. Alvidrez, J.H. and Johnson, J. (2017). Parametric PV grid-support function characterization for simulation environments. IEEE 44th Photovoltaic Specialist Conference (PVSC), pp. 2153–2158. doi: 10.1109/PVSC.2017.8366295.

Index

T

U

V

W